Mathematical Evolutionary Theory

MATHEMATICAL EVOLUTIONARY THEORY

Edited by Marcus W. Feldman

PRINCETON UNIVERSITY PRESS

PRINCETON, NEW JERSEY

Randall Library UNC-W

Copyright © 1989 by Princeton University Press

Published by Princeton University Press, 41 William Street,
Princeton, New Jersey 08540
In the United Kingdom: Princeton University Press,
Guildford, Surrey

All Rights Reserved

Library of Congress Cataloging-in-Publication Data

Mathematical evolutionary theory / edited by Marcus W. Feldman.
   p. cm.
   Collection of papers presented in honor of Samuel Karlin.
   Includes bibliographies and indexes.
   ISBN 0-691-08502-1 (alk. paper)   ISBN 0-691-08503-X
   (pbk. : alk. paper)
     1. Evolution—Mathematics.   2. Karlin, Samuel, 1923–
   I. Feldman, Marcus W.   II. Karlin, Samuel, 1923–
   QH371.M295 1988   575.01′51—dc19   88-15591

This book has been composed in Lasercomp Times Roman
by Syntax International Private Limited

Clothbound editions of Princeton University Press books
are printed on acid-free paper, and binding materials are
chosen for strength and durability. Paperbacks, although satisfactory
for personal collections, are not usually suitable for library rebinding

Printed in the United States of America by Princeton University Press,
Princeton, New Jersey

Designed by Laury A. Egan

QH
371
.M295
1989

CONTENTS

DEDICATION

The **papers** in this volume celebrate Samuel Karlin's contributions to mathematical evolutionary theory. His earliest contributions dealt with genetic drift, the stochastic phenomena induced by the finiteness of a population's size. In the late 1960s and 1970s his work addressed the interaction between linkage, selection, and the mating system in a deterministic context. This was succeeded by papers advancing the statistical analysis of data for gene frequencies and familial aggregation. Karlin's most recent studies in the mathematical evolutionary theory have concerned the evolution of behavior and the development of numerical algorithms for comparing and interpreting DNA sequences.

The authors of the papers collected here have all been influenced by Sam Karlin, either as a mentor, collaborator, constructive critic, or through extended stays at Stanford. The papers span the wide range of topics in evolutionary theory to which Sam has contributed. This is our opportunity to acknowledge the profound effect that he has had on the direction and quality of interdisciplinary study in mathematical biology, and on our own research.

PREFACE

Mathematical theory has been central to the study of evolutionary biology since the rediscovery of Mendel's work. Will Provine's masterful biography of Sewall Wright* amply documents the profound effect that this theory has had on the leading natural historians of this century. The mathematical theory of evolution was established by Wright, R. A. Fisher, and J.B.S. Haldane in order to explain observations about variability within and between populations and species. The need to explain the origin and maintenance of this variability led to the dynamical theories of population genetics and, to some extent, to ecology. The need to describe the extant patterns of variation led to far-reaching developments in mathematical statistics.

In the past ten years, application of advances in biotechnology to the study of populations has resulted in the exposure of previously unexpected extensive genetic variability in nature. At the same time, more advanced mathematical technology has been applied to the models designed to describe the origin and maintenance of this variability. Samuel Karlin has, to a great extent, orchestrated this mathematical advance.

Mathematical evolutionary theory spans a range of subjects similar to that covered by population biology as a whole. What are the roles of population size and subdivision on the rate of evolution? Does it make a difference to our picture of the evolutionary process if genes are linked or unlinked? Can the evolutionary effect of departures from random mating such as inbreeding, assortative mating, and sexual selection be quantified? What are the advantages of sex, recombination, and dispersal? Is there a natural framework within which to simultaneously study behaviors and genes that affect these behaviors? What aspects of the environment are most important for the evolution of life-history patterns? What kinds of inference about organismal evolution can legitimately be made from molecular genetic variability? The papers here contribute to our understanding of some of these issues by suggesting new mathematical models, by more extensive analysis of standard models, or by comparing data and theory.

The papers have been divided, somewhat arbitrarily, into two groups. Part I addresses general problems in evolutionary genetics; these do not refer to a specific biological situation, organism, or behavior but take a

---

* W. B. Provine, *Sewall Wright and Evolutionary Biology* (Chicago: University of Chicago Press, 1986).

general perspective on the mathematical consequences of small population size, mutation, gene linkage, population subdivision, and gene flow. In Part II the theory is developed for specific biological situations, some of which have been associated with general theory, as is the case with kin selection. Both approaches to theory have been important to the modern study of evolution.

Stanford, California
December 1987

Mathematical Evolutionary Theory

PART 1

Stochastic and Deterministic
Genetic Theory

$\mathbf{T}$he foundation stone of population genetic theory with finite population size is the Wright-Fisher sampling model. It is described in the paper by Ewens and underlies much of the discussion in the papers by Gillespie, Watterson, and Kaplan and Hudson. It also forms the background for part of Tavaré's paper. The binomial (or multinomial) sampling scheme produces a Markov chain that describes the change in the genetic constitution of the population over time. The eigenvectors of this Markov chain have not been found in a useful form, and Wright in 1931 used a diffusion approximation to the discrete time stochastic process in order to obtain information about the transient properties of the gene frequencies

The eigenvalues of the Wright-Fisher Markov chain determine the rate of evolution of the system. The leading nonunit eigenvalue in the simplest case, for example, with N diploid individuals, is $1 - 1/2N$, and this is the rate of approach to homozygosity. When the population is partitioned into two sexes, or undergoes inbreeding, the rate-determining eigenvalue is more complicated, and is often expressed in terms of N_e, the effective population size. Other definitions of effective population size have involved analogies with different properties of the stochastic model.

In his paper (Chapter 1), Ewens discusses the various definitions of effective population size when a population is subdivided into demes. At every generation some of these demes become extinct and are recolonized. The manner of recolonization, fast or gradual, is shown greatly to affect the various effective population sizes and their relative magnitudes.

Much of the huge volume of population genetic data accumulated during the past twenty years has been interpreted in terms of the interaction between drift and mutation. One of the most widely studied genetic models with finite population size includes a constant mutation rate, μ, to novel alleles. This is the infinite alleles model about which so much has been written in the context of the selection-neutrality controversy. The parameter $\theta = 4N\mu$, with N the population size, has been known for many years to govern the distribution of $\{\beta_i\}$, the number of alleles having i representatives $(i = 1, 2, \ldots, n)$ is a sample of size n from such a population.

The magnitude of the compound parameter θ has been a source of controversy between selectionists and neutralists since the mid-1960s. Watterson (Chapter 2) studies the distribution of $\{\beta_i\}$ when θ is not constant over time, analytically and by numerical simulation. In one simulation the mutation rate is held constant and the population is allowed to cycle from one size to another. In particular, it is shown that the simulated sample mean of the homozygosity agrees well with the expected population homozygosity derived earlier by Maruyama and Fuerst (1985b; see Chapter 2). The variance of the sample mean, however, can be uncomfortably high.

Tavaré (Chapter 3) studies a birth and death process subject to immigration at the time points of a Poisson process. Starting from an initial immigrant propagule, the new arrivals initiate lineages that evolve independently from one another. The stochastic process of interest keeps track of the sizes of the families in the order of their appearance. When conditioned on the population size, the age-ordered family sizes have a joint distribution that is the size-biased version of that derived by Ewens (1972; see Chapter 2) for the number of alleles with i representatives in the infinitely-many neutral alleles model (the β_i of Watterson's paper). The special case of the birth process with immigration provides a simple way to study the genealogical structure of the (age-ordered) stationary infinite alleles model. For example, the asymptotic fractions of the population accounted for by the different families have, when written in decreasing order, the Poisson-Dirichlet distribution that plays so central a role in the mathematical theory of neutral evolution.

The next paper, by Gillespie (Chapter 4), further examines the Wright diffusion approximation. For Wright's diffusion approximation to be mathematically legitimate, parameters such as the rates of mutation, rates of migration, and fitness differences between genotypes must be of the same order of magnitude as the reciprocal of the population size, N, as N increases. In his paper, Gillespie surveys what is known about the properties of limiting processes that emerge when other assumptions are made about the relationship between these key parameters of evolution and the population size.

In the remaining articles of this section the evolutionary forces under study are genotypic selection, recombination, and migration. The interaction of these has been a major focus for population genetic theory since 1970. Linkage disequilibrium is the widely used term for the degree of association between the alleles at a pair of loci. With multiple loci an adequate description of the gamete frequency distribution must involve higher-order associations, that is, associations among three or more genes. Christiansen (Chapter 5) suggests that useful measures of association can be generated from the powers of a single matrix that has proven effective in earlier studies of selection on multiple linked loci (see his formula 4). He calls these *linear interactions*, and shows that they possess desirable properties under iteration in the absence of selection.

In earlier studies of subdivided populations, the linkage disequilibrium between a pair of genes had been shown to depend on the variation in gene frequencies among the subpopulations. Christiansen extends this principle to multiple loci and to some simple migration arrangements among the subpopulations. It is shown, for example, that the measure of association among a given set of genes depends only on the frequencies of alleles at that set of genes.

Weir and Cockerham (Chapter 6) have been concerned for a number of years with the statistical and stochastic sampling properties of linkage disequilibrium in populations that are sampled with respect to the genotype at two or more loci. They use "descent measures," measures of the probability that specific alleles (or combinations of alleles) in the gametes that unite to form a zygote are identical by descent. In the present paper, they express the frequencies of the ten possible genotypes at two diallelic loci in terms of the two gene frequencies and a set of disequilibrium measures. The latter are defined as specific sums of the genotype frequencies and can be expressed in terms of the descent measures. Some of these disequilibrium measures depend on three or four positions in the two-locus genotype and may not have the same time-dependent behavior as the usual linkage disequilibrium.

With the assumption of multinomial sampling of individuals from a population, maximum likelihood estimates of the various disequilibria and variances of these estimates are computed, although some of the algebra is daunting. These estimates are then used to construct statistics for testing whether specific disequilibria are zero. The test statistics have chi-square asymptotic distributions.

The validity of this procedure was tested numerically with 10,000 replicate samples of size 100. When all disequilibria were zero, the fit of the test statistic to the chi-square distribution was very good. The presence of disequilibria in the underlying population did not have enough of an effect on the testing procedure to invalidate it.

Liberman, Feldman, and Holsinger (Chapter 7) address the notion of evolutionary optimality. Their approach extends the notion of Evolutionary Genetic Stability, originally used in the study of the sex ratio, to neutral genetic modifiers of the rate of migration between two populations. The genetic model is diploid and consists of two loci, one of which has two alleles and is called "major" because it is subject to viability selection on the genotypes. The population is divided into two habitats with arbitrary viability regimes in each. The second locus, the modifier, has an arbitrary number of alleles, and its function is to control genetically the level of migration between the demes. The linkage between the genes is arbitrary.

The first part of the paper develops a class of equilibria at which the migration-modifying alleles have frequencies equivalent to those that would emerge from a one-locus viability model with the migration rates playing the roles of the genotypic viabilities. It is then shown that at such an equilibrium only those alleles that initially reduce the average migration rate in the population can invade. This suggests that *zero* migration has the property of EGS.

In the Appendix to Chapter 7, internal stability of the equilibrium is analyzed numerically. In view of the authors' earlier work on mutation modification, the equilibrium structure here is surprisingly complicated. In particular, the observation of several instances of cyclic behavior of the genotype frequencies poses interesting questions about the mathematical structure of migration models in population genetics.

CHAPTER ONE

The Effective Population Sizes in
the Presence of Catastrophes

Warren J. Ewens

Introduction

My long association with Sam Karlin through our work in mathematical population genetics has been a most memorable and enjoyable experience for me. It started in 1964 when I was a postdoctoral student at Stanford, has continued to this day, and my aim in this paper is to continue it even further. A simplified description of one aspect of our association is as follows: I would become interested in some problem and partially develop its mathematical properties, but would eventually be defeated by some aspect of the mathematical analysis, or not see the full generality of the theory, whereupon I would take the problem to Sam, who would solve the mathematical problem, or generalize the theory (or do both). I do not wish to imply that more than a small fraction of Sam's work in population genetics was initiated in this way, but much of mine was finished along these lines. I remember problems relating to the distribution of the number of alleles maintained in the infinitely many alleles model (Ewens 1963, 1964; Karlin and McGregor 1967), the fundamental theorem of natural selection (Ewens 1969a,b; Karlin and Feldman 1970a and subsequent papers), the unexpected complexities of two-locus systems (Ewens 1968; Karlin and Feldman 1970b and subsequent papers), the evolution of dominance (Ewens 1967; Feldman and Karlin 1971), and in particular the eigenvalues (Ewens and Kirby 1975; Karlin and Avni 1975) and sampling properties (Ewens 1972; Karlin and McGregor 1972) of the infinitely many alleles model, as well as problems of heterozygosity and the effective population size (all the above references). I can never thank Sam enough for the inspiration of this association, and I here present him with a set of further problems, evocative of the those just described, and again concerning eigenvalues, sampling theory, heterozygosity, and effective population sizes, as well as other matters recognizable to the initiated.

So, play it again, Sam: What do you make of this one?

Subdivided Populations and Catastrophes

One of the main concerns of theoretical population genetics is the analysis of the degree and nature of genetic variation in natural populations. The degree of variation is, of course, also of interest in practice to those who are concerned with the loss of genetic variation through random genetic drift, particularly in specific or unusual situations. One circumstance, discussed at length at a recent conference on minimum viable population sizes (Soulé 1987), is that of a large population divided into comparatively small subpopulations that are liable to complete extinction through catastrophic events, with the niche previously occupied by an extinct subpopulation being taken over by migrants from another subpopulation. We discuss here the rate of loss of genetic variation in such cases and the amount of genetic variation maintained when mutation is present, extending the work of Maruyama and Kimura (1980). We find that accepted ideas in this situation do not necessarily hold. This occurs largely because the very subdivision of the large population has an effect on genetic variation often counterbalancing the effects caused by the extinction process.

Many reasonable models of a "catastrophe and recolonization" process may be formed, and any conclusion drawn should not be largely an artefact of the particular model chosen. We consider two models here but it is clear that others are possible. Further, we sometimes reach different conclusions in the two models we do consider. A complete investigation of the process of modeling for this problem is needed and is far from straightforward.

The rate of loss of genetic variation in a population is often measured by calculating one or another concept of the effective population size. These are defined relative to the simple Wright-Fisher model and we therefore start by introducing this model and discussing the effective population sizes to which it leads.

The Wright-Fisher Model and the Effective Population Sizes

Although most populations of interest to us are diploid, our analysis here is more easily carried out, and the main features are not lost, in the haploid case, which we (and others, e.g., Maruyama and Kimura 1980) therefore use throughout. The "simple" Wright-Fisher model (i.e., allowing no selection, mutation, population subdivision, etc.) considers a population of M individuals (or genes) in any generation, each of which is of allelic type A_1 or A_2. It is assumed that random sampling of individuals with re-

placement occurs in choosing the parents forming any daughter generation, so that if there are i A_1 genes in any generation, the probability p_{ij}
that in the next generation there will be j A_1 genes is given by

$$p_{ij} = \binom{M}{j}\left(\frac{i}{M}\right)^j\left(1 - \frac{i}{M}\right)^{M-j}, \qquad i, j = 0, 1, 2, \ldots, M. \qquad (1.1)$$

Clearly $\mathbf{P} = \{p_{ij}\}$ is the transition matrix of a Markov chain with absorbing states at 0 and M. Many properties of this model are known, in
particular

1. $\lambda_{\max}$ = largest nonunit eigenvalue of $\mathbf{P} = 1 - M^{-1}$,   (1.2)
2. π = prob (two individuals taken at random have
        same parent) = M^{-1},   (1.3)
3. $\text{var}(x_{t-1}\,|\,x_t) = x_t(1 - x_t)M^{-1}$, where x_t is the fraction of
   individuals in generation t who are A_1.   (1.4)

Thus in this model,

$$M = (1 - \lambda_{\max})^{-1}, \qquad (1.5)$$

$$M = \pi^{-1}, \qquad (1.6)$$

$$M = x_t(1 - x_t)/\text{var}(X_{t+1}\,|\,x_t). \qquad (1.7)$$

Suppose in a more complicated model, allowing, say, for geographical
subdivision, the largest nonunit eigenvalue of the appropriate transition
matrix is λ^*. Then the (eigenvalue) effective population size for this model
is defined, using (1.5), as

$$N_e^{(e)} = (1 - \lambda^*)^{-1}, \qquad (1.8)$$

and the interpretation of this is that insofar as the leading eigenvalue
is concerned, the complicated model behaves as a simple Wright-Fisher
model of size $N_e^{(e)}$. Similarly, if π^* is the probability that two individuals
have the same parent, the (inbreeding) effective population size is defined
as

$$N_e^{(i)} = (\pi^*)^{-1}. \qquad (1.9)$$

Finally, the variance effective population size $N_e^{(v)}$ (if it exists) is defined,
using (1.7), as

$$N_e^{(v)} = x_t(1 - x_t)/\text{var}(x_{t+1}\,|\,x_t). \qquad (1.10)$$

These three quantities are not necessarily equal, so the expression "effective population size" should not be used without a further adjective
describing which of the three is intended. Further, in complicated models
they can be very difficult to calculate, and indeed $N_e^{(v)}$ might not even be

well defined, in that there might be no scalar Markovian variable x such that the right-hand side in (1.10) is a constant independent of x_t.

Finally, a fourth definition of effective population size, which we call here the mutation effective population size $N_e^{(m)}$, has been introduced by Maruyama and Kimura (1980), and this new concept should be of considerable value in describing the likely degree of genetic variation in populations subject to mutation. If genes mutate, at rate u, in such a way that every mutant is of an entirely novel allelic type, then genetic variation is maintained in the population. We now quickly summarize the standard theory (Kimura and Crow 1964) assessing this degree of variation for the simple Wright-Fisher model for purposes of comparison with later calculations. If P_{t+1} is the probability that two individuals chosen at random in generation $t + 1$ are of the same allelic type, then neither can be a mutant [probability $(1 - u)^2$] and they are either both descended from the same parent (probability M^{-1}) or different parents who are of the same allelic type [probability $(1 - M^{-1})P_t$]. Thus

$$P_{t+1} = (1 - u)^2[M^{-1} + (1 - M^{-1})P_t]. \qquad (1.11)$$

The stationary probability P is then, exactly,

$$P = (1 - u)^2[M - (M - 1)(1 - u)^2]^{-1}. \qquad (1.12)$$

If, in a more complicated model, the stationary value of this probability is P^*, we use (1.12) to define the mutation effective population size exactly as the solution for y of the equation

$$P^* = (1 - u)^2[y - (y - 1)(1 - u)^2]^{-1}. \qquad (1.13)$$

$N_e^{(m)}$ is not necessarily equal to $N_e^{(e)}$, $N_e^{(i)}$, or $N_e^{(v)}$, and is to be interpreted simply as the size of a simple Wright-Fisher population having the same stationary value of P as the more complicated model. Unfortunately, $N_e^{(m)}$ is not necessarily independent of the mutation rate u, although as we note later, it will often be rather insensitive to the value of u and also have a well-defined limit as $u \to 0$. It is therefore a potentially valuable parameter in describing an important aspect of the complicated model.

We now consider two models of catastrophes and the values of the various effective population sizes that they define.

Model 1: Total Replacement

Theory

We consider a total population of Mn individuals consisting of n subpopulations with M individuals in each. In each generation, k of the subpopulations become extinct, due to catastrophes, and the niche occupied by an extinct subpopulation is refilled by randomly choosing one of the surviving subpopulations to produce (apart from its "normal" reproduction of M individuals) a further M to fill the niche. These choices are independent, so that a surviving subpopulation can fill more than one niche. Superimposed on this "demographic" process is a simple Wright-Fisher process within any subpopulation, as described above. Random changes in allele frequency thus arise at two levels, the demographic or "between subpopulations" changes (through the random choice of subpopulations to become extinct), and genetic or "within subpopulations" changes (through the Wright-Fisher process).

A fundamental parameter in this model is the probability q that two individuals chosen at random from distinct subpopulations have parents in the same subpopulation. We find

$$q = k(2n - k - 1)/[n(n - 1)(n - k)]. \qquad (1.14)$$

To find the eigenvalue population size for this model, we must first find a Markovian variable having a transition matrix generalizing $\mathbf{P}$. The appropriate variable is the vector $(a_1, a_2, \ldots, a_n)$, where $a_1, \ldots, a_n$ are the numbers of A_1 genes in the various subpopulations. Note that $a_1, \ldots, a_n$ are interchangeable: the two states $(a_1, a_2, \ldots, a_n)$ and $(a_2, a_1, \ldots, a_n)$, for example, are regarded as being identical. There are then $R = \begin{pmatrix} M + n \\ M \end{pmatrix}$ states for the process, two of which $[(0, 0, \ldots, 0)$ and $(M, M, \ldots, M)]$ are absorbing, with the rest transient. Thus the matrix of transition probabilities allows two unit eigenvalues and $R - 2$ eigenvalues less than unity in modulus. The method described below for finding these is reminiscent of that used by Ewens and Kirby (1975); is there an approach reminiscent of Karlin and Avni (1975)?

We first write $a_1, \ldots, a_n$ in increasing order $a_1 \leq a_2 \leq a_3 \leq \cdots \leq a_n$, and then list the possible values in dictionary order, as in Table 1.1.

The first two states are now temporarily ignored, and if we also ignore all zeroes, we can define "sample configurations"

$$\{2\}, \{3\}, \ldots, \{M\}, \{1, 1\}, \{1, 2\}, \ldots, \{M, M, \ldots, M\} \qquad (1.15)$$

in conformity with the vectors listed in Table 1.1. We write a typical sample configuration $\underline{x} = (x_1, \ldots, x_s)$ and seek the probability $P_{t+1}(\underline{x}) = P_{t+1}(x_1, \ldots, x_s)$ that in generation $t + 1$, a sample of x_1 genes from one subpopulation, x_2 from another, $\ldots$, x_s from an sth subpopulation, are all of the same allelic type. Since a sample of genes from s subpopulations can have parents in at most s subpopulations and since also a sample of x_i genes in any one subpopulation can have at most x_i parents (all, of course, in the same subpopulation), it follows by arguments parallel

Table 1.1
Values of the vector $(a_1, a_2, \ldots, a_n)$.

a_1	a_2	a_3	$\cdots$	a_{n-1}	a_n
0	0	0	$\cdots$	0	0
0	0	0	$\cdots$	0	1
0	0	0	$\cdots$	0	2
.	.	.	.	.	.
0	0	0	$\cdots$	0	M
0	0	0	$\cdots$	1	1
0	0	0	$\cdots$	1	2
.	.	.	.	.	.
M	M	M	$\cdots$	M	M

to those leading to (1.11) that, apart from an additive constant, $P_{t+1}(x_1, \ldots, x_s)$ is a linear combination of probabilities of the form $P_t(y_1, \ldots, y_u)$, where $(y_1, \ldots, y_u)$ precedes $(x_1, \ldots, x_s)$ in the ordering (1.15). Thus if we form a vector $\underset{\sim}{P}_{t+1}$ from the $P_{t+1}(\underset{\sim}{x})$ values, with the $\underset{\sim}{x}$ values ordered as in (1.15),

$$\underset{\sim}{P}_{t+1} = \mathbf{D}\underset{\sim}{P}_{t}, \tag{1.16}$$

where $\mathbf{D}$ is a triangular matrix. The eigenvalues of $\mathbf{D}$ are the nonunit eigenvalues we seek, and these are its diagonal elements, that is, typically the coefficient of $P_t(\underset{\sim}{x})$ in $P_{t+1}(\underset{\sim}{x})$. This is easily seen to be

$$g(\underset{\sim}{x}) = \left[\prod_{j=1}^{s} \{1 - M^{-1}\}\{1 - 2M^{-1}\}\{1 - 3M^{-1}\} \cdots \{1 - (x_j - 1)M^{-1}\} \right]$$

$$\times \frac{\binom{n-k}{s}}{\binom{n}{s}} \sum_{j=0}^{s} \binom{s}{j} \frac{k!}{(k-s+j)!} \frac{1}{(n-k)^{s-j}}. \tag{1.17}$$

The latter term in this expression is the probability that s daughter subpopulations have s different parent subpopulations (i.e., is the "demographic" contribution to the eigenvalue), whereas the initial term comes from standard (Feller 1951) eigenvalues for the Wright-Fisher model (i.e., is the "genetic" contribution). The largest eigenvalue depends on the relative contributions from these two sources, being

$$\lambda^* = \max(1 - M^{-1}, 1 - q), \tag{1.18}$$

and thus

$$N_e^{(e)} = \max(M, q^{-1}). \tag{1.19}$$

The ultimate rate of loss of genetic variation is thus decided entirely either by within population (genetic) factors, as measured by M, or by between population (demographic) factors as measured by q^{-1}, and not by any combination of the two.

It is interesting to confirm the eigenvalues (1.17) in the special case $M = 1$, $k = 1$. The model reduces in this case to the well-known Moran model (1958) of genetics where at unit time points a single individual dies and is replaced, with the dying individual not being a possible parent of the new individual. (This model first interested Karlin [Karlin and McGregor 1962] in genetics.) The largest nonunit eigenvalue in this model is well known to be $1 - 2/n(n - 1)$, and this is what $N_e^{(e)}$ reduces to with $M = k = 1$.

It is usually accepted that a subdivided population subject to extinction of subpopulations will lose genetic variation more rapidly than an equally large random-mating population, or equivalently that it has a smaller eigenvalue effective population size. The above shows that this is not necessarily true: for example, when $M = 10^3$, $n = 10^4$, $k = 4$, the eigenvalue effective population size in the subdivided case is 1.25×10^7, whereas the actual population size is only 10^7. We will see later that when mutation exists, the subdivided population can maintain more genetic variation, on the average, than a random-mating population of the same size, again against accepted views.

We find $N_e^{(i)}$ by calculating the probability π^* that two different individuals chosen at random have the same parent. Elementary arguments show that

$$\pi^* = [M - 1 + qM(n - 1)]/(Mn - 1)M],$$

so that

$$N_e^{(i)} = M(Mn - 1)/[M - 1 + qM(n - 1)]. \tag{1.20}$$

It is interesting to compare the numerical value of $N_e^{(i)}$ with the actual population size Mn. Elementary calculations show that $N_e^{(i)} > Mn$ if and only if

$$qMn < 1. \tag{1.21}$$

This turns out to be a fundamental inequality, which we will return to several times later.

There appears to be no well-defined expression for $N_e^{(v)}$, since there is no scalar Markovian variable in this model. We therefore do not discuss $N_e^{(v)}$ further, other than to remark that the only hope for a reasonable definition of $N_e^{(v)}$ is through the quasi-Markovian variable concept of Norman (1975).

We turn finally to $N_e^{(m)}$, which can be calculated, using (1.13), once we have an expression for P^*. Now two individuals taken from the same subpopulation will be of the same allelic type with probability P given by (1.12). Two individuals taken from different subpopulations will have parents in the same subpopulation (possibly the same parent) with probability q. Thus, if P_d is the probability that two individuals from different subpopulations are of the same allelic type, the identity

$$P_d = (1 - u)^2 [(1 - q)P_d + q\{M^{-1} + (1 - M^{-1})P\}]$$

must hold. This gives

$$P_d = qP[1 - (1 - u)^2(1 - q)]^{-1}, \tag{1.22}$$

and hence

$$P^* = [(M - 1)P + M(n - 1)P_d]/(Mn - 1) \tag{1.23}$$

$$= P[M - 1 + M(n - 1)q\{1 - (1 - u)^2(1 - q)\}^{-1}]/(Mn - 1). \tag{1.24}$$

This is an exact expression and from it we find, exactly,

$$N_e^{(m)} = M\beta + (\beta - 1)/\alpha, \tag{1.25}$$

where

$$\alpha = (1 - u)^{-2} - 1$$

$$\beta = (Mn - 1)[M - 1 + M(n - 1)q\{1 - (1 - u)^2(1 - q)\}^{-1}]^{-1}.$$

It is necessary to use these exact calculations to develop various properties of $N_e^{(m)}$, which we now do.

First, $N_e^{(m)}$ is dependent on u, but as we note later, this dependence is often rather weak and $N_e^{(m)}$ thus provides a useful parameter for measuring the expected degree of variation in the subdivided population. The limit of $N_e^{(m)}$ as the mutation rate u approaches zero is well defined, being

$$\bar{N}_e^{(m)} = M + M(n - 1)(1 - q)/[q(Mn - 1)]. \tag{1.26}$$

This is close to an expression given by Maruyama and Kimura (1980) in a similar model. Next, the behavior of $N_e^{(m)}$ as u approaches zero is of some interest. Careful handling of small-order terms shows that $N_e^{(m)}$ approaches its limiting value from *below* when (1.21) holds, and otherwise from *above*. We will return to this property later.

Finally, it is interesting to compare $N_e^{(m)}$ with $N_e^{(e)}$ and $N_e^{(i)}$. Elementary calculations show that $N_e^{(e)} > \bar{N}_e^{(m)}$ if and only if (1.21) holds. So far as the comparison of $N_e^{(m)}$ and $N_e^{(i)}$ is concerned, a simple (but incorrect) argument would make these identical. This argument claims that two randomly chosen individuals will be of identical genetic type if neither is a mutant [probability $(1 - u)^2$], and they are either descended from the same parent [probability $(N_e^{(i)})^{-1}$], or different parents [probability $1 - (N_e^{(i)})^{-1}$] who are of identical allelic type (probability P^*). This argument would lead to

$$P^* = (1 - u)^2[(N_e^{(i)})^{-1} + \{1 - (N_e^{(i)})^{-1}\}P^*], \qquad (1.27)$$

so that, from (1.13), $N_e^{(i)} = N_e^{(m)}$. However, this cannot be the case, since $N_e^{(m)}$, unlike $N_e^{(i)}$, is a function of u, and the fallacy in the above argument (pointed out to me by R. C. Griffiths) is that the probability that the two (different) parents of two randomly chosen individuals are of the same allelic type is not the same as the corresponding probability for the individuals themselves. Note that this fallacy does not arise in the argument leading to (1.25) and (1.26). Thus the final P^* in (1.27) should be replaced by P^* (parents), the probability that the parents of two randomly chosen individuals who are themselves different individuals are of the same allelic type.

This fallacious argument is nevertheless of use, since it shows that $N_e^{(i)} < N_e^{(m)}$ if and only if P^*(parents) $< P^*$. Now

$$P^*(\text{parents}) = P\gamma + P_d(1 - \gamma), \qquad (1.28)$$

where $\gamma = (N - 1)/(N_e^{(i)} - 1)$ is the probability that two randomly chosen individuals have different parents who are nevertheless in the same subpopulation. Using (1.23) and (1.28), we eventually find that $N_e^{(i)} < N_e^{(m)}$ if and only if (1.21) holds. We have thus shown, by piecing together various of the above conclusions, that if (1.21) holds,

$$Mn < N_e^{(i)} < N_e^{(m)} < \bar{N}_e^{(m)} < N_e^{(e)}, \qquad (1.29a)$$

while if (1.21) does not hold,

$$Mn > N_e^{(i)} > N_e^{(m)} > \bar{N}_e^{(m)} > N_e^{(e)}. \qquad (1.29b)$$

When $qMn = 1$, all effective population sizes equal the actual population size. These inequalities highlight the importance of the parameter qMn in assessing the likely genetic properties of the subdivided population. They also raise a curious point: $N_e^{(i)}$ and the actual population size are non-genetic concepts, $N_e^{(m)}$ is a totally genetic concept and $N_e^{(e)}$ is a partially genetic concept. Despite this, the various effective population sizes can be ordered, as in (1.28) and (1.29), according to the value of the single parameter qMn.

NUMERICAL EXAMPLE

Table 1.2 provides numerical examples of the values of the actual population size $N_a = Mn$, together with the various effective population sizes defined above, for various n and k combinations. There are many points of interest in the table, the most striking being the wide variation possible in the various values. Usually N_a and $N_e^{(i)}$ are quite close, as are $N_e^{(e)}$ and

Table 1.2

Values of the actual population size (N_a), the inbreeding effective size $N_e^{(i)}$, the mutation effective size for $u = 10^{-5}$(a) and 10^{-6}(b), and the limiting value $\bar{N}_e^{(m)}$, together with the eigenvalues effective size $N_e^{(e)}$, for various (n, k) combinations. ($M = 1000$)

Value of n		Value of k		
		1	10	100
10^3	N_a	10^6	10^6	10^6
	$N_e^{(i)}$	10^6	0.98×10^6	0.83×10^6
	$N_e^{(m)}$(a)	0.51×10^6	0.51×10^6	0.58×10^4
	$N_e^{(m)}$(b)	0.50×10^6	0.50×10^6	0.57×10^4
	$\bar{N}_e^{(m)}$	0.50×10^6	0.50×10^6	0.57×10^4
	$N_e^{(e)}$	0.50×10^6	0.50×10^6	0.47×10^4
10^4	N_a	10^7	10^7	10^7
	$N_e^{(i)}$	10^7	10^7	0.98×10^7
	$N_e^{(m)}$(a)	0.45×10^8	0.50×10^7	0.51×10^6
	$N_e^{(m)}$(b)	0.50×10^8	0.50×10^7	0.50×10^6
	$\bar{N}_e^{(m)}$	0.50×10^8	0.50×10^7	0.50×10^6
	$N_e^{(e)}$	0.50×10^8	0.50×10^7	0.50×10^6
10^5	N_a	10^8	10^8	10^8
	$N_e^{(i)}$	10^8	10^8	10^8
	$N_e^{(m)}$(a)	0.25×10^{10}	0.46×10^9	0.50×10^8
	$N_e^{(m)}$(b)	0.45×10^{10}	0.50×10^9	0.50×10^8
	$\bar{N}_e^{(m)}$	0.50×10^{10}	0.50×10^9	0.50×10^8
	$N_e^{(e)}$	0.50×10^{10}	0.50×10^9	0.50×10^8

$N_e^{(m)}$, but the former two vary from being some two hundred times larger than the latter two to being fifty times smaller. This shows that great care must be taken in assessing whether the subdivided population tends to maintain more variation, or preserve variation longer, than an undivided population. The reason why such large differences between the N_e's are possible (and this is largely an artefact of modeling of the situation) is that the very fact of subdivision tends to preserve genetic variation, and the loss of variation through the random loss of subpopulations might, or might not, be strong enough to offset this. What is perhaps needed is a catastrophe model that does not rely so strongly on subdivision into isolated populations. Migration is a factor that should also be taken into account.

Returning to the above model, the observation that the various effective population sizes can differ radically from each other, together with the two sets of inequalities (1.28) and (1.29) and the condition (1.21), highlights the importance of the parameter qMn in this model. This leads to the first of the problems this model suggests, namely, why is the parameter qMn of such significance, and what is important about the equation $qMn = 1$?

A second problem follows from this. It is well known (Ewens 1979) that for so-called exchangeable models (Cannings 1974), the equations

$$N_e^{(v)} = N_e^{(i)} = N_e^{(e)}$$

are true (assuming $N_e^{(v)}$ is well defined). Is it then true that the above model is exchangeable when $qMn = 1$? Is there a well-defined $N_e^{(v)}$ in this case?

A third problem is to construct a realistic catastrophe model that is not dependent, as is the above model, on population subdivision, so that the effects of catastrophes can be disentangled from the effects of subdivision. A possible model is to assume that some fixed number M^* of individuals dies in a catastrophe in each generation and that all individuals in the daughter generation are descended from the surviving individuals. (Simple models along these lines are exchangeable, so that the various effective population sizes are equal.)

As a final problem, it is unclear to what extent the sampling theory for random mating populations (Ewens 1972; Karlin and McGregor 1972), referred to in several papers in this volume, holds in the subdivided population model: the calculations above give no clues on this question.

Model 2: Gradual Replacement

This model has many of the features of the previous one. There exists a fixed number n of subpopulations, and in each generation k of this number become extinct because of catastrophes. However, the niche previously occupied by an extinct subpopulation is now filled by one single offspring individual, drawn at random from the surviving individuals. At the same time, the number of individuals in each surviving subpopulation increases by unity. This model is very close to that analyzed by Maruyama and Kimura (1980).

We consider first the nongenetic properties of the model, focusing on the sizes of the various subpopulations. When $k = 1$, these sizes may be written in increasing order,

$$1 = M_1 < M_2 < M_3 < \cdots < M_n. \tag{1.30}$$

Apart from M_1, these sizes are random variables and it can be shown that, at stationarity,

$$
\begin{aligned}
P(M_i) = \binom{n-1}{i-1}&\left[\left(\frac{i-1}{n}\right)^{M_i-1} - \binom{i-1}{1}\left(\frac{i-2}{n}\right)^{M_i-1}\right.\\
&+ \binom{i-1}{2}\left(\frac{i-3}{n}\right)^{M_i-1} \cdots\\
&\left.\pm \binom{i-1}{i-2}\left(\frac{1}{n}\right)^{M_i-1}\right],
\end{aligned}
\tag{1.31}
$$

$$M_i = i, i+1, i+2, \ldots,$$

$$
\begin{aligned}
P(M_1, \ldots, M_i) = \binom{n-1}{i-1}&\left[\left(\frac{1}{2}\right)^{M_2-1}\left(\frac{2}{3}\right)^{M_3-1}\left(\frac{3}{4}\right)^{M_4-1}\right.\\
&\left.\times \left(\frac{i-2}{i-1}\right)^{M_{i-1}-1}\left(\frac{i-1}{n}\right)^{M_i-1}\right],
\end{aligned}
\tag{1.32}
$$

$$1 = M_1 < M_2 < M_3 < M_4 < \cdots < M_i,$$

and in particular

$$
P(M_1, \ldots, M_n) = \left[\left(\frac{1}{2}\right)^{M_2-1}\left(\frac{2}{3}\right)^{M_3-1} \cdots \left(\frac{n-1}{n}\right)^{M_n-1}\right],
$$

$$1 = M_1 < M_2 < M_3 < M_4 < \cdots < M_n. \tag{1.33}$$

Any randomly chosen subpopulation has size given by the geometric distribution, so that for general k,

$$P(M) = (k/n)(1 - k/n)^{M-1}, \qquad M = 1, 2, 3, \ldots. \tag{1.34}$$

Thus

$$E(M) = n/k, \qquad \mathrm{Var}(M) = n(n-k)/k^2, \tag{1.35}$$

and the mean total population size is n^2/k. From this, we get an approximation to the probability that two individuals drawn at random are from the same subpopulation:

$$[2n^2(n - k)/k^2] \div (n^4/k^2) \simeq 2(n - k)/n^2. \tag{1.36}$$

This is approximately twice the value arising in the previous model and occurs because the two individuals are more likely to be chosen from the larger subpopulations. It is also possible to derive an expression for the probability distribution of the total population size. However, this calculation is complicated, and we note here only that the mean total population size is (as noted above) n^2/k, with variance (computed by G. A. Watterson) of

$$n^2(n - k)^2/\{k^2(2n - k - 1)\}. \tag{1.37}$$

Note that this result is less than the sums of the variances of the individual population sizes, due to a correlation of $-(2n - k - 1)^{-1}$ between individual population sizes.

We now add a genetical component to the model. First, suppose any individual is of allelic type A_1 or A_2, and the allelic type of any parent is passed without mutation to each offspring. The population will eventually consist of entirely A_1 or A_2 individuals, and the rate at which this occurs will be measured by the eigenvalue effective size of the model. To calculate this, we must first describe the state of the process (generalizing the vector $\mathbf{a}$ in Model 1) by a vector of pairs,

$$[(a_1, M_1), (a_2, M_2) \dots, (a_n M_n)], \tag{1.38}$$

where a_i is the number (out of M_i) of individuals in subpopulation i who are A_1. We next impose a Wright-Fisher model in each subpopulation: if the subpopulation survives the catastrophe at any given time, then in an obvious notation,

$$P[(a_i, M_i) \longrightarrow (b_i, M_i + 1)] = \binom{M_i + 1}{b_i} \left(\frac{a_i}{M_i}\right)^{b_i} \left(1 - \frac{a_i}{M_i}\right)^{M_i + 1 - b_i}. \tag{1.39}$$

The eigenstructure of this process appears to be very difficult, and I can make no progress toward finding $N_e^{(e)}$. (There are, of course, infinitely many eigenvalues.)

An intuitive argument giving an approximation for the inbreeding effective population size is as follows. If at any time the subpopulation sizes are $M_1, M_2, \dots, M_n$, with $\sum M_i = M$, the probability that two individuals taken at random are from a population size M_i is $M_i(M_i - 1)/M(M - 1)$. The probability that they then have the same parent is $1/(M_i - 1)$, and so the overall probability that two individuals have the same parent is

$$\sum_i M_i(M_i - 1)(M_i - 1)^{-1}/M(M - 1) = (M - 1)^{-1}.$$

The expected value of this quantity is approximately $(n^2/k)^{-1}$, and hence the inbreeding effective population size should be approximately the mean actual population size, n^2/k. It would be desirable to have a more precise argument leading to a more accurate value for $N_e^{(i)}$, using the complete distribution of population size.

To discuss the mutation-effective population size, we now allow mutation at rate u, with all mutants being of novel allelic type. We first calculate P_M, the probability that two individuals taken from a subpopulation of size M are of the same allelic type. Clearly $P_1 = 1$. For $M > 1$, an argument similar to that leading to (1.12) shows that

$$P_M = (1 - u)^2[(M - 1)^{-1} + (M - 2)(M - 1)^{-1}P_{M-1}]. \qquad (1.40)$$

G. A. Watterson has pointed out to me that putting $Q_M = (M - 1)P_M$ leads to a very rapid solution of this recurrence relation:

$$P_M = Q[1 - (1 - u)^{2M-2}](M - 1)^{-1}, \qquad M = 2, 3, 4, \ldots, \qquad (1.41)$$

where

$$Q = (1 - u)^2/[1 - (1 - u)^2].$$

Since any subpopulation size actually assumes the value M with probability given by (1.34), we may calculate the mean probability $\bar{P}$ that two individuals taken at random from the same subpopulation are of the same allelic type as

$$\bar{P} = \sum P_M(k/n)(1 - k/m)^{M-1} \simeq \alpha \log(1 + \alpha^{-1}), \qquad (1.42)$$

where

$$\alpha = k/(2un). \qquad (1.43)$$

A numerical check for (1.42) is available. Maruyama and Kimura (1980) considered a model almost identical to the above, the only difference being that catastrophes and population size increases occur in continuous time rather than at discrete time points (as here), with the probability of extinction of a subpopulation in time $(t, t + \delta t)$ being $\lambda\,\delta t$. To compare their simulations with (1.48), it is necessary to put $\lambda = k/n$ (and also their v equal to u). Table 1.3 compares their simulated values of $\bar{P}$ with those calculated from (1.48). We see that the two sets of values are quite close.

It seems much more difficult to arrive at an expression for $\bar{P}_d$, the probability that two individuals from different populations are of the same

Table 1.3

Simulation (Maruyama and Kimura 1980) and approximating
theoretical values [from (42) and (45)] of $\bar{P}$ and $P_d/\bar{P}$.

		$\bar{P}$		$P_d/\bar{P}$	
u	λ	Sim.	Theoret.	Sim.	Theoret.
0.05	0.1	0.599	0.693	0.154	0.166
0.02	0.1	0.599	0.841	0.295	0.333
0.01	0.1	0.872	0.912	0.507	0.500
0.001	0.1	0.987	0.990	0.853	0.909
0.002	0.05	0.958	0.962	0.735	0.714
0.005	0.05	0.907	0.912	0.506	0.500

allelic type. The argument leading to (1.22) does not appear to carry over
to this model. We note from (1.22) that in Model 1, if q and u are both
small,

$$P_d/P \approx 1/(1 + 2u/q) \approx 1/(1 + n^2 u/k), \qquad (1.44)$$

which for $n = 10$ (the value used in the Maruyama and Kimura simulations) gives

$$P_d/P \approx 1/(1 + 10u/\lambda). \qquad (1.45)$$

We also present, in Table 1.3, the simulation values of P_d/P and the values
calculated from (1.45). Again, "theoretical" and simulation values are close,
although here we have little justification for using (1.45) as it derives from
a model different from the present one. Despite this, and encouraged by
the simulation results, we now form an approximation to P^*, the probability that two individuals drawn at random are of the same allelic type.
We have from (1.36) that

$$P^* \approx 2n^{-1}\bar{P} + (n - 2)n^{-1}P_d,$$

and this is approximately

$$\alpha(\alpha + 1)(\alpha + \tfrac{1}{2}n)^{-1} \log(1 + \alpha^{-1}). \qquad (1.46)$$

Letting $u \to 0$ and using (1.13) leads to a suggested approximate limiting
mutation-effective population size of $n^2/2k$, about half the mean actual
population size. In the case $n = 10^5$, $k = 100$, discussed by Muruyama

and Kimura (1980), this is 5×10^7, in exact agreement with the value they calculate, and half of the mean actual population size (10^8). The expression (1.37) suggests that the standard deviation in the actual population size is about 2.2×10^5, so that the population size will seldom exceed 1.005×10^8. In other words, the actual population size and the mutation-effective population size will seldom differ by more than a factor of 2.01. The conjecture by Maruyama and Kimura that the actual population size might often be of order 10^{20} (so that there would be an enormous difference between actual and mutation-effective population sizes) seems to be without foundation.

Note that if the mutation-effective population size in this model is about half the actual population size, we reach conflicting conclusions between Models 1 and 2 on the relative values of these quantities when, in Model 1, $qMn < 1$. This emphasizes strongly a point made above and returned to below, that the conclusions we draw should not be artefacts of particular models and that we need a more general approach to the catastrophe problem.

It is clear that in Model 1, explicit expressions can be found for many quantities of interest. However, for Model 2 the best that has been found above, for most quantities of interest, is a set of approximations, sometimes based on tenuous arguments. Some specific problems for Model 2, some of which are very difficult, are the following. First, find the complete set of eigenvalues and hence the eigenvalue effective population size. Second, find the probability that two individuals have the same parent and then find the inbreeding population size. Third, find an exact expression for P^* and thus find the mutation-effective population size. Fourth, is there a parameter (analogous to qMn in Model 1) that determines the relative magnitudes of these effective population sizes and the mean actual population size? Fifth, is there ever a well-defined concept of a variance-effective population size? Sixth, is there a more realistic catastrophe model analogous to Model 2 but not so dependent on population structure? Seventh, can the whole concept of modeling the catastrophe process be considered so that the conclusions drawn are not artefacts of particular models? Finally, can one derive a reasonable sampling theory for Model 2, generalizing the one we found together many years ago for the random mating population case?

References

Cannings, C. C. 1974. The latent roots of certain Markov chains arising in genetics: A new approach, I. Haploid models. *Adv. Appl. Prob.* 6: 260–290.

Ewens, W. J. 1963. The diffusion equation and a pseudo-distribution in genetics. *J. Roy. Stat. Soc. B.* 25: 405–412.

Ewens, W. J. 1964. The maintenance of alleles by mutation. *Genetics* 50: 891–898.

Ewens, W. J. 1967. A note on the mathematical theory of the evolution of dominance. *Amer. Nat.* 101: 35–40.

Ewens, W. J. 1968. A genetic model having complex linkage behaviour. *Theoret. Appl. Genet.* 38: 140–143.

Ewens, W. J. 1969a. Mean fitness increases when fitnesses are additive. *Nature* 221: 1076.

Ewens, W. J. 1969b. A generalized fundamental theorem of natural selection. *Genetics* 63: 531–537.

Ewens, W. J. 1972. The sampling theory of selectively neutral alleles. *Theor. Pop. Biol.* 3: 87–112.

Ewens, W. J. 1979. *Mathematical Population Genetics*. Springer-Verlag, Berlin and NewYork.

Ewens, W. J., and Kirby, K. 1975. The eigenvalues of the neutral alleles process. *Theor. Pop. Biol.* 7: 212–220.

Feldman, M., and Karlin, S. 1971. The evolution of dominance: A direct approach through the theory of linkage and selection. *Theor. Pop. Biol.* 2: 482–492.

Feller, W. 1951. Diffusion processes in genetics. In *Proc. Second Berkeley Symposium on Mathematical Statistics and Probability*, pp. 227–246. Ed. J. Neyman. University of California Press, Berkeley.

Karlin, S., and Avni, H. 1975. Derivation of the eigenvalues of the configuration process induced by a labeled direct product branching process. *Theor. Pop. Biol.* 7: 221–228.

Karlin, S., and Feldman, M. 1970a. Convergence to equilibrium of the two-locus additive viability model. *J. Appl. Prob.* 7: 262–271.

Karlin, S., and Feldman, M. 1970b. Linkage and selection: Two-locus symmetric viability model. *Theor. Pop. Biol.* 1: 39–71.

Karlin, S., and McGregor, J. L. 1962. On a genetic model of Moran. *Proc. Camb. Phil. Soc.* 58: 299–311.

Karlin, S., and McGregor. J. L. 1967. The number of mutant forms maintained in a population. In *Proc. Fifth Berkeley Symposium on Mathematical Statistics and Probability*, pp. 415–438. Ed. L. Le Cam and J. Neyman. University of California Press, Berkeley.

Karlin, S., and McGregor, J. L. 1972. Addendum to a paper of W. Ewens. *Theor. Pop. Biol.* 3: 113–116.

Kimura, M., and Crow, J. F. 1964. The number of alleles that can be maintained in a finite population. *Genetics* 49: 725–738.

Maruyama, T., and Kimura, M. 1980. Genetics variability and effective population size when local extinction and recolonization of subpopulations are frequent. *Proc. Nat. Acad. Sci. USA* 77: 6710–6714.

Moran, P.A.P. 1958. Random processes in genetics. *Proc. Camb. Phil. Soc.* 54: 60–71.

Norman, M. F. 1975. Diffusion approximation of non-Markovian processes. *Ann. Prob.* 3: 358–364.

Soulé, M. E. 1987. *Viable Populations for Conservation*. Cambridge University Press, New York.

CHAPTER TWO

The Neutral Alleles Model
with Bottlenecks

Geoffrey A. Watterson

2.1. Introduction

If **a sample** of n genes is chosen at random from a large population (say of size N diploids) then Ewens' (1972) sampling distribution describes the probability that the sample will contain a certain number of alleles, at various frequencies for those alleles. Perhaps the neatest way to describe the distribution is first to introduce the sample "frequency spectrum" $\beta_1, \beta_2, \ldots, \beta_n$, where

β_i = number of alleles having i representative genes in the sample.

Then Ewens' formula for the probability of getting a particular set of values for the spectrum is

$$P(\beta_1, \beta_2, \ldots, \beta_n) = \frac{n!}{(\theta)_{(n)}} \prod_{i=1}^{n} [(\theta/i)^{\beta_i}/\beta_i!] \qquad (2.1.1)$$

where $\beta_1, \beta_2, \ldots, \beta_n$ are nonnegative integers such that

$$\sum_{i=1}^{n} i\beta_i = n, \qquad (2.1.2)$$

and where

$$(\theta)_{(n)} = \theta(\theta + 1) \cdots (\theta + n - 1).$$

The parameter θ is a mutation parameter, as explained below. It is assumed, in deriving (2.1.1), that evolution has been proceeding for an infinitely long period before the sample was taken. Throughout that time, the population size N is assumed to have remained constant, and the mutation rate, per gene per generation, has always been u ($0 \leq u \leq 1$). All mutants are of different allelic type from previously existing alleles, and all alleles and genotypes are selectively neutral. Mating is at random, or, at least, N denotes the effective population size. Then the scaled mutation rate,

$$\theta = 4Nu, \tag{2.1.3}$$

is the only parameter appearing in (2.1.1).

Notice that the number of alleles, K, in the sample is given by the random quantity

$$K = \sum_{i=1}^{n} \beta_i. \tag{2.1.4}$$

The distribution (2.1.1) was proved by Karlin and McGregor (1972) as the limiting distribution in the case of sampling from the Wright-Fisher model as $N \to \infty$, $u \to 0$, with θ fixed. The distribution is exact for sampling without replacement from the Moran model, as was shown by Kelly (1977), where, however, θ is then defined as $\theta = 2Nu/(1 - u)$, $2N$ being the actual number of haploids in the population.

The probability generating function (p.g.f.) corresponding to (2.1.1) is

$$E\left(\prod_{i=1}^{n} s_i^{\beta_i}\right) = \frac{n!}{\theta_{(n)}} \text{ coefficient of } \phi^n \text{ in } \exp\left[\theta \sum_{i=1}^{n} s_i \phi^i / i\right]. \tag{2.1.5}$$

See (2.10) in Watterson (1974).

The p.g.f. for the number of alleles, K, and its probability distribution are

$$E(s^K) = (\theta s)_{(n)} / (\theta)_{(n)} \tag{2.1.6}$$

and

$$P(K = k) = \theta^k |S_n^{(k)}| / (\theta)_{(n)}, \qquad k = 1, 2, \ldots, n \tag{2.1.7}$$

[see Ewens 1972 (27), (28)]. In (2.1.7), $S_n^{(k)}$ is a Stirling number of the first kind.

In this chapter, we generalize the distribution (2.1.1) to cases when the parameter θ has *not* remained constant in the generations preceding the sample. Two special cases have been discussed already (see Watterson 1984a), in which θ has taken two or three different values. We shall here allow θ to take an arbitrary number of values in the past, including, for instance, the periodic bottleneck case when θ varies alternately between two values (e.g., see Maruyama and Fuerst 1985b). The variation in θ could be due to (1) the variation of the population size over time, with the mutation rate u remaining constant; (2) the variation in mutation rate u, with constant N; or (3) the variation of both N and u over time. Whichever is the case, we will assume that N remains large compared to the sample size n (especially for values of N in the recent past) so that the derived formulas are good approximations whether the sample was from a Wright-Fisher, Moran, or similar model.

Because the formulas are complicated, we also give a method to simulate samples, and illustrate its use in periodic bottleneck cases.

2.2. The Sample Distribution

Suppose that the sample is taken at the present generation, labeled "0". For the previous t_1 generations, the mutation parameter θ took the value θ_1, corresponding to $N = N_1$ and $u = u_1$. For the previous t_2 generations (i.e., from t_1 to $t_1 + t_2$ generations ago) the value was θ_2, corresponding to $N = N_2$ and $u = u_2$. Similarly, for earlier parameter values $\theta_3 = 4N_3 u_3$, $\theta_4 = 4N_4 u_4, \ldots$.

$$\begin{array}{ccccc} & \theta_1 = 4N_1 u_1 & \theta_2 = 4N_2 u_2 & \theta_3 = 4N_3 u_3 & \\ \vdash & \vdash & \vdash & \vdash & \longrightarrow \text{ generations} \\ 0 & t_1 & t_1 + t_2 & t_1 + t_2 + t_3 & \text{ago.} \\ \text{Present} & & & & \end{array}$$

We define a Markov chain $\{L_0, L_1, L_2, \ldots\}$ as follows. Let $L_0 = n$, the initial sample size. L_1 denotes the number of genes in the population at time t_1 ago which have nonmutant descendant genes in the sample. That is, L_1 is the number of "lines of descent," without mutation, that are involved in the sample, starting from the ancestral generation t_1 ago (see Griffiths 1980). Similarly, define $L_2, L_3, \ldots$ as the numbers of nonmutant lines of descent in the sample, from ancestors at generations $t_1 + t_2$, $t_1 + t_2 + t_3, \ldots$ generations ago. Of course $(L_k, k = 0, 1, 2, \ldots)$ is a "pure-death" process with $L_0 \geq L_1 \geq L_2 \geq L_3, \ldots$, because our sample genes have fewer and fewer ancestors as time recedes into the past. The transition probabilities are, for $k = 1, 2, 3, \ldots, i, j = 0, 1, 2, \ldots$,

$$P(L_k = j \mid L_{k-1} = i) = \begin{cases} p_k(i, j), & j \leq i \\ 0, & j > i, \end{cases}$$

where

$$p_k(i, j) = \sum_{l=j}^{i} e^{-l(l + \theta_k - 1)t_k/4N_k} (-1)^{l-j} (2l + \theta_k - 1) \frac{(j + \theta_k)_{(l-1)} (i)_{[l]}}{j!(l - j)!(i + \theta_k)_{(l)}}, \quad (2.2.1)$$

and $(i)_{[l]} = i(i - 1) \cdots (i - l + 1)$. [See Tavaré 1984, (5.2).] Notice that (2.2.1) is not time-homogeneous; it depends on k.

It is of interest later that, given $L_{k-1} = i$, L_k has the following representation:

$$L_k = \sup \left\{ j: \sum_{l=j}^{i} T_l > t_k \right\}, \quad (2.2.2)$$

where $T_1, T_2, \ldots, T_i$ are independent exponentially distributed variables having means

$$E(T_l) = 4N_k/[l(l + \theta_k - 1)],$$

while $T_0 = \infty$. [See Kingman 1982, (1.7), for the $\theta_k = 0$ case.]

Let $\mathbf{L} = (L_0, L_1, L_2, \ldots)$ be the lines-of-descent process, and $\mathbf{l} = (l_0, l_1, l_2, \ldots)$ a particular realization of it. We will assume that

$$n = l_0 \geq l_1 \geq l_2 \geq \cdots \geq l_{r-1} > l_r = 0, \tag{2.2.3}$$

so that, if $\mathbf{L} = \mathbf{l}$, then L_r is the *first* value having $L_r = 0$ (in backwards time), and all the sample alleles arose by mutation less than $t_1 + t_2 + \cdots + t_r$ generations ago.

In what follows, we will first condition on $\mathbf{L} = \mathbf{l}$, and discuss the allelic composition of the sample in terms of the allelic compositions of the l_1 ancestors at t_1, the l_2 ancestors at $t_1 + t_2$, etc. Later, we will average the conditional results over all possible realizations $\mathbf{l}$.

We write $\beta_{k,j}$, $j = 1, 2, \ldots, l_k$, for the number of alleles having j representative genes among the sample's ancestors at time $t_1 + t_2 + \cdots + t_k$ ago. As the total number of such ancestors is l_k, we must have

$$\sum_{j=1}^{l_k} j\beta_{kj} = l_k.$$

Assuming that $L_r = 0$ but $L_{r-1} = l_{r-1} > 0$, we note that the frequency spectrum $(\beta_{r-1,1}, \beta_{r-1,2}, \ldots, \beta_{r-1,l_{r-1}})$ of the most remote ancestors consists solely of alleles arising as a result of mutations during a period in which the mutation parameter was θ_r. This spectrum therefore has the Ewens distribution (2.1.1), with n replaced by l_{r-1} and θ replaced by θ_r. Its p.g.f. is, from (2.1.5),

$$E\left(\prod_{j=1}^{l_{r-1}} s_j^{\beta_{r-1,j}} \bigg| \mathbf{L} = \mathbf{l}\right) = \frac{l_{r-1}!}{(\theta_r)_{(l_{r-1})}} \text{ coefficient of } \phi_{r-1}^{l_{r-1}} \text{ in}$$

$$\exp\left[\theta_r \sum_{j=1}^{l_{r-1}} s_j \phi_{r-1}^j / j\right]. \tag{2.2.4}$$

We trace these alleles in the l_{r-1} ancestor sample down through their descendants and into the eventual final sample. This requires us to study how one set of ancestors, say those at $t_1 + t_2 + \cdots + t_k$, inherits alleles from the ancestors at $t_1 + t_2 + \cdots + t_k + t_{k+1}$, together with any new mutants that arise between those two times. This transition has been investigated by Watterson (1984a) where in (2.7) of that paper, the p.g.f. for the "old" alleles and the new mutant alleles was obtained. Here, we modify that equation to yield the p.g.f. for the combined (old and new) frequency spectrum $(\beta_{k,1}, \beta_{k,2}, \ldots, \beta_{k,l_k})$, conditional on the frequency spectrum $\beta_{k+1} = (\beta_{k+1,1}, \beta_{k+1,2}, \ldots, \beta_{k+1,l_{k+1}})$, and conditional on $L_k = l_k$, $L_{k+1} = l_{k+1}$. We find

$$E\left(\prod_{j=1}^{l_k} s_j^{\beta_{k,j}} \Big| \beta_{k+1}, \mathbf{L} = \mathbf{l}\right) = a_k(\mathbf{l}) \text{ coefficient of } \phi_k^{l_k} \text{ in}$$

$$\prod_{j=1}^{l_{k+1}} \left[\sum_{l=j}^{l_k} s_l \phi_k^l (l-1)_{[j-1]}/(j-1)!\right]^{\beta_{k+1,j}} \exp\left[\theta_{k+1} \sum_{l=1}^{l_k} s_l \phi_k^l / l\right], \quad (2.2.5)$$

where the normalizing constant is

$$a_k(\mathbf{l}) = (l_k - l_{k+1})!/(l_{k+1} + \theta_{k+1})_{(l_k - l_{k+1})}. \quad (2.2.6)$$

Equation (2.2.5) summarizes the transition from one frequency spectrum to the next. As the right-hand side of (2.2.5) involves the β_{k+1} spectrum as powers, it is possible to take expectations with respect to β_{k+1} to obtain the following recurrence relation for successive p.g.f.'s:

$$E\left(\prod_{j=1}^{l_k} s_j^{\beta_{k,j}} \Big| \mathbf{L} = \mathbf{l}\right) = a_k(\mathbf{l}) \text{ coefficient of } \phi_k^{l_k} \text{ in}$$

$$E\left(\prod_{j=1}^{l_{k+1}} v_j^{\beta_{k+1,j}} \Big| \mathbf{L} = \mathbf{l}\right) \exp\left[\theta_{k+1} \sum_{l=1}^{l_k} s_l \phi_k^l / l\right], \quad (2.2.7)$$

where

$$v_j = \sum_{l=j}^{l_k} s_l \phi_k^l (l-1)_{[j-1]}/(j-1)!.$$

It is now possible to find explicitly the solutions for the p.g.f.'s of $\{\beta_k\}$ in succession, starting with the known function (2.2.4) when $k = r - 1$, and applying the recurrence (2.2.7) successively for $k = r - 2, r - 3, r - 4, \ldots,$ 2, 1, 0. We find that the last step in this recursion yields

$$E\left(\prod_{j=1}^{l_0} s_j^{\beta_{0,j}} \Big| \mathbf{L} = \mathbf{l}\right) = \left[\prod_{k=0}^{r-1} a_k(\mathbf{l})\right] \text{ coefficient of } \prod_{k=0}^{r-1} \phi_k^{l_k - l_{k+1}} \text{ in}$$

$$\exp\left[\sum_{l=1}^{l_0} s_l g_l(\phi_0, \phi_1, \phi_2, \ldots, \phi_{r-1})/l\right], \quad (2.2.8)$$

where

$$g_l(\phi_0, \phi_1, \phi_2, \ldots, \phi_{r-1}) = (\theta_1 - \theta_2)\phi_0^l + (\theta_2 - \theta_3)(\phi_0 + \phi_1)^l$$
$$+ (\theta_3 - \theta_4)(\phi_0 + \phi_1 + \phi_2)^l + \cdots$$
$$+ (\theta_r - \theta_{r+1})(\phi_0 + \phi_1 + \phi_2 + \cdots + \phi_{r-1})^l. \quad (2.2.9)$$

For purposes of symmetry in the expressions (2.2.6), (2.2.8) and (2.2.9), we have used the conventions $\theta_{r+1} = 0$, $l_0 = n$ and $l_r = 0$. The proof of (2.2.8) is fairly straightforward by induction, and involves the identity

$$\sum_{j=1}^{l} [(l-1)_{[j-1]}/(j-1)!][\phi^j/j] = [(1+\phi)^l - 1]/l.$$

It remains to average (2.2.8) over the probabilities (2.2.1) of the realizations $\{L_k\}$. There seems to be no great simplification possible; the final expression for the sample frequency spectrum's p.g.f. is (suppressing the subscript 0 in the notation $\beta_{0,j}$):

$$E\left(\prod_{j=1}^{l_0} s_j^{\beta_j}\right) = \sum_{r=2}^{\infty} \sum_{l_1=1}^{l_0} \sum_{l_2=1}^{l_1} \cdots \sum_{l_{r-1}=1}^{l_{r-2}} p_1(l_0, l_1)p_2(l_1, l_2) \cdots p_r(l_{r-1}, 0)$$

$$\times \left[\prod_{k=0}^{r-1} a_k(\mathbf{l})\right] \text{coefficient of } \prod_{k=0}^{r-1} \phi_k^{l_k - l_{k+1}} \text{ in}$$

$$\exp\left[\sum_{l=1}^{l_0} s_l g_l(\phi_0, \phi_1, \phi_2, \ldots, \phi_{r-1})/l\right]$$

$$+ p_1(l_0, 0) \frac{l_0!}{(\theta_1)_{(l_0)}} \text{ coefficient of } \phi_0^{l_0} \text{ in}$$

$$\exp\left[\theta_1 \sum_{l=1}^{l_0} s_l \phi_0^l/l\right], \tag{2.2.10}$$

where the second contribution comes from the case when $l_0 = n$ but $l_1 = 0$, so that $r = 1$.

While (2.2.10) is rather intimidating, it is possible to use it to obtain moments (which we will do in the next section) and the probability distribution it generates. We find the generalized Ewens distribution for the sample's spectrum to be

$$P(\beta_1, \beta_2, \ldots, \beta_n) = \sum_{r=2}^{\infty} \sum_{l_1=1}^{l_0} \sum_{l_2=1}^{l_1} \cdots \sum_{l_{r-1}=1}^{l_{r-2}} p_1(l_0, l_1)p_2(l_1, l_2) \cdots p_r(l_{r-1}, 0)$$

$$\times \left[\prod_{k=0}^{r-1} a_k(\mathbf{l})\right] \text{coefficient of } \prod_{k=0}^{r-1} \phi_k^{l_k - l_{k+1}} \text{ in}$$

$$\prod_{l=1}^{l_0} \{[g_l(\phi_0, \phi_1, \ldots, \phi_{r-1})/l]^{\beta_l}/\beta_l!\}$$

$$+ p_1(l_0, 0) \frac{l_0!}{(\theta_1)_{(l_0)}} \text{ coefficient of } \phi_0^{l_0} \text{ in}$$

$$\prod_{l=1}^{l_0} [(\theta_1 \phi_0^l/l)^{\beta_l}/\beta_l!].$$

As g_l is of degree l in the ϕ's, the product $\prod_{l=1}^{l_0} g_l^{\beta_l}$ is of degree $\sum_l l\beta_l \equiv l_0 \equiv n$ in the ϕ's. So too is the term $\prod_{k=0}^{r-1} \phi_k^{l_k - l_{k+1}}$, whose coefficient needs to be found. We can therefore replace one of the dummy variables, say ϕ_0, by 1 to get

$$P(\beta_1, \beta_2, \ldots, \beta_n) = \sum_{r=2}^{\infty} \sum_{l_1=1}^{l_0} \sum_{l_2=1}^{l_1} \cdots \sum_{l_{r-1}=1}^{l_{r-2}} p_1(l_0, l_1) p_2(l_1, l_2) \cdots p_r(l_{r-1}, 0)$$

$$\times \left[\prod_{k=0}^{r-1} a_k(\mathbf{l}) \right] \text{coefficient of } \prod_{k=1}^{r-1} \phi_k^{l_k - l_{k+1}} \text{ in}$$

$$\prod_{l=1}^{l_0} \{ [g_l(1, \phi_1, \phi_2, \ldots, \phi_{r-1})/l]^{\beta_l}/\beta_l! \}$$

$$+ p_1(l_0, 0) \frac{l_0!}{(\theta_1)_{(l_0)}} \prod_{l=1}^{l_0} [(\theta_1/l)^{\beta_l}/\beta_l!]. \tag{2.2.11}$$

If there are only two stages in our process, with θ_1 applying over t_1 generations and θ_2 applying over $t_2 = \infty$ generations, then only the $r = 1$ and $r = 2$ terms in the above are relevant, and (2.2.11) reduces to the one-step bottleneck result in Watterson (1984a, [2.13]), although the notation there is very different. Similarly, for a two-step bottleneck, when the parameters prior to the sample were θ_1 for t_1 generations, θ_2 for t_2 generations, and θ_3 for $t_3 = \infty$ generations, only the $r = 0, 1$, and 2 terms in (2.2.10) and (2.2.11) are needed, and the expression (2.2.10) is equivalent to (4.1) in Watterson (1984a), although (2.2.10) is in a slightly simplified form.

For subsequent reference, we note that putting $s_1 = s_2 = \cdots = s_{l_0} = 1$ in (2.2.8) yields the identity

$$\text{coefficient of } \prod_{k=0}^{r-1} \phi_k^{l_k - l_{k+1}} \text{ in } \exp\left[\sum_{l=1}^{l_0} g_l(\phi_0, \phi_1, \ldots, \phi_{r-1})/l \right]$$

$$= 1 \bigg/ \left[\prod_{k=0}^{r-1} a_k(\mathbf{l}) \right], \tag{2.2.12}$$

which may be proved directly, of course.

2.3. Expected Frequency Spectrum

To save writing complicated equations, we shall discuss the expected frequency spectrum conditional on $\mathbf{L} = \mathbf{l}$ as per (2.2.3). Unconditional results are obtained by averaging over the distribution of $\mathbf{L}$. The (conditional) expected number of alleles having frequency j in the sample is $E(\beta_j | \mathbf{L} = \mathbf{l})$, for $j = 1, 2, \ldots, n$. This we can find by differentiating (2.2.8) with respect to s_j, at $s_1 = s_2 = \cdots = s_n = 1$. This yields

$$E(\beta_j | \mathbf{L} = \mathbf{l}) = \left[\prod_{k=0}^{r-1} a_k(\mathbf{l}) \right] \times \text{coefficient of } \prod_{k=0}^{r-1} \phi_k^{l_k - l_{k+1}} \text{ in } j^{-1}$$

$$\times g_j(\phi_0, \phi_1, \ldots, \phi_{r-1}) \exp \left[\sum_{l=1}^{l_0} g_l(\phi_0, \phi_1, \ldots, \phi_{r-1})/l \right].$$

Now, using (2.2.9) and the multinomial theorem, we have that

$$\text{coefficient of } \prod_{k=0}^{r-1} \phi_k^{n_k} \text{ in } g_j(\phi_0, \phi_1, \ldots, \phi_{r-1})$$

$$= \frac{j!}{n_0! n_1! \cdots n_{r-1}!} \sum_{k=0}^{r-1} (\theta_{k+1} - \theta_{k+2}) \left(\prod_{i=k+1}^{r-1} \delta_{n_i, 0} \right)$$

$$= b_j(\mathbf{n}), \text{ say,} \tag{2.3.1}$$

where $\mathbf{n} = (n_0, n_1, \ldots, n_{r-1})$ consists of nonnegative integers with

$$n_0 + n_1 + \cdots + n_{r-1} = j, \quad \text{and} \quad n_k \le l_k - l_{k+1}$$

for each k, and where $\delta_{n,0}$ is Kronecker's delta. Thus

$$E(\beta_j | \mathbf{L} = \mathbf{l}) = j^{-1} \left[\prod_{k=0}^{r-1} a_k(\mathbf{l}) \right] \sum_{\mathbf{n}} b_j(\mathbf{n}) \times \text{coefficient of } \prod_{k=0}^{r-1} \phi_k^{l_k - l_{k+1} - n_k} \text{ in}$$

$$\exp \left[\sum_{l=1}^{l_0} g_l(\phi_0, \phi_1, \ldots, \phi_{r-1})/l \right]$$

$$= j^{-1} \left[\prod_{k=0}^{r-1} a_k(\mathbf{l}) \right] \sum_{\mathbf{n}} b_j(\mathbf{n}) \left[\prod_{k=0}^{r-1} a_k^{-1}(\mathbf{l}, \mathbf{n}) \right] \tag{2.3.2}$$

where

$$a_k^{-1}(\mathbf{l}, \mathbf{n}) = a_k^{-1}(l_0 - n_0 - n_1 - \cdots - n_{r-1}, l_1 - n_1 - n_2 - \cdots - n_{r-1},$$

$$\ldots, l_{r-1} - n_{r-1})$$

$$= \frac{(l_{k+1} - n_{k+1} - n_{k+2} - \cdots - n_{r-1} + \theta_{k+1})_{(l_k - l_{k+1} - n_k)}}{(l_k - l_{k+1} - n_k)!} \tag{2.3.3}$$

by (2.2.6) and (2.2.12).

While (2.3.2) is fully explicit, subject to the definitions (2.2.6), (2.3.1), and (2.3.3), it is certainly rather unwieldy in general. Three special cases deserve attention, $j = 1$, $j = 2$, and $j = n$. These correspond respectively to "singly present" alleles, "doubly present" alleles (in the terminology of Maruyama and Fuerst 1985a), and alleles "fixed in the sample" (of which there could be either 0 or 1).

When $j = 1$, the sum over $\mathbf{n}$ in (2.3.2) can include only cases of the form $n_0 = 0, n_1 = 0, \ldots, n_{i-1} = 0, n_i = 1, n_{i+1} = 0, \ldots, n_{r-1} = 0,$ for $i = 0, 1, \ldots, r - 1$. For the ith such case,

$$b_1(\mathbf{n}) = \sum_{k=i}^{r-1} (\theta_{k+1} - \theta_{k+2}) = \theta_{i+1},$$

and

$$a_k^{-1}(\mathbf{l}, \mathbf{n}) = 1/a_k(\mathbf{l}) \qquad\qquad\qquad \text{for } k > i$$
$$= (l_i - l_{i+1})/[(l_i + \theta_{i+1} - 1)a_i(\mathbf{l})] \qquad \text{for } k = i$$
$$= (l_{k+1} + \theta_{k+1} - 1)/[(l_k + \theta_{k+1} - 1)a_k(\mathbf{l})] \qquad \text{for } k < i.$$

So

$$E(\beta_1 \mid \mathbf{L} = \mathbf{l}) = \sum_{i=0}^{r-1} \theta_{i+1}(l_i - l_{i+1})$$
$$\left(\prod_{k=0}^{i-1} \frac{l_{k+1} + \theta_{k+1} - 1}{l_k + \theta_{k+1} - 1}\right) \Big/ (l_i + \theta_{i+1} - 1). \qquad (2.3.4)$$

Similarly it may be shown that

$$E(\beta_2 \mid \mathbf{L} = \mathbf{l}) = \sum_{i=0}^{r-1} \theta_{i+1}(l_i - l_{i+1})_{[2]}$$
$$\left[\prod_{k=0}^{i-1} \frac{(l_{k+1} + \theta_{k+1} - 1)_{[2]}}{(l_k + \theta_{k+1} - 1)_{[2]}}\right] \Big/ (l_i + \theta_{i+1} - 1)_{[2]}$$
$$+ 2 \sum_{i<j}\sum \theta_{j+1}(l_i - l_{i+1})(l_j - l_{j+1})(l_{i+1} + \theta_{i+1} - 1)$$
$$\times \left[\prod_{k=0}^{i-1} \frac{(l_{k+1} + \theta_{k+1} - 1)_{[2]}}{(l_k + \theta_{k+1} - 1)_{[2]}}\right]\left[\prod_{k=i+1}^{j-1} \frac{(l_{k+1} + \theta_{k+1} - 1)}{(l_k + \theta_{k+1} - 1)}\right] \Big/$$
$$[(l_i + \theta_{i+1} - 1)_{[2]}(l_j + \theta_{j+1} - 1)], \qquad (2.3.5)$$

and that

$$E(\beta_n \mid \mathbf{L} = \mathbf{l}) = n^{-1}\left[\prod_{k=0}^{r-1} a_k(\mathbf{l})\right]\theta_r \frac{n!}{(l_0 - l_1)!(l_1 - l_2)! \cdots l_{r-1}!}$$
$$= \theta_r(n - 1)! \prod_{k=0}^{r-1} [1/(l_{k+1} + \theta_{k+1})_{(l_k - l_{k+1})}]. \qquad (2.3.6)$$

In (2.3.4), (2.3.5), and (2.3.6), any empty products (e.g., $\prod_{k=0}^{-1}$) are interpreted as 1.

For unconditional moments, (2.3.3)–(2.3.6) should be averaged over the distribution of $\mathbf{L}$. When this is done, (2.3.6) becomes equivalent to the special case $\beta_1 = \beta_2 = \cdots = \beta_{n-1} = 0$, $\beta_n = 1$ of (2.2.11), because $E(\beta_n) = P(\beta_n = 1)$, the probability that the sample contains one allele only. This allele must have been present in all of the most remote ancestors at $t_1 + t_2 + \cdots + t_{r-1}$ ago. Special cases of (2.3.3)–(2.3.6) are consistent with results found in Watterson (1984a) for sampling a population after one-step or two-step bottlenecks.

The total number, $K = \sum_{j=1}^{n} \beta_j$, of alleles in the sample has conditional p.g.f.,

$$E(s^K | \mathbf{L} = \mathbf{l}) = \left[\prod_{i=0}^{r-1} a_i(\mathbf{l}) \right] \text{coefficient of} \prod_{i=0}^{r-1} \phi_i^{l_i - l_{i+1}} \text{ in}$$

$$\exp\left[s \sum_{l=1}^{l_0} g_l(\phi_0, \phi_1, \ldots, \phi_{r-1})/l \right]$$

$$= \prod_{i=0}^{r-1} \left[(l_{i+1} + \theta_{i+1}s)_{(l_i - l_{i+1})}/(l_{i+1} + \theta_{i+1})_{(l_i - l_{i+1})} \right], \quad (2.3.7)$$

by (2.2.6) and (2.2.12), which generalizes (4.4) in Watterson (1984a), and (2.1.6) above. Recall that we assume $l_r = 0$.

The probability distribution for K is

$$P(K = k | \mathbf{L} = \mathbf{l}) = \left[\prod_{i=0}^{r-1} a_i(\mathbf{l}) \right] \text{coefficient of} \prod_{i=0}^{r-1} \phi_i^{l_i - l_{i+1}} \text{ in}$$

$$\left[\sum_{l=1}^{l_0} g_l(\phi_0, \phi_1, \ldots, \phi_{r-1})/l \right]^k \Big/ k!. \quad (2.3.8)$$

This expression is not particulary attractive, although it simplifies to (2.1.7) if $r = 1$, and to Watterson [1984a, (3.5)] for $r = 2$.

The conditional expected number of alleles in the sample is, from (2.3.7),

$$E(K | \mathbf{L} = \mathbf{l}) = \sum_{i=0}^{r-1} \sum_{j=l_{i+1}+1}^{l_i} \frac{\theta_{i+1}}{\theta_{i+1} + j - 1}. \quad (2.3.9)$$

Again, this generalizes (4.7) in Watterson (1984a).

The conditional homozygosity, calculated as $E(\sum_{j=1}^{n} j^2 \beta_j | \mathbf{L} = \mathbf{l})$ could, of course, be written down from (2.3.2). However, it is probably easier in practice to find such quantities by simulation, as we now describe.

2.4. Simulation of Samples

To simulate the allelic composition of a sample of $l_0 = n$ genes, we first simulate a realization $\mathbf{l}$ of $\mathbf{L}$. This can be achieved in various ways. Perhaps the simplest method to program (if not the quickest to run on a computer) is to use the representation (2.2.2) in succession. If $l_0, l_1, l_2, \ldots, l_{k-1}$ have already been found, with $l_{k-1} > 0$, the next value l_k can be computed as

$$l_k = \sup\left\{j: \sum_{l=j}^{l_{k-1}} T_l > t_k\right\},$$

where $T_0 = \infty$ and, for $l = 1, 2, \ldots, l_{k-1}$,

$$T_l = -4N_k \ln(U_l)/[l(l + \theta_k - 1)],$$

in which $U_j, U_{j+1}, \ldots, U_{l_{k-1}}$ are independent random variables uniformly distributed on $[0, 1]$. Should it turn out that $l_k = 0$ (so that $r = k$ in our previous notation), no further values need be computed.

Having found **l**, as in (2.2.3) for instance, we now proceed to simulate the sample alleles. These will be chosen from among alleles labeled A_1, A_2, $A_3, \ldots$, starting with A_1 as oldest, but whenever a new mutant is called for, the next unused label along the list is chosen. For the first gene in the sample, we assign allelic type A_1. For the next $l_{r-1} - 1$ genes (if any), we proceed in succession. Suppose $j - 1$ of them have already had their allelic types allocated. Then for the jth gene, it will either be a new mutant [with probability $\theta_r/(\theta_r + j - 1)$] or it will be of allelic type possessed by a randomly chosen one of the $j - 1$ genes already allocated [each having probability $1/(\theta_r + j - 1)$ of being chosen]. This procedure applies for genes $j = 2, 3, \ldots, l_{r-1}$.

For genes $(l_{r-1} + 1), \ldots, l_{r-2}$, the scheme is continued except that for the jth gene, it will be a new mutant [with probability $\theta_{r-1}/(\theta_{r-1} + j - 1)$] or will copy the allelic type of one of the previous genes, $1, 2, \ldots, j - 1$ [each having probability $1/(\theta_{r-1} + j - 1)$ of being copied].

In general, for genes $j = (l_k + 1), \ldots, l_{k-1}$, in the sample, the jth gene will be a new mutant [with probability $\theta_k/(\theta_k + j - 1)$] or will copy the allelic type of one of the previous $j - 1$ genes [each having probability $1/(\theta_k + j - 1)$ of being copied].

And finally, the genes $j = (l_1 + 1), (l_1 + 2), \ldots, l_0$ making up the remainder of the sample of size l_0 $(= n)$ are such that the jth is a new mutant [with probability $\theta_1/(\theta_1 + j - 1)$], or it copies the type of one of the $j - 1$ previous genes [each with probability $1/(\theta_1 + j - 1)$].

Once the alleles have been allocated to the $l_0 = n$ genes, it is then trivial to count up the number of alleles, β_j, having j representatives in the sample, for $j = 1, 2, \ldots, n$. These β_j have probability distribution (2.2.11), in which the age-order of occurrence of alleles in the sample, and the precise allocation of the alleles to the frequencies is disregarded.

The justification for the above simulation method is briefly as follows. The first l_{r-1} genes in the sample can be thought of as being nonmutant descendants from the l_{r-1} ancestors at time $t_1 + t_2 + \cdots + t_{r-1}$ ago. These ancestors have their allelic types consistent with Ewens' distribution (2.2.4). It is known (see Hoppe 1984 and Watterson 1984b) that the scheme proposed above is correct for such a sample. In general, for genes $j = (l_{k+1} + 1), (l_{k+1} + 2), \ldots, l_k$, these can be thought of as nonmutant descendants from $l_k - l_{k+1}$ of the l_k ancestors of the sample at time $t_1 + t_2 + \cdots + t_k$. (The other l_{k+1} of those ancestors can be thought of as having types already allocated, with respect to the l_{k+1} ancestors at time $t_1 + \cdots + t_{k+1}$). The proposed method of simulating allelic types for these extra $l_k - l_{k+1}$ genes uses the mutation parameter θ_{k+1}, which is current between times $t_1 + t_2 + \cdots + t_k$ and $t_1 + t_2 + \cdots + t_{k+1}$ ago. And the method can be shown to be consistent with the transition probabilities generated by (2.2.5). We will omit the detailed proof of this assertion. However, it is clear that the method is at least consistent with (2.3.9) for the expected number of different alleles in the sample, and indeed consistent with (2.3.7), which has an interpretation as a convolution of Bernoulli trials for the production of new mutants in the sample; cf. Watterson [1984a, (3.3) and (4.5)].

2.5. Periodic Bottlenecks

As an example of the use of the simulation method to calculate moments for sample statistics, we discuss the case of a population that has gone through a periodic succession of bottlenecks. The population size has varied between two values N_1 and N_2, for respective lengths of time t_1 and t_2 generations. In terms of our previous notation, we will assume that for "odd" phases of the cycle we have

$$N_1 = N_3 = N_5 = \ldots, t_1 = t_3 = t_5 = \ldots, \theta_1 = \theta_3 = \theta_5 = \ldots,$$

and for "even" phases,

$$N_2 = N_4 = N_6 = \ldots, t_2 = t_4 = t_6 = \ldots, \theta_2 = \theta_4 = \theta_6 = \ldots.$$

The mutation rates $u_1, u_2, u_3, \ldots$ are assumed to be all equal, say, to u.

Maruyama and Fuerst [1985b, (6)] give the following result for the expected homozygosity in a *population* at the end of an "odd" phase:

$$J = \frac{1}{1 - e^{-\lambda_1 t_1 - \lambda_2 t_2}} \left[\frac{1}{(\theta_2 + 1)} (1 - e^{-\lambda_2 t_2}) e^{-\lambda_1 t_1} + \frac{1}{(\theta_1 + 1)} (1 - e^{-\lambda_1 t_1}) \right],$$

(2.5.1)

where $\lambda_1 = (\theta_1 + 1)/2N_1$, $\lambda_2 = (\theta_2 + 1)/2N_2$.

For purposes of illustration, we take one hundred replicate samples each of size $n = 100$ genes, simulated according to the scheme in Section

2.4, and calculate for each sample the homozygosity

$$\hat{F} = \sum_{j=1}^{n} j^2 \beta_j / n^2, \qquad (2.5.2)$$

the number of singly present alleles $\beta(1)$, and the total number of alleles

$$K = \sum_{j=1}^{n} j\beta_j. \qquad (2.5.3)$$

For comparison with J, it is better to modify $\hat{F}$ to be the probability that two genes, chosen at random from the sample *without replacement*, are of the same allelic type. We denote this statistic by

$$\tilde{F} = \sum_{j} j(j-1)\beta(j) / [n(n-1)]$$
$$= (n\hat{F} - 1)/(n-1). \qquad (2.5.4)$$

We also record the number, r, such that the most remote ancestor of our sample arose as a mutant between generations $t_1 + t_2 + \cdots + t_{r-1}$ and $t_1 + t_2 + \cdots + t_{r-1} + t_r$ ago.

The means and standard deviations of $\tilde{F}$, $\beta(1)$, K, and r are given in Table 2.1, estimated from the one hundred replicates, under sixteen different sets of time parameters. Following Maruyama and Fuerst (1985b, Fig. 1), we take $u = 10^{-6}$, $N_1 = 5 \times 10^7$ and $N_2 = 5 \times 10^3$, so that $\theta_1 = 200$ and $\theta_2 = 0.02$, for all sixteen cases. However, we consider the various combinations of values $\lambda_1 t_1$, $\lambda_2 t_2 = 0.01, 0.1, 1.0, 10$. Also in Table 2.1, we show values for J as in (2.5.1), for comparison with the mean $\tilde{F}$ values.

From Table 2.1 we might expect that because θ_1 is very high, the homozygosity in the sample would be low. While this is correct (for both $\tilde{F}$ and J) when $\lambda_1 t_1$ is large relative to $\lambda_2 t_2$, the homozygosity is high in the opposite case when $\lambda_1 t_1$ is small relative to $\lambda_2 t_2$. This latter effect is because the high homozygosity at time t_1, the end of the low θ_2 phase, has not had time to dissipate by time $t = 0$, in spite of the high θ_1 value. If we had interchanged the values θ_1 and θ_2 so that the sample was taken at the end of the low θ phase, our conclusions would have been reversed; see Maruyama and Fuerst [1985b, (5)].

In Table 2.1, we see that there is very good agreement between sample means for $\tilde{F}$ and the population value J. However, with samples of one hundred genes, there is still a considerable variation in $\tilde{F}$, with its standard deviation being proportionately greatest for low $\tilde{F}$ values.

As is to be expected, the number of singly present alleles and the total number of alleles are high when the recent high mutation parameter θ_1 has been in effect for large $\lambda_1 t_1$ values.

Finally, in Table 2.1 we see that the oldest allelic types in the sample can be traced back through many cycles when $\lambda_1 t_1 = \lambda_2 t_2 = 0.01$, but to

Table 2.1

Sample characteristics after periodic bottlenecks.

$\lambda_1 t_1$	$\lambda_2 t_2$	$\tilde{F}$	(s.d.)	J	$\beta(1)$	(s.d.)	K	(s.d.)	r	(s.d.)
		Homozygosity			**Singly present alleles**		**Total alleles**		**Most remote ancestor**	
0.01	0.01	0.478	(0.223)	0.490	1.2	(1.0)	5.5	(1.9)	553.6	(286.1)
	0.1	0.857	(0.196)	0.887	0.6	(0.7)	2.1	(0.9)	284.7	(207.5)
	1.0	0.969	(0.086)	0.965	0.5	(0.7)	1.6	(0.7)	117.7	(99.0)
	10.0	0.969	(0.085)	0.971	0.6	(0.7)	1.6	(0.7)	24.7	(25.6)
0.1	0.01	0.090	(0.044)	0.089	11.8	(3.6)	26.4	(4.3)	118.2	(45.6)
	0.1	0.476	(0.194)	0.468	4.7	(2.2)	8.5	(2.6)	63.3	(36.4)
	1.0	0.852	(0.127)	0.841	4.5	(2.0)	5.8	(2.0)	34.7	(29.2)
	10.0	0.899	(0.067)	0.888	4.5	(2.1)	5.6	(2.2)	12.4	(9.9)
1.0	0.01	0.011	(0.003)	0.011	52.2	(6.5)	70.1	(4.5)	18.1	(5.6)
	0.1	0.055	(0.023)	0.056	35.4	(5.1)	47.7	(4.9)	13.1	(4.5)
	1.0	0.269	(0.094)	0.267	33.0	(5.0)	37.8	(4.9)	7.7	(4.0)
	10.0	0.361	(0.076)	0.364	32.5	(5.1)	36.7	(5.0)	5.0	(2.7)
10.0	0.01	0.0051	(0.0016)	0.0050	66.5	(6.9)	80.9	(4.1)	1.7	(1.0)
	0.1	0.0051	(0.0015)	0.0050	66.5	(6.2)	81.0	(4.0)	1.8	(1.1)
	1.0	0.0051	(0.0016)	0.0050	67.0	(7.0)	81.2	(4.3)	1.8	(1.0)
	10.0	0.0050	(0.0016)	0.0050	66.6	(7.4)	81.1	(4.5)	1.7	(1.0)

less than one complete period when $\lambda_1 t_1 = \lambda_2 t_2 = 10.00$. However, when $N_1 = 5 \times 10^7$ and $N_2 = 5 \times 10^3$, these inferences are deceptive. In fact, in Table 2.1, the first value, $r = 553.6$, corresponds to about $0.01(277\lambda_1^{-1} + 276.6\lambda_2^{-1}) = 1{,}405{,}227$ generations, whereas the last value, $r = 1.7$, corresponds to $10(\lambda_1^{-1} + 0.7\lambda_2^{-1}) = 5{,}043{,}752$ generations.

From Table 2.1 it seems that, provided the mutation parameter $\theta_1 = 200$ was in effect for a $\lambda_1 t_1 = 10$ time interval, it matters hardly at all what happened prior to that phase. The previous bottlenecks do not seem to influence $\tilde{F}$, $\beta(1)$, K, or r.

References

Ewens, W. J. 1972. The sampling theory of selectively neutral alleles. *Theor. Pop. Biol.* 3: 87–112.

Griffiths, R. C. 1980. Lines of descent in the diffusion approximation of neutral Wright-Fisher models. *Theor. Pop. Biol.* 17: 37–50.

Hoppe, F. M. 1984. Pólya-like urns and the Ewens' sampling formula. *J. Math. Biol.* 20: 91–94.

Karlin, S., and McGregor, J. 1972. Addendum to a paper of W. Ewens. *Theor. Pop. Biol.* 3: 113–116.

Kelly, F. P. 1977. Exact results for the Moran neutral allele model. *Adv. Appl. Prob.* 9: 197–201.

Kingman, J.F.C. 1982. The coalescent. *Stoch. Proc. & Applns.* 13: 235–248.

Maruyama, T., and Fuerst, P. A. 1985a. Population bottlenecks and nonequilibrium models in population genetics. II. Number of alleles in a small population derived from a large steady-state population by means of a bottleneck. *Genetics* 111: 675–689.

Maruyama, T., and Fuerst, P. A. 1985b. Population bottlenecks and nonequilibrium models in population genetics. III. Genic homozygosity in populations which experience periodic bottlenecks. *Genetics* 111: 691–703.

Tavaré, S. 1984. Line-of-descent and genealogical processes, and their applications in population genetics models. *Theor. Pop. Biol.* 26: 119–164.

Watterson, G. A. 1974. The sampling theory of selectively neutral alleles. *Adv. Appl. Prob.* 6: 463—488.

Watterson, G. A. 1984a. Allele frequencies after a bottleneck. *Theor. Pop. Biol.* 26: 387–407.

Watterson, G. A. 1984b. Estimating the divergence time of two species. Statistical Research Report 94, Monash University, Australia.

CHAPTER THREE

The Genealogy of the Birth, Death, and Immigration Process

Simon Tavaré

3.1. Introduction

It is indeed a pleasure for me to contribute to this dedicatory volume for Professor Samuel Karlin. Among Karlin's many contributions that address mathematical or statistical issues in the broad area of biology is a collection devoted to the analysis of a variety of stochastic processes that arise in the mathematical theory of population genetics. This theory is the most developed (and the most elegant) in the setting of the infinitely many neutral alleles models, and it is to such problems that this paper is addressed.

In a seminal paper, Karlin and McGregor (1967) describe the following model. Imagine families of individuals each initiated by a single individual at the time points of a renewal process. The size of each family fluctuates in time according to the probabilistic laws of a given stochastic process, different families evolving independently of each other. If we also interpret the families as novel mutant alleles, the model then describes the evolution of the genetic composition of a population of varying size. If all families are given distinct allelic labels, the process may be viewed as a version of the infinitely many neutral alleles model, neutrality here corresponding to the fact that each family is assumed to evolve with identical probabilistic structure.

A detailed description of such a process might include a study of the joint distribution of the numbers of families of different sizes at a given time t, and its asymptotic behavior as t increases. Not surprisingly, the most explicit results are available when the renewal process is Poisson, and the families fluctuate according to a linear birth-and-death process. The model just described is then a detailed version of the classical birth-and-death process with immigration. This particular model is one of the generating mechanisms for the well-known Ewens Sampling Formula.

This distribution, which has been the object of detailed study, was derived originally by Ewens (1972) in the context of a population model of fixed but large size; see also Karlin and McGregor (1972). In our variable population size setting, it arises as the joint distribution of the family size statistics, conditional on the total population size; see Watterson (1974) and Kendall (1975).

In fact, I will describe a richer class of processes which keep track of the sizes of families (or number of representatives of each distinct allele, in the population genetics setting) in the order of their appearance in the population. For studying many aspects of the age structure of the families, this approach seems to have some advantages over the unlabeled process described earlier. One consequence of this representation is an age-ordered version of the Ewens Sampling Formula that is a variable population-size analogue of an earlier result of Donnelly and Tavaré (1986).

In Section 3.2 are recorded the basic properties of the birth, death, and immigration process. Section 3.3 provides the probabilistic structure of the age-ordered family size process, and details the connection with size-biasing. Section 3.4 studies distributions of this process that are conditional on the total population size, and describes some connections with reversibility. The final section specializes to the case of the birth process with immigration. Its jump chain is intimately connected to the genealogical structure of the infinitely many neutral alleles model; see Watterson (1984), Hoppe (1984), Donnelly (1986). Its representation via the birth process with immigration provides a simple way to study its asymptotic properties.

3.2. The BDI Process

We begin with a brief discussion of the simple (linear) birth-and-death (BD) process. This is a time-homogeneous Markov process whose states are labeled by $0, 1, 2, \ldots$, and whose behavior is specified by two non-negative parameters:

λ, the birth rate per head per unit time

μ, the death rate per head per unit time.

The (stable, conservative, regular) Q-matrix of the process has elements determined by

$$q_{i,i+1} = i\lambda, \qquad i = 0, 1, 2, \ldots$$
$$q_{i,i-1} = i\mu, \qquad i = 1, 2, 3, \ldots,$$

the other off-diagonal elements being zero. The number of individuals $N(t)$ alive in the population at time t has a well-known distribution (cf. Kendall 1949), and we recall here that

$$P[N(t) = n \mid N(0) = 1] \equiv g_n(t) = \begin{cases} (1 - a_t)(1 - b_t)b_t^{n-1}, & n \geq 1 \\ a_t, & n = 0, \end{cases} \quad (3.1a)$$

where

$$b_t = \frac{\lambda \exp[(\lambda - \mu)t] - \lambda}{\lambda \exp[(\lambda - \mu)t] - \mu}$$

$$a_t = \mu b_t/\lambda, \qquad \text{if} \quad \lambda \neq \mu \qquad (3.1b)$$

and

$$a_t = b_t = \frac{\lambda t}{1 + \lambda t}, \qquad \text{if} \quad \lambda = \mu. \qquad (3.1c)$$

We will be interested in the birth-and-death process with immigration (BDI), which can be constructed from a sequence of independent BD processes as follows. Initially, we suppose there are no individuals in the population. At the time points $T_1, T_2, \ldots$ of a homogeneous Poisson process of rate θ, we initiate immigrant families, each starting from a single individual, and evolving independently as a BD process. As the BDI process evolves, families appear, fluctuate in size, and possibly become extinct. Often, we will not distinguish among families of the same size, and so most useful information is contained in the process

$$\{[\xi_0(t), \xi_1(t), \ldots], t \geq 0\},$$

in which $\xi_n(t)$ is the number of families which have n members at time t; $n = 1, 2, \ldots$, and $\xi_0(t)$ is the number of extinct families at time t if $t \geq T_1$, and 0 if $t < T_1$. Much of the stochastic structure of $\xi(t)$ is contained in Theorem 1 below.

THEOREM 1 (Karlin and McGregor 1967):

$$E \prod_{n=0}^{\infty} s_n^{\xi_n(t)} = \prod_{n=0}^{\infty} \exp\left\{ -\theta(1 - s_n) \int_0^t g_n(u)\, du \right\},$$

for $|s_n| \leq 1$, $n = 0, 1, 2, \ldots$ *. That is, the $\xi_n(t)$ are independent Poisson-distributed random variables with means*

$$E\xi_n(t) = \theta \int_0^t g_n(u)\, du.$$

It is convenient in the following to scale time so that $\lambda = 1$. Then the immigration rate θ and the death rate μ are the only free parameters, and it follows from (3.1) that

$$E\xi_n(t) = \theta b_t^n/n, \qquad n = 1, 2, \ldots.$$

Hence from THEOREM 1 we obtain

$$P[\xi_i(t) = m_i, i = 1, 2, \ldots] = (1 - b_t)^\theta b_t^{\Sigma rm_r} \prod_{r=1}^\infty \frac{1}{m_r!} \left(\frac{\theta}{r}\right)^{m_r}. \qquad (3.2)$$

There are a number of other statistics of interest in a BDI process. We mention

$$M(t) = \sum_{j=0}^\infty \xi_j(t), \text{ the total number of immigrations up to time } t;$$

$$F(t) = \sum_{j=1}^\infty \xi_j(t), \text{ the total number of families alive at time } t;$$

$$I(t) = \sum_{j=1}^\infty j\xi_j(t), \text{ the total number of individuals alive at time } t.$$

It follows directly from THEOREM 1 that

$$Es^{F(t)}w^{I(t)} = (1 - b_t)^\theta(1 - wb_t)^{-\theta s},$$

so that $I(t)$ has a negative binomial distribution with

$$P[I(t) = n] = \binom{\theta + n - 1}{n}(1 - b_t)^\theta b_t^n, \qquad n = 0, 1, \ldots, \qquad (3.3)$$

while $F(t)$ has a Poisson distribution with mean $-\theta \ln(1 - b_t)$.

Further properties of the random variables $\{\xi_n(t)\}$ may be found in the papers of Karlin and McGregor (1967), Watterson (1974), and Kendall (1975). Kendall's work focuses on the reversibility of the process $\xi(.) \equiv \{[\xi_1(t), \xi_2(t), \ldots], t \geq 0\}$ in the case $\mu > 1$; see Section 3.4 for further details. We turn first, though, to a study of a related process that studies some aspects of the age structure of the families.

3.3. An Age-Ordered BDI Process

In this section we extend the analysis of the process $\xi(t)$ of Section 3.2 by keeping track of the size of extant families *in the order of their appearance*. Recalling that new families arise at the points $T_1, T_2, \ldots$ of a Poisson process of rate θ, we say that a family that originated at time T_r with at least one member alive at time t is older than a family extant at time t that originated at time T_s if $r < s$. We will also keep track of the number $F(t)$ of families that survive at time t.

The states z of our new time-homogeneous Markov process $\{A(t), t \geq 0\}$ will be either of the form

$z = (0)$, if no families survive at time t [i.e., $F(t) = 0$]; or

$z = (l; \mu_1, \ldots, \mu_l)$, if l families survive at time t, and the oldest family has μ_1 members, the second oldest μ_2 members, . . . , the youngest μ_l members. It is implicit that $\mu_1 \geq 1, \ldots, \mu_l \geq 1$ for such a z.

It is a simple matter to compute the transition rates q_{zw} from state z to state w:

Immigrations
(a) If $z = (l; \eta_1, \ldots, \eta_l)$, $w = (l + 1; \eta_1, \ldots, \eta_l, 1)$, and $l \geq 1$, then $q_{zw} = \theta$.
(b) If $z = (0)$, and $w = (1; 1)$, then $q_{zw} = \theta$.

Births
If $z = (l; \eta_1, \ldots, \eta_l)$, $w = (l; \eta_1, \ldots, \eta_{i-1}, \eta_i + 1, \eta_{i+1}, \ldots, \eta_l)$, and $l \geq 1$, then $q_{zw} = \eta_i \lambda$, $i = 1, \ldots, l$.

Deaths
(a) If $z = (l; \eta_1, \ldots, \eta_l)$, $w = (l; \eta_1, \ldots, \eta_{i-1}, \eta_i - 1, \eta_{i+1}, \ldots, \eta_l)$, and $\eta_i > 1$, then $q_{zw} = \eta_i \mu$.
(b) If $z = (l; \eta_1, \ldots, \eta_{i-1}, 1, \eta_{i+1}, \ldots, \eta_l)$, $w = (l - 1; \eta_1, \ldots, \eta_{i-1}, \eta_{i+1}, \ldots, \eta_l)$, then $q_{zw} = \mu$.
(c) If $z = (1; 1)$, $w = (0)$, then $q_{zw} = \mu$.

In this treatment we are not interested in keeping track of families that have existed but are now extinct, although this may be of interest elsewhere. We also note that off-diagonal elements of the Q-matrix not specified above are zero, and the diagonal elements are determined by the requirement that the process be conservative. It also follows by an obvious modification of Kendall's argument (1975, p. 336) that this Q-matrix is also regular; it is clearly irreducible.

What is interesting is that the distribution of $A(t)$ can be found explicitly.

THEOREM 2 *With time standardized so that* $\lambda = 1$, *and* $A(0) = (0)$,

$$P[A(t) = (0)] = (1 - b_t)^\theta,$$

while for $l \geq 1, \eta_1 \geq 1, \ldots, \eta_l \geq 1,$

$$P[A(t) = (l; \eta_1, \ldots, \eta_l)] = \frac{(1 - b_t)^\theta b_t \sum \eta_i \theta^l}{\eta_l(\eta_l + \eta_{l-1}) \cdots (\eta_l + \cdots + \eta_1)}. \quad (3.4)$$

Proof: The Q-matrix of $A(.)$ is conservative, stable, and regular, so that the forward equations have a unique solution that specifies a probability distribution. We will therefore verify the equations

$$\frac{d}{dt} p_w(t) = \sum_z p_z(t) q_{zw}, \quad (3.5)$$

where $p_w(t) = P[A(t) = w]$. Suppose first that $w = (l; \eta_1, \ldots, \eta_l)$ and $l > 0$. The left-hand side of (3.5) is

$$(b_t\mu - 1)(\theta b_t + r(1 - b_t))b_t^{-1} p_w(t), \quad (3.6)$$

where $\eta_1 + \cdots + \eta_l = r$. Now consider all the paths that lead from z to w in a single change of state. There are two cases to consider.

CASE 1: $\mu_l > 1$

The possible transitions are

(a) $z = (l + 1; \eta_1, \ldots, \eta_i, 1, \eta_{i+1}, \ldots, \eta_l)$, for which $q_{zw} = \mu$, $i = 1, \ldots, l$,

and

$z = (l + 1; 1, \eta_1, \ldots, \eta_l)$, for which $q_{zw} = \mu$.

(b) $z = (l; \eta_1, \ldots, \eta_i + 1, \ldots, \eta_l)$, for which $q_{zw} = (\eta_i + 1)\mu$, $i = 1, \ldots, l$.

(c) $z = (l; \eta_1, \ldots, \eta_i - 1, \ldots, \eta_l)$, for which $q_{zw} = (\eta_i - 1)$, $i = 1, \ldots, l$.

Using (3.4), we see that

$$\sum_{z \neq w} p_z(t) q_{zw}$$

$$= (1 - b_t)^\theta b_t^{r+1} \theta^{l+1} \mu \left\{ \frac{1}{(1 + \eta_l)(1 + \eta_l + \eta_{l-1}) \cdots (1 + \eta_l + \cdots + \eta_1)} \right.$$

$$+ \frac{1}{\eta_l(\eta_l + 1)(\eta_l + \eta_{l-1} + 1) \cdots (1 + \eta_l + \cdots + \eta_1)} + \cdots$$

$$+ \frac{1}{\eta_l(\eta_l + \eta_{l-1}) \cdots (\eta_l + \cdots + \eta_1)(1 + \eta_l + \cdots + \eta_1)} \Bigg\}$$

$$+ (1 - b_t)^\theta b_t^{r+1} \theta^l \mu \times \left\{ \sum_{i=1}^l \frac{1}{\substack{\eta_l(\eta_l + \eta_{l-1}) \cdots (\eta_l + \cdots + \eta_{i+1}) \\ \times (\eta_l + \cdots + \eta_i + 1) \cdots (\eta_l + \cdots + \eta_1 + 1)}} \right\}$$

$$+ (1 - b_t)^\theta b_t^{r-1} \theta^l \times \left\{ \sum_{i=1}^l \frac{\eta_i - 1}{\substack{\eta_l(\eta_l + \eta_{l-1}) \cdots (\eta_l + \cdots + \eta_{i+1}) \\ \times (\eta_l + \cdots + \eta_i - 1) \cdots (\eta_l + \cdots + \eta_1 - 1)}} \right\}.$$

A little algebra reduces this to

$$= (1 - b_t)^\theta b_t^{r+1} \theta^{l+1} \mu \times \frac{1}{\eta_l(\eta_l + \eta_{l-1}) \cdots (\eta_l + \cdots + \eta_1)}$$

$$+ (1 - b_t)^\theta b_t^{r+1} \theta^l \mu \times \frac{r}{\eta_l(\eta_l + \eta_{l-1}) \cdots (\eta_l + \cdots + \eta_1)}$$

$$+ (1 - b_t)^\theta b_t^{r-1} \theta^l \times \frac{r}{\eta_l(\eta_l + \eta_{l-1}) \cdots (\eta_l + \cdots + \eta_1)}$$

$$= p_w(t)[\mu b_t(\theta + r) + r/b_t].$$

Since $q_{ww} = -r(1 + \mu) - \theta$, the right-hand side of (3.5) becomes

$$p_w(t)[\mu b_t(\theta + r) + r/b_t - r(1 + \mu) - \theta]$$
$$= p_w(t)b_t^{-1}[\mu b_t^2(\theta + r) + r - rb_t(1 + \mu) - \theta b_t]$$
$$= p_w(t)b_t^{-1}(\theta b_t + r(1 - b_t))(b_t \mu - 1)$$
$$= \text{left-hand side of (3.5), and we are done.}$$

Case 2: $\mu_l = 1$

In this case the possible states and their transition rates are given by (a) and (b) above, but (c) must be replaced by

(c') $z = (l; \eta_1, \ldots, \eta_i - 1, \ldots, \eta_{l-1}, 1)$, for which $q_{zw} = (\eta_i - 1)$, $i = 1, \ldots, l - 1$,

and

$z = (l - 1; \eta_1, \ldots, \eta_{l-1})$, for which $q_{zw} = \theta$.

But notice that

$$\sum_{z \text{ of type (c')}} p_z(t)q_{zw}$$

$$
= (1 - b_t)^\theta b_t^{r-1} \theta^l \sum_{i=1}^{l-1} \frac{\eta_i - 1}{1(1 + \eta_{l-1}) \cdots (1 + \eta_{l-1} + \cdots + \eta_{i+1})}
$$
$$
\times (\eta_{l-1} + \cdots + \eta_i) \cdots (\eta_{l-1} + \cdots + \eta_1)
$$

$$
+ (1 - b_t)^\theta b_t^{r-1} \theta^{l-1} \times \frac{1}{\eta_{l-1}(\eta_{l-1} + \eta_{l-2}) \cdots (\eta_{l-1} + \cdots + \eta_1)} \times \theta
$$

$$
= (1 - b_t)^\theta b_t^{r-1} \theta^l \times \left\{ \frac{1}{\eta_{l-1}(\eta_{l-1} + \eta_{l-2}) \cdots (\eta_{l-1} + \cdots + \eta_1)} \right.
$$

$$
+ \frac{\eta_{l-1} - 1}{\eta_{l-1}(\eta_{l-1} + \eta_{l-2}) \cdots (\eta_{l-1} + \cdots + \eta_1)}
$$

$$
+ \frac{\eta_{l-2} - 1}{(1 + \eta_{l-1})(\eta_{l-1} + \eta_{l-2}) \cdots (\eta_{l-1} + \cdots + \eta_1)} + \cdots
$$

$$
\left. + \frac{\eta_1 - 1}{(1 + \eta_{l-1})(1 + \eta_{l-1} + \eta_{l-2}) \cdots (1 + \eta_{l-1} + \cdots + \eta_2)(\eta_{l-1} + \cdots + \eta_1)} \right\}
$$

$$
= (1 - b_t)^\theta b_t^{r-1} \theta^l \times \left\{ \frac{1}{(\eta_{l-1} + \eta_{l-2}) \cdots (\eta_{l-1} + \cdots + \eta_1)} \right.
$$

$$
+ \frac{\eta_{l-2} - 1}{(1 + \eta_{l-1})(\eta_{l-1} + \eta_{l-2}) \cdots (\eta_{l-1} + \cdots + \eta_1)} + \cdots
$$

$$
\left. + \frac{\eta_1 - 1}{(1 + \eta_{l-1})(1 + \eta_{l-1} + \eta_{l-2}) \cdots (\eta_{l-1} + \cdots + \eta_1)} \right\}
$$

$$
= (1 - b_t)^\theta b_t^{r-1} \theta^l
$$

$$
\times \left\{ \frac{1}{(1 + \eta_{l-1})(\eta_{l-1} + \eta_{l-2} + \eta_{l-3}) \cdots (\eta_{l-1} + \cdots + \eta_1)} + \cdots \right.
$$

$$
\left. + \frac{1}{(1 + \eta_{l-1})(1 + \eta_{l-1} + \eta_{l-2}) \cdots (\eta_{l-1} + \cdots + \eta_1)} \right\} = \cdots
$$

$$
= (1 - b_t)^\theta b_t^{r-1} \theta^l \times \frac{r}{(1 + \eta_{l-1})(1 + \eta_{l-1} + \eta_{l-2}) \cdots (1 + \eta_{l-1} + \cdots + \eta_1)}
$$

$$
= p_w(t) r / b_t.
$$

The remainder of the verification of (3.5) now proceeds just as in Case 1. The proof may be completed by checking the remaining case, which occurs when $l = 0$. This is easy, and the details are omitted.

It is worthwhile at this stage to comment on the relationship between the distribution of $A(t)$ specified by (3.4) and that of $\xi(t)$ given by (3.2).

Assume then that for a given $t > 0$, the extant family sizes $\eta_1, \ldots, \eta_l$ correspond to family size statistics $m_1, m_2, \ldots$ given by

$$m_i = \text{card}\{j: \eta_j = i\}, \qquad i = 1, 2, \ldots, \tag{3.7a}$$

satisfying

$$m_1 + m_2 + \cdots = l, \tag{3.7b}$$

and

$$m_1 + 2m_2 + 3m_3 + \cdots = \eta_1 + \eta_2 + \cdots + \eta_l. \tag{3.7c}$$

The distribution of $\xi(t)$ follows from that of $A(t)$ by a collapsibility argument. For $m_1, m_2, \ldots$ satisfying the conditions of (3.7), we see that

$$P[\xi_i(t) = m_i, i = 1, 2, \ldots] = \sum P[A(t) = (l; \eta_1, \eta_2, \ldots, \eta_l)],$$

the sum ranging over all $\eta_1, \ldots, \eta_l$ satisfying (3.7a). But this last is

$$\left(\prod_{r \geq 1} m_r!\right)^{-1} \sum P[A(t) = (l; \eta_{\pi(1)}, \ldots, \eta_{\pi(l)})],$$

where π denotes a permutation of $\{1, 2, \ldots, l\}$, and the combinatorial quantity arises because the sum over all permutations π counts the required terms $\prod m_r!$ times. Observing that

$$\sum_\pi \frac{1}{\eta_{\pi(l)}(\eta_{\pi(l)} + \eta_{\pi(l-1)}) \cdots (\eta_{\pi(1)} + \cdots + \eta_{\pi(l)})} = \frac{1}{\eta_1 \eta_2 \cdots \eta_l},$$

we obtain

$$P[\xi_i(t) = m_i, i \geq 1] = \frac{(1 - b_t)^\theta \theta^l b_t^{\Sigma \eta_i}}{\eta_1 \eta_2 \cdots \eta_l} \times \frac{1}{\prod_{i \geq 1} m_i!}$$

$$= (1 - b_t)^\theta b_t^{\Sigma r m_r} \prod_{r \geq 1} \left(\frac{\theta}{r}\right)^{m_r} \frac{1}{m_r!},$$

this last following because $\eta_1 \cdots \eta_l = 1^{m_1} 2^{m_2} \ldots$, and the relationships in (3.7) hold.

Conversely, there is a simple operational description of how the age-ordering arises from the process $\xi(.)$. This relies on size-biasing in the following way. Imagine that $\xi_i(t) = m_i, i \geq 1$, that $I(t) = n$, and that (3.7) holds. Randomly select an individual, and remove him and all members of his family. These are assigned the label 1. Next, randomly choose one of the remaining individuals, and remove him and all his family. This group is given the label 2. Continue in this way until the remaining family is labeled l. This assignment produces η_1 individuals with label 1, η_2 with label 2, ..., η_l with label l with conditional probability (given the family sizes m_i) of

$$\left(\prod_r m_r!\right) \frac{\eta_1}{n} \times \frac{\eta_2}{n - \eta_1} \times \cdots \times \frac{\eta_{l-1}}{\eta_l + \eta_{l-1}}$$

$$= \left(\prod_r m_r!\right) \frac{\eta_1 \eta_2 \cdots \eta_l}{n(\eta_2 + \cdots + \eta_l) \cdots (\eta_l + \eta_{l-1})\eta_l}.$$

Hence the unconditional probability of getting η_1 individuals with label 1, ..., η_l with label l is

$$(1 - b_t)^\theta b_t^{\sum r m_r} \frac{\theta^{\sum m_r}}{\left(\prod_r m_r!\right)\left(\prod_r r^{m_r}\right)} \left(\prod_r m_r!\right)$$

$$\times \frac{\eta_1 \cdots \eta_l}{\eta_l(\eta_l + \eta_{l-1}) \cdots (\eta_l + \cdots + \eta_1)},$$

which reduces to the distribution in (3.4) after simplification. Thus we see that the joint distribution of the number of individuals in the oldest family, ..., the youngest family is the same as that of the number of individuals labeled 1, 2, ... in the size-biased permutation.

3.4. Conditional Distributions and Reversibility

This section is divided into two parts, the first devoted to the structure of the processes $A(.)$ and $\xi(.)$ conditional on the total population size $I(.)$, and the second to reversibility.

3.4.1 CONDITIONAL DISTRIBUTIONS

In the sequel, we will let $P_n(.)$ denote conditional probabilities given that $I(t) = n$. It follows from (3.2) and (3.3) that

$$P_n[\xi_i(t) = m_i, i = 1, 2, \ldots, n] = \binom{\theta + n - 1}{n}^{-1} \prod_{j=1}^n \left(\frac{\theta}{j}\right)^{m_j} \frac{1}{m_j!}. \quad (3.8)$$

Notice that this distribution is independent of t (and μ), so that supression of the t in $P_n(.)$ is justified. The distribution (3.8) is known in the population genetics literature as the Ewens Sampling Formula (Ewens 1972). It was derived in the present context by Watterson (1974), and Kendall (1975).

The age-ordered version follows from (3.3) and (3.4):

$$P_n[A(t) = (l; n_1, \ldots, n_l)]$$

$$= \binom{\theta + n - 1}{n}^{-1} \frac{\theta^l}{n_l(n_l + n_{l-1}) \cdots (n_l + \cdots + n_1)}. \quad (3.9)$$

Donnelly and Tavaré (1986) showed that (3.9) arises as the distribution of an age-ordered sample of size n from a stationary Moran model of constant size, and also in the limit of large population size in a wide variety of other models.

Of course, (3.8) and (3.9) may be related by the same size-biasing argument described in the last section. This same idea can be applied in several other ways also. For example, conditional on $I(t) = n$, the probability that a given family of size i is the oldest is i/n, since this is the probability that this family of size i is the first chosen in the size-biased sample. Here is another example.

Let O denote the number of individuals in the oldest family. Then from (3.9) we obtain

$$P_n[F(t) = l; O = j] = \sum \frac{1}{(\eta_2 + \cdots + \eta_l) \cdots (\eta_{l-1} + \eta_l)\eta_l} \times \frac{\theta^l}{\theta_{(n)}} \frac{n!}{n},$$

where $\theta_{(n)} = \theta(\theta + 1) \cdots (\theta + n - 1)$, and the summation is over

$\eta_2, \ldots \eta_l \in \{1, 2, \ldots n - j\}$ with $\eta_2 + \cdots + \eta_l = n - j$. This last sum is easily effected, revealing that

$$P_n[F(t) = l; O = j] = \frac{\theta n! |S_{n-j}^{(l-1)}|}{n\theta_{(n)}(n - j)!}, \quad (3.10)$$

where $S_n^{(j)}$ is a Stirling number of the first kind. For given values of j and n, $F(t)$ may take any value between 1 and $n - j + 1$. Hence summing (3.10) over $l = 1, 2, \ldots, n - j + 1$ gives the marginal distribution of O:

$$P_n[O = j] = \frac{\theta n! \theta_{(n-j)}}{n\theta_{(n)}(n - j)!}; \quad (3.11)$$

See Kelly (1977). By size-biasing, it is clear that

$$P_n[O = j] = \frac{jE[\xi_j(t) | I(t) = n]}{n},$$

whence from (3.11), we obtain

$$E[\xi_j(t) | I(t) = n] = \frac{\theta n! \theta_{(n-j)}}{j\theta_{(n)}(n - j)!},$$

as found by Karlin and McGregor (1967), equation (4.13).

3.4.2 Limit Distributions and Reversibility

In this section I will describe some of the connections between the age-ordered BDI process and reversibility. Let's assume that the death rate μ is greater than the birth rate λ ($\equiv 1$ in the following). Any family that arises in the BDI must, with probability one, become extinct in finite time. It follows immediately from (3.4) that for any state $z = (l; \eta_1, \ldots, \eta_l)$,

$$\lim_{t \to \infty} P[A(t) = z \mid A(0) = (0)] \equiv \pi(z) = \frac{\left(1 - \dfrac{1}{\mu}\right)^{\theta}\left(\dfrac{1}{\mu}\right)^{\sum \eta_i} \theta^l}{\eta_l(\eta_l + \eta_{l-1}) \cdots (\eta_l + \cdots + \eta_1)}.$$

(3.12)

Since the state space is irreducible, it also follows that

$$\lim_{t \to \infty} P[A(t) = z \mid A(0) = w] = \pi(z)$$

for any w. The distribution $\{\pi(z)\}$ is also the invariant measure of the process $A(.)$, and we can if we wish arrange to extend the definition of the process to the whole time axis $(-\infty, \infty)$ in such a way that the homogeneity is preserved, and such that $P[A(t) = z] = \pi(z)$ for all times t and states z. Assume this has been done. The process $A(.)$ provides a rather detailed description of the appearance and disappearance of families through time. Keeping track of the numbers of families of different sizes (rather than their age order) can be achieved by grouping collections of states $z = (l; \eta_1, \ldots, \eta_l)$ of the $A(.)$ process which have the same family-size statistics $(m_1, m_2, \ldots)$ [cf. (3.7)]. This lumped process is precisely the stationary "FS-process" $\xi(.)$ discussed so eloquently by Kendall (1975).

Recall that an irreducible process with (regular, stable, and conservative) Q-matrix $\{q_{zw}\}$ is symmetrically reversible with respect to the measure $\pi(.)$ if and only if

$$\pi(z)q_{zw} = \pi(w)q_{wz} \qquad \text{for all } z, w.$$

(3.13)

Kendall showed *inter alia* that the process $\xi(.)$ is symmetrically reversible with respect to the measure π determined by letting $t \to \infty$ in (3.2). It is tempting to address questions about the ages of families (rather than just their age ordering) by exploiting this reversibility; the past history of a family should be stochastically similar to its future. However, from a realization of the family-size process *alone* this is not possible, since from it one cannot follow the progress of a given family from birth to extinction. What is needed is a more sophisticated, labeled process that allows families to be distinguished; cf. Watterson (1976) and Kelly (1979, p. 152) in the context of the Moran model. While the process $A(.)$ does give some information about age-ordering, there seems to be no direct way to use it to study the ages themselves. This process is clearly not reversible with respect to the measure π in (3.12) [take, for example, $z = (2; 2, 2)$ and $w = (2; 1, 2)$; the ratio of the two sides of (3.13) is 3/4] and its time reversal has a very complicated structure.

3.5. The Birth Process with Immigration

One special case of the BDI model is the linear birth process with immigration. The simplicity of the structure of this process allows the development of a much richer asymptotic theory than is presently known for the case allowing death. Details and further ramifications may be found elsewhere (Tavaré 1987). We will content ourselves here with an overview of the results.

Formally, the birth process with immigration is the special case of the BDI in which the death rate $\mu = 0$. Notice that since death is impossible, a family grows without limit as time increases and, in particular, can never go extinct. Since a Poisson process in reversed time is still a Poisson process (with the same rate, θ in this case), all questions about the ages themselves of families in the BDI process can be answered immediately. Here interest focuses on the asymptotic behavior of the sizes of these families as time increases.

Again, we will use the notation $A(t)$ to denote the value of the (age-ordered) process at time t, and we assume $A(0) = (0)$. We will assume in the sequel that time has been scaled so that the birth rate is $\lambda = 1$. From (3.4) with $\mu = 0$ it follows that

$$P[A(t) = 0] = e^{-\theta t},$$

$$P[A(t) = (l; \eta_1, \ldots, \eta_l)] = \frac{e^{-\theta t}(1 - e^{-t})^{\sum \eta_i} \theta^l}{\eta_l(\eta_l + \eta_{l-1}) \cdots (\eta_l + \cdots + \eta_1)}. \qquad (3.14)$$

Notice that the first term in (3.14) is just the probability that no events have occurred in the Poisson process in time t. We will also require the distribution of $I(t)$, the total number of individuals in the population at time t [and $I(0) = 0$]. This follows in the same way from (3.3) as

$$P[I(t) = n] = \binom{\theta + n - 1}{n} e^{-\theta t}(1 - e^{-t})^n, \qquad n = 0, 1, \ldots. \qquad (3.15)$$

We will now decompose the structure of the process $\{A(t), t \geq 0\}$ into its jump chain $\{J_n, n \geq 0\}$ and its time-scale process. To this end, define $\tau_0 = 0$, and let τ_n be the time of the nth change of state of the process. It will be convenient to let

$$S_n = \{(l; \mu_1, \ldots, \mu_l): \mu_1 + \cdots + \mu_l = n\}$$

be the subset of the state space of $A(.)$ with a total of n individuals, for $n = 0, 1, 2, \ldots$. The jump chain $J_0 = (0)$, $J_n = A(\tau_n +)$, $n = 1, 2, \ldots$ has one-step transition probabilities determined by

$$P[J_n = (l; \eta_1, \ldots, \eta_{i-1}, \eta_i + 1, \eta_{i+1}, \ldots, \eta_l) | J_{n-1} = (l; \eta_1, \ldots, \eta_l)]$$

$$= \frac{\eta_i}{n - 1 + \theta}, \qquad i = 1, 2, \ldots, l. \tag{3.16}$$

$$P[J_n = (l + 1; \eta_1, \ldots, \eta_l, 1) | J_{n-1} = (l; \eta_1, \ldots, \eta_l)] = \frac{\theta}{n - 1 + \theta},$$

if $(l; \eta_1, \ldots, \eta_l) \in S_{n-1}$.

It can be shown that $\{J_n, n \geq 0\}$ and $\{I(t), t \geq 0\}$ are independent stochastic processes, and since $A(t) = J_{I(t)}$, $t \geq 0$, it follows that

$$P[A(t) = (l; \eta_1, \ldots, \eta_l)] = \sum_{n \geq 0} P[J_n = (l; \eta_1, \ldots, \eta_l)] P[I(t) = n].$$

Hence,

$$P[J_n = (l; \eta_1, \ldots, \eta_l)] = \binom{\theta + n - 1}{n}^{-1} \frac{\theta^l}{\eta_l(\eta_l + \eta_{l-1}) \cdots (\eta_l + \cdots + \eta_1)} \tag{3.17}$$

if $(l; \eta_1, \ldots, \eta_l) \in S_n$.

The jump chain $\{J_n, n \geq 0\}$ is a Markov chain that arises in the study of the genealogy of the stationary infinitely many neutral alleles model. Hoppe (1984) and Watterson (1984) describe a Pólya-like urn with a transition mechanism similar to that of $\{J_n, n = 0, 1, \ldots\}$. They focus on the fact that the urn model gives rise to the Ewens sampling formula. Connections between their process and results on age ordering are developed and exploited in Donnelly (1986) and Donnelly and Tavaré (1986). See also Aldous (1985) for a related model.

Our attention is now directed to the asymptotic behavior of the family sizes in $A(t)$ as $t \to \infty$. To study this question recall that the new families arise at the points $T_1, T_2, \ldots$ of a homogeneous Poisson process of rate θ. We will change notation slightly, and define

$$\zeta_n(t) = \begin{cases} 0, & \text{if } T_n > t \\ \text{size of family initiated at time } T_n, & \text{if } T_n \leq t. \end{cases}$$

Now recall for a moment how a typical family evolves. A family initiated at time 0, say, with a single individual grows according to a linear birth process $[N(t), t \geq 0]$ of rate 1. It is a well-known fact about such a process that $e^{-t}N(t)$ converges almost surely as $t \to \infty$ to a random variable having an exponential distribution with mean 1. It follows from this structure that if $E_1, E_2, \ldots$ are independent and identically distributed random variables having the exponential distribution with mean 1, then

$$e^{-t}[\zeta_1(t), \zeta_2(t), \ldots] \longrightarrow (e^{-T_1}E_1, e^{-T_2}E_2, \ldots)$$

almost surely as $t \to \infty$.

There are several interesting consequences that flow from this result once the structure of the limit vector is uncovered. As an example, notice that since $T_1, T_2, \ldots$ are the points of a Poisson process of rate θ, and the $\{E_i\}$ are i.i.d., the collection $\{(T_i, E_i), i = 1, 2, \ldots\}$ may be identified as the points of a marked Poisson process. From this it can be shown that the points $\{E_i \exp(-T_i), i = 1, 2, \ldots\}$ may be viewed as the points (in some order) of a Poisson process on $(0, \infty)$ with mean measure density $\theta e^{-x}/x, x > 0$. Hence it follows that, almost surely as $t \to \infty$, $I(t)^{-1}[\zeta_1(t), \zeta_2(t), \ldots] \to (P_1, P_2, \ldots)$, say, and a calculation establishes that the random variables $\{P_i\}$ have the representation

$$P_i = (1 - Z_1)(1 - Z_2) \cdots (1 - Z_{i-1})Z_i, \qquad i = 1, 2, \ldots, \quad (3.17)$$

where the $\{Z_i\}$ are independent and identically distributed random variables with density $\theta(1 - x)^{\theta - 1}, x \in (0, 1)$. P_i is the (asymptotic) fraction of the population that belongs to the ith oldest family.

The decreasing-order statistics of the random variables $\{P_i\}$ have the Poisson-Dirichlet distribution with parameter θ; see Kingman (1975). Such random vectors arise not only in the population genetics setting (see Kingman 1980), but also in the context of species abundance models (McCloskey 1965 and Engen 1978, for example), and (when $\theta = 1$) in the theory of random permutations (Vershik and Shmidt 1977).

Acknowledgments

I would like to thank Dr. Peter Donnelly and Professor A. G. Pakes for critical reviews of earlier drafts of this paper. This research was supported in part by N.S.F. grant DMS 86-08857.

References

Aldous, D. J. 1985. Exchangeability and related topics. *Lecture Notes in Mathematics* 1117: 1–198. Springer-Verlag, New York.

Donnelly, P. J. 1986. Partition structures, Pólya urns, the Ewens sampling formula and the ages of alleles. *Theor. Pop. Biol.* 30: 271–288.

Donnelly, P. J., and Tavaré, S. 1986. The ages of alleles and a coalescent. *Adv. Appl. Prob.* 18: 1–19.

Engen, S. 1978. *Stochastic abundance models.* Halsted Press, New York.

Ewens, W. J. 1972. The sampling theory of selectively neutral alleles. *Theor. Pop. Biol.* 3: 87–112.

Hoppe, F. M. 1984. Pólya-like urns and the Ewens sampling formula. *J. Math. Biol.* 20: 91–94.

Karlin, S., and McGregor, J. 1967. The number of mutant forms maintained in a population. In *Proc. Fifth Berkeley Symposium on Mathematical Statistics and Probability*, ed. L. LeCam and J. Neyman, pp. 415–438. University of California Press, Berkeley.

Karlin, S., and McGregor, J. 1972. Addendum to a paper of W. Ewens. *Theor. Pop. Biol.* 3: 113–116.

Kelly, F. P. 1977. Exact results for the Moran neutral allele model. *Adv. Appl. Prob.* 9: 197–201.

Kelly, F. P. 1979. *Reversibility and Stochastic Networks.* Wiley, New York.

Kendall, D. G. 1949. Stochastic processes and population growth. *J. Roy. Statist. Soc. B* 11: 230–264.

Kendall, D. G. 1975. Some problems in mathematical genealogy. In *Perspectives in Probability and Statistics: Papers in honour of M. S. Bartlett*, pp. 325–345. Ed. J. Gani. Applied Probability Trust, Sheffield. Academic Press, London.

Kingman, J.F.C. 1975. Random discrete distributions. *J. Roy. Statist. Soc. B* 37: 1–22.

Kingman, J.F.C. 1977. The population structure associated with the Ewens sampling formula. *Theor. Pop. Biol.* 11: 274–283.

Kingman, J.F.C. 1980. *Mathematics of Genetic Diversity.* CBMS Regional Conference Series in Applied Math. 34. S.I.A.M., New York.

Kingman, J.F.C. 1982. The coalescent. *Stoch. Proc. Appl.* 13: 235–248.

McCloskey, J. W. 1965. A model for the distribution of individuals by species in an environment. Ph.D. thesis, Michigan State University.

Tavaré, S. 1987. The birth process with immigration, and the genealogical structure of large populations. *J. Math. Biol.* 25: 161–168.

Vershik, A. M., and Shmidt, A. A. 1977. Limit measures arising in the asymptotic theory of symmetric groups. I. *Theor. Prob. Applns.* 22: 70–85.

Watterson, G. A. 1974. The sampling theory of selectively neutral alleles. *Adv. Appl. Prob.* 6: 463–488.

Watterson, G. A. 1976. Reversibility and the age of an allele. 1. Moran's infinitely many neutral alleles model. *Theor. Pop. Biol.* 10: 239–253.

Watterson, G. A. 1984. Estimating the divergence time of two species. *Statistical Research Report* 94, Monash University, Australia.

CHAPTER FOUR

When Not to Use Diffusion Processes in Population Genetics

John H. Gillespie

Introduction

The use of diffusion processes to approximate the dynamics of population genetics models has yielded insights that would be unapproachable by exact methods. Wright (1931, 1945) laid the foundations with techniques that he developed *de novo* to obtain stationary densities and leading eigenvalues of processes that arise in population genetics. Building on Wright's results, Kimura (1955, 1964) pioneered work on the transient properties of diffusions through the use of the forward equation to obtain transient densities and through the backward equation to obtain fixation probabilities. The Australian group (Moran 1962; Watterson 1962; Ewens 1964, 1965) did the most important early work on waiting time problems and the assessment of the accuracy of diffusion approximations. Altogether, the successes of the diffusion approach are impressive, both in terms of the breadth of problems that have succumbed to a diffusion analysis and in the accuracy of the approximations.

There are, however, certain problems with the diffusion approach. The general aim of diffusion methodology in population genetics is to trade accuracy for tractability by approximating the dynamics of discrete time stochastic processes with continuous time diffusion processes. The approximations are usually very good when the genetic processes change little from one generation to the next. To assure this behavior, the parameters of the models must be delicately balanced. In certain cases, such as the strictly neutral model, this balance of parameters is very natural to the biological underpinnings of the model. In other cases, particularly for models involving natural selection or fluctuating parameters, the balance of parameters may not be a natural part of the underlying biology. In these instances the use of diffusion approximations may not be the best approach.

In this paper some particular problems with diffusion approximations in population genetics will be discussed to call attention to these problems rather than to provide solutions. Nonetheless, various preliminary observations may convince the reader that there are a plethora of stochastic models of real biological significance that are in need of the attention of mathematicians who are jaded by diffusions.

The Problem with Unequal Parameters

Karlin and McGregor (1964) were among the first to recognize clearly that diffusion processes are but a small subset of the stochastic processes that naturally appear when one attempts to approximate population genetics models. They pointed out that in using a diffusion process to approximate the dynamics of a stochastic model, one is implicitly making a limiting argument that involves certain parameters of the model approaching zero and a rescaling of time. The limiting argument is very sensitive to the rate with which the various parameters approach zero. If the parameters are properly balanced, then the limiting process will be a diffusion process.

For example, in a haploid population of size n with two alleles that mutate back and forth at the same rate, u, the mean and variance of the change in the frequency of one of the alleles is

$$\mathbf{E}\{\Delta x\} = 2u(1/2 - x)$$
$$\mathbf{Var}\{\Delta x^2\} \approx x(1 - x)/n.$$

(4.1)

In order to obtain a diffusion approximation of this model one usually assumes that u and $1/n$ are small and of the same order of magnitude and that time is measured in units of n generations. This yields the diffusion

$$\mathbf{E}\{dx\} = 2u(1/2 - x)\,dt$$
$$\mathbf{E}\{dx^2\} = x(1 - x)\,dt.$$

(4.2)

Karlin and McGregor (1964), among others (e.g., Watterson 1962; Moran 1962), noted that this diffusion can only be obtained rigorously as a limit of a sequence of processes if we assume that $u = O(1/n)$, measure time in units of n generations, and let $n \rightarrow \infty$.

The assumption that $u = O(1/n)$ seems strange at first glance. As pointed out by Sewall Wright in the discussion following the Karlin and McGregor (1964) paper, there is no biological reason why the mutation rate should depend on the population size. Karlin and McGregor's response was the obvious one: the question they were addressing was mathematical rather than biological. Mathematically we require $u = O(1/n)$ in order to obtain the limiting diffusion process (4.2). The biological message is clear: if the population that we are modeling does not have the property that $u \approx 1/n$, then the diffusion approximation may not be accurate. Karlin and McGregor pointed out that there are other assumptions that could be made about the magnitudes of u, $1/n$, and the scaling of time, and that these assumptions lead to limiting processes that are either not diffusions, or are diffusions that are other than (4.2). For example, if the mutation rate is fixed $[u = O(1)]$ and if time is not scaled, but the allele frequency is transformed by

$$y(t) = \sqrt{n}(x(t) - 1/2), \qquad (4.3)$$

then as $n \to \infty$, $y(t)$ approaches the Ornstein-Uhlenbeck process

$$\begin{aligned} \mathbf{E}\{dy\} &= -2uy\,dt \\ \mathbf{E}\{dy^2\} &= (1/4)\,dt. \end{aligned} \qquad (4.4)$$

Clearly this diffusion is very different than (4.2) and would apply to a population whose size is so large than $1/n \ll u$. Bacteria populations may well fit into this category even when u is the nucleotide mutation rate. (By recalling that there are 10^{10} pine trees in New Jersey alone [Harshberger 1970], we can see that other species may fit this as well.)

If we recognize that there are two candidate diffusions for one biological model, this opens the more general question as to the appropriateness of the "standard" diffusion processes of population genetics as models of real populations. The parameters that are used in diffusion models usually include

$1/n$—reciprocal of the population size
u—mutation rate
m—migration rate
s—selection coefficient
r—recombination rate between pairs of alleles
α—selfing rate

The common practice is to combine several of these parameters in a single model. For example, in the diffusion processes of the nearly neutral allele theory, the parameters $1/n$, u, and s are used. In so doing, several implicit assumptions are being made, the most critical of which is that $1/n$, s, and u are all small and of the same order of magnitude.

To think that in biological populations all of these parameters are of similar orders of magnitude is patently absurd. Thus many workers have explored models in which some parameters are much larger or smaller than others. For example, Slatkin (1981) has considered cases where $m \ll 1/n$, while Nagylaki (1980) has examined the case where $m \gg 1/n$. Papers that consider models where the disequilibrium is assumed to be zero may be assuming that $r \gg (s, 1/n)$. Gillespie (1983a,b) has examined strong-selection, weak-mutation limits, $u \ll 1/n \ll s$. Some of the limiting processes obtained with these assumptions are diffusions, the Nagylaki large migration limit, while others are not, for example, the strong-selection, weak-mutation limit. To my knowledge, no one has attempted a systematic survey of all of the limiting processes that are possible when one removes the assumption that all of the parameters are of similar orders of magnitude. At this point we do not even have a clear idea as to when the limiting processes will be diffusions and when they will be other processes. In this paper I will argue that for many situations of biological importance, diffusion limits may not be the appropriate ones.

There is another compelling reason to investigate some of the limiting processes obtained when different scalings are used, which is the issue of tractability. Whereas the diffusion processes that have been solved in population genetics have given us tremendous insights, it is fair to say that many of the most interesting evolutionary problems have remained refractory to the diffusion approach. Foremost among these are multiple allele and multiple locus problems with selection, mutation, and genetic drift. Thus, for example, the theory of molecular evolution by natural selection is still in its infancy when compared to the neutral allele theory simply because of the intractability of the mathematics. The preeminent position of the neutral allele theory as an explanation for molecular evolution may be due more to the simplicity of its mathematics than to its biological underpinnings.

Some progress has been achieved with nondiffusion limits under the strong-selection, weak-mutation (SSWM) limit. Here we assume that we have a standard k allele, constant fitness population genetics model. If $n \to \infty$, $s \to 0$, and $u \to 0$ such that $ns \to \infty$ and $nu \to 0$, then the limiting process appears to be a continuous time finite Markov chain where each state corresponds to an equilibrium (either stable or unstable) of the underlying deterministic process (Gillespie 1983b). From this Markov chain all of the waiting time properties that are required to describe certain models of molecular evolution may be obtained. This class of limiting processes seems to be particularly interesting from a biological point of view. Consider that the mutation rate to a particular allele will be the nucleotide mutation rate, which is of the order of 10^{-9}. Thus in all but the largest of populations one would reasonably expect that $u \ll 1/n$. On the other hand, for selection to dominate drift, $s \gg 1/n$ is required. Even for modest populations this means that selection coefficients could be as small as 10^{-5} to 10^{-4} and still have selection dominate drift.

SSWM limit processes have been most useful to describe fixation processes for multiple alleles. To date, two different models have been examined. In the first model we assume that there are m alleles at a particular haploid locus, and that the mutation rate from allele A_i to A_j is the same for all i and j. The fitnesses of the alleles are assigned at random from some arbitrary probability distribution and remain fixed through-out the process. The alleles are labeled such that allele A_1 is the most fit, followed by A_2, and so on to allele A_m, which is the least fit. In the case of strong selection and weak mutation, the population will be nearly fixed for one of the alleles except for relatively brief periods when a new allele is in the process of becoming fixed in the population. Thus we can declare that the population is in state k when the kth allele is the most frequent. Furthermore, the only fixations that are likely to occur are ones that replace a more fit allele with a less fit allele. Once the A_1 allele becomes fixed in the population, no further fixations are likely to occur. If the population begins with allele A_k as the fixed allele, then a random number of fixations will occur until allele A_1 finally becomes fixed. A question of some biological interest concerns the distribution of the number of fixations leading to the fixation of allele A_1.

This problem is completely intractable using diffusion approximations, but can be solved using the SSWM limit. The distribution of the number of fixations will be the same as the distribution of the number of iterations of the following algorithm.

Algorithm A1

1. Generate m positive IID random variables, $\mathbf{X}_i$, from some probability distribution that is in the domain of attraction of the Gumbel extreme value distribution. Rank the random variables such that $\mathbf{X}_1 > \mathbf{X}_2 > \cdots > \mathbf{X}_m$.
2. Set the initial state of the process, $(\mathbf{Z}, n)$, to $(Z_k, 0)$, with $k > 1$.
3. Among the random variables that are larger than $\mathbf{Z}$, choose one at random according to the probabilities,

$$\text{Prob.}\{\text{choosing } \mathbf{X}_j\} = \frac{(\mathbf{X}_j - \mathbf{Z})}{\sum_{\mathbf{X}_i > \mathbf{Z}} (\mathbf{X}_i - \mathbf{Z})},$$

If j is chosen, set $\mathbf{Z} = \mathbf{X}_j$ and $n = n + 1$.
4. If $\mathbf{Z} = \mathbf{X}_1$ stop, or else go to step 3.

As $m \to \infty$, the mean number of iterations of this algorithm approaches

$$E\{n\} \sim \frac{1}{2} + \frac{1}{k} + \sum_{i=2}^{k-1} \frac{i+3}{2i(i+1)} \tag{4.5}$$

(Gillespie 1983a). The remarkable fact that this mean does not depend on the probability distribution that was used to assign the fitnesses is due to the role of extreme value theory. As $m \to \infty$, only the top k alleles are relevant to the process. The joint distribution of these alleles is the multi-variate extreme value distribution regardless of the distribution that was used to assign the fitnesses. Note also that the mutation rate and popula-tion size do not enter into this result. They are relevant to the time required to complete the process, but not to the number of steps. Unfortunately, the distribution of n is not known. Perhaps Sam Karlin can find it.

This problem is only of limited biological importance since the assump-tion that all alleles can mutate to all others at the same rate will not hold for real species. Rather, the structure of DNA suggests that with each mutation that is fixed in the population, an entirely new set of mutations will appear. These alleles will be two mutational steps away from the initial sequence. For this model the algorithm is as follows.

ALGORITHM A2

1. Generate m positive IID random variables, X_i^0, from some proba-bility distribution that is in the domain of attraction of the Gumbel extreme value distribution. Rank the random variables such that $X_1^0 > X_2^0 > \cdots X_m^0$.
2. Set the initial state of the process, (Z, n), to $(X_k^0, 0)$, with $k > 1$.
3. Generate m new IID random variables, X_i^{n+1}. If all of the new ran-dom variables are less than Z, then stop. If one or more of these random variables is greater than Z, choose one at random according to the probabilities,

$$\text{Prob.}\{\text{choosing } X_j^{n+1}\} = \frac{(X_j^{n+1} - Z)}{\displaystyle\sum_{X_i^{n+1} > Z} (X_i^{n+1} - Z)}.$$

If j is chosen, set $Z = X_j^{n+1}$. Set $n = n + 1$. Repeat 3.

I have been unable to find the mean number of steps until this process stops. Figure 4.1 shows the results of a Monte Carlo simulation of this algorithm and compares it with the previous case. Once again, extreme value theory plays a role and makes the process independent of the dis-tribution used to assign fitnesses. This case corresponds to the "molecular landscape" described in Gillespie (1984), where it was argued that this is a model that could be used to describe molecular evolution.

The simple structure of these algorithms might suggest that they are essentially combinatoric in nature and that their properties could be dis-covered by a ranking argument. This does not appear to be the case be-cause of the random choice involved in each step. A purely combinatoric algorithm that can be solved is as follows.

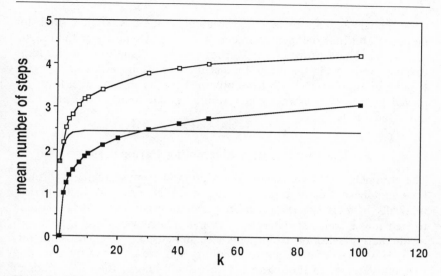

Figure 4.1. The mean number of steps from the three algorithms, A1 (solid squares), A2 (open squares), and A3 (no squares).

ALGORITHM A3

1. Generate m positive IID random variables, X_i^0, from some probability distribution. Rank the random variables such that $X_1^0 > X_2^0 > \cdots X_m^0$.
2. Set the initial state of the process, (Z, n), to $(X_k^0, 0)$, with $k > 1$.
3. Generate m new IID random variables, X_i^{n+1}. If all of the new random variables are less than Z, stop. If one or more of these random variables is greater than Z, set $Z = \max\{X_i^{n+1}\}$. Set $n = n + 1$. Repeat step 3.

For this case, it is easy to show by a purely combinatoric argument that the mean number of steps before the process stops is

$$E\{n\} = 1 + 2(1 - (1/2)^k)(e - 2). \tag{4.6}$$

This curve is also shown in Figure 4.1. Note that when $k = 1$ the mean number of steps for algorithm A2 is very close to $e - 1$ as obtained for the combinatoric model, A3. However, by $k = 3$ the two models diverge significantly. This emphasizes the role of the extreme value theory in the original problem.

The use of SSWM limits to generate these algorithms suggests that a number of interesting models of evolution may be studied without having to resort to multidimensional diffusion models. A common property of these models appears to be that the number of steps is fairly small for modest k and that the mean number of steps grows no faster than $\log(k)$ as $k \to \infty$. These properties may provide an explanation for the large variance observed in the rates of molecular evolution (see Gillespie 1986). Kauffman and Levin (1987) have obtained results that parallel those reported here as well as many new results for landscapes that are much more general.

The Problem with Fluctuating Parameters

The parameters that are used in diffusion models must fluctuate through time and space. For example, population size is known to fluctuate dramatically in most species, yet most population genetic models continue to view it as a constant. There is, of course, a fairly well developed theory of fluctuating selection coefficients (see Felsenstein 1976). Most of this theory involves diffusion approximations that are obtained by assuming that the variance of the parameters are small and of the same order of magnitude as the mean. Karlin and Levikson (1974) and Seno and Shiga (1984) have provided the appropriate theory to show the convergence to diffusions.

In the case of fluctuating parameters, the use of diffusion approximations seems on shakier ground than in the case of models with fixed parameters. Recall that genetic drift is a process for which the stochastic element is due to binomial sampling. For large n, the binomial distribution converges to a normal distribution making the process fit the "locally Brownian" behavior of diffusions. Binomial sampling has the property that the higher-order moments converge to zero faster than the variance. Thus,

$$\mathbf{E}\{dx^r\} = o(1/n)\,dt, \qquad r > 2, \tag{4.7}$$

for such processes.

There is no similar a priori justification for assuming that the higher-order moments of models with fluctuating parameters converge to zero faster than the variance. For example, the innocent-appearing random variable $\mathbf{X}$ with density

$$\varepsilon x^{\varepsilon - 1}, \qquad 0 < x < 1 \tag{4.8}$$

has the property that $\mathbf{E}\{\mathbf{X}^r\} = \varepsilon/(r + \varepsilon)$. As $\varepsilon \to 0$, *all* of the moments approach zero at the same rate. If this random variable were used in a random environment model, the limiting process would not be a diffusion process. It is worthwhile discussing this case in more detail, but the reader is forewarned that no results will be presented. This section is presented as a challenge to occupy Sam Karlin until his 70th birthday volume.

Consider an infinite diploid population with two alleles at a particular locus. Let the fitnesses of the three genotypes A_1A_1, A_1A_2, and A_2A_2 be $1 + U_t$, $1 + (U_t + V_t)/2$, and $1 + V_t$ respectively, where $\{U_t\}$ and $\{V_t\}$ are stochastic processes. The stochastic difference equation to describe the change of the frequency of the A_1 allele, x, is the familiar

$$\Delta x = x(1 - x)(U_t - V_t)/[2(1 + xU_t + (1 - x)V_t)]. \qquad (4.9)$$

It is well known that the conditions for the existence of a stable polymorphism (i.e., a nontrivial stationary distribution) in the case where the collections $\{U_t\}$ and $\{V_t\}$ are made up of independent, identically distributed random variables are

$$\mathbf{E}\{\log[1 + (U_t + V_t)/2]\} > \mathbf{E}\{\log(1 + U_t)\},$$
$$\mathbf{E}\{\log[1 + (U_t + V_t)/2]\} > \mathbf{E}\{\log(1 + V_t)\} \qquad (4.10)$$

(Gillespie 1973, Karlin and Liberman 1974, and Norman, 1975). These conditions are not very informative as they stand since they are not in terms of fitness moments. This can be achieved if we make a diffusion-type assumption, namely that

$$\mathbf{E}\{U_t\} = \mu_1\varepsilon, \qquad \mathbf{E}\{V_t\} = \mu_2\varepsilon$$
$$\mathrm{Var}\{U_t\} = \mathrm{Var}\{V_t\} = \sigma^2\varepsilon \qquad (4.11)$$
$$\mathbf{E}\{U_t^r\} = o(\varepsilon), \qquad \mathbf{E}\{V_t^r\} = o(\varepsilon), \qquad r > 2.$$

As $\varepsilon \to 0$,

$$\mathbf{E}\{\log(1 + U_t)\}/\varepsilon \sim \mu_1 - \sigma^2/2 \qquad (4.12)$$

so the conditions for stationarity become

$$|\mu_1 - \mu_2| < \sigma^2/2. \qquad (4.13)$$

Note that this result is true for any distribution of U_t and V_t for which the moment assumptions hold. This invariance has quite possibly given a false sense of security about the biological relevance of diffusions for random environment models.

A nondiffusion approach might assume that U_t and V_t may be represented as

$$U_t = \mu_1\varepsilon + \xi_t, \qquad V_t = \mu_2\varepsilon + \zeta_t,$$

where ξ_t and ζ_t are IID random variables distributed according to the density (4.8). In this case

$$\mathbf{E}\{\log(1 + U_t)\}/\varepsilon \sim \mu_1 - \pi^2/12. \qquad (4.14)$$

In comparing this result to (4.12), the most important thing to notice is that the term $\pi^2/12$ is due to the particular form of the density (4.8). Had we used another density, say,

$$\Gamma(\varepsilon)^{-1}x^{\varepsilon-1}e^{-x}, \tag{4.15}$$

the result would have been different. Thus the limiting process will not exhibit the independence of the distribution of the fitnesses that characterizes diffusion processes.

I have been totally unsuccessful in obtaining any properties of these nondiffusion limits. It is worthwhile, nonetheless, to make a few comments on the nature of the processes. We begin by illustrating in Figure 4.2 the sample path behavior for three processes, one with U_t and V_t being normally distributed, one with the power distribution (4.8), and one with the gamma distribution (4.15). For each of the cases the variance of U_t and V_t is 0.01. The wildly different sample path behavior for the three cases is obvious. The two nondiffusion type processes exhibit very little change over several generations followed by large jumps, while the diffusion type process shows lots of "jitter."

The explanation for this behavior is easy to see. For the power density, we might ask: what is the probability that the fitnesses exceed some fixed value α, and what is the distribution of the fitness, given that it exceeds α?

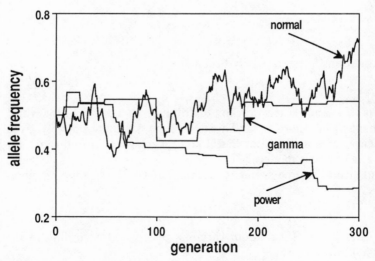

Figure 4.2. Sample paths from simulations of the selection model given by Eq. 4.9 for three different distributions of environmental noise.

The answers are, as $\varepsilon \to 0$,

$$\mathbf{Pr}(\mathbf{U}_t > \alpha) \sim -\varepsilon \log(\alpha),$$
$$\mathbf{Pr}(\mathbf{U}_t < x \,|\, \mathbf{U}_t > \alpha) \sim x^{-1}/-\log(\alpha), \qquad \alpha < x < 1. \tag{4.16}$$

From this it is obvious that the selection coefficient, $\mathbf{U}_t$, tends to be very close to zero most of the time (in an asymptotic sense, as $\varepsilon \to 0$), but when it does take on a value that is not close to zero, that value is (asymptotically) independent of ε. If the limiting process is obtained by rescaling time to be measured in units of $1/\varepsilon$ generations, then, at the limit, the number of times that $\mathbf{U}_t$ exceeds α in an interval of length t is Poisson distributed with mean $-\log(\alpha)t$. Thus the sample path for the power process exhibited in Figure 4.2 could be approximated by a process that remains unchanged for an exponentially distributed length of time with mean $-1/2 \log(\alpha)$ (the division by two is to take into account that either $\mathbf{U}_t$ or $\mathbf{V}_t$ could take on a substantial value), and then takes a jump whose displacement is determined by (4.9), randomized by the conditional density (4.16). The randomization follows this sequence: Given that either $\mathbf{U}_t$ or $\mathbf{V}_t$ exceeds α, the probability that it is $\mathbf{U}_t$ that exceeds α rather than $\mathbf{V}_t$ is 1/2. Given that it is $\mathbf{U}_t$, the probability that the allele frequency in the next generation is x' is given by the density of the transformation

$$x' = x[1 + (1 + x)\mathbf{U}/2]/(1 + x\mathbf{U}), \tag{4.17}$$

where $\mathbf{U}$ has the density $-1/[u \log(\alpha)]$, $\alpha < u < 1$. With probability 1/2, x' has the density

$$x' = x[1 + (1 - x)\mathbf{U}/2]/[1 + (1 - x)\mathbf{U}], \tag{4.18}$$

where $\mathbf{U}$ has the same density as before. The accuracy of this approximating process will improve as $\alpha \to 0$.

This view of the process illustrates why it appears to be impossible to make analytic progress. The jumps that occur do not tend to diminish as the limiting operation ($\varepsilon \to 0$) is carried out. Thus the difference operator, (4.9), cannot be approximated as done in the diffusion limit. The difficulty of the problem seems to be essentially the same as for the original discrete-time stochastic difference equation.

There is an intriguing direction that this line of inquiry might be pushed in hopes of obtaining some solvability, or at least some interesting mathematics. We have made a natural progression from assuming $\mathbf{E}\{\mathbf{U}_t^r\} = o(\varepsilon)$, $r > 2$, to $\mathbf{E}\{\mathbf{U}_t^r\} = O(\varepsilon)$, $r > 1$, so why not continue and assume that the moments are infinite? For example, we might assume that $1 + \mathbf{U}_t$ and $1 + \mathbf{V}_t$ are distributed according to a Levy Stable distribution on $(0, \infty)$

with characteristic function

$$\exp(-bz^\alpha), \qquad 0 < \alpha < 1, \tag{4.19}$$

for which all moments are infinite. An initial question to ask of this family concerns the existence of a stationary distribution. It is not obvious how to approach this question because the proof of the conditions (4.10) requires finite moments. To add to the computational complexities, the densities of the members of the stable family cannot be written down in a finite number of terms except for the case $\alpha = 1/2$. For this case the density is

$$\sqrt{a/\pi}\,\exp(-a/x)x^{-3/2}. \tag{4.20}$$

The distribution of the fitness of the heterozygote, $1 + (U_t + V_t)/2$, is a stable distribution with the same exponent but with a replaced by $2a$. Thus by using the fact that the expectation of the logarithm of a random variable with this distribution is

$$\gamma + 2\log(2) + \log(a), \tag{4.21}$$

(γ is Euler's constant) it is easy to show that the conditions (4.10) will be met. Whether this implies that a stationary density exists for the allele frequency remains an open question.

In many ways the stable distributions are a natural extension of the normal distribution. Perhaps the fact that normality leads to tractability for diffusions will have a parallel in the stable laws leading to tractability for processes with infinite moments. If so, then these processes may prove to be good approximations to processes of real biological significance that are driven by noise processes such as the power density, (4.8).

The problems introduced in this section are not purely of mathematical interest. Distributions like the power distribution (4.8) may be better representations of the nature of environmental fluctuations than is the normal distribution. This distribution is closer to a situation where in most years the environment is not particularly harsh so that fitness differences between alleles are minimal, and on occasional years the environment is very harsh causing large fitness differences to be expressed. By comparison, the normal model has moderate fitness differences each generation. This theory may help in one unsatisfying aspect of the use of diffusion models for randomly fluctuating environments. In the symmetric SAS-CFF model the stationary density of the infinite allele case is a Poisson-Dirichlet distribution just as it is for the neutral allele model (Gillespie 1977a). When the data are compared to the predictions of the Poisson-Dirichlet distribution, it is generally observed that there are too many rare alleles (Ohta 1976). The strictly neutral allele theory may be modified to account for this by the additional assumption of mildly deleterious alleles. There has been no equally natural way to bring the SAS-CFF model in line with the data. Since the limiting nondiffusions discussed in this section carry with them unmistakable vestiges of the distribution of the environmental fluctuations, it may well turn out that the stationary distribution can be made to deviate from the Poisson-Dirichlet in the same direction as the data.

Another area of relevance concerns natural selection for variances. The claim has been made that if two genotypes differ only in the variance of their offspring number, then natural selection will always favor the genotype with the lower variance (Gillespie 1977b). This result was obtained by assuming that the third and higher-order moments are negligible. For our current models this cannot be assumed. In the case of fluctuating fitness, the best measure of overall fitness is the expectation

$$\mathbf{E}\{\log(1 + \mathbf{U}_t)\} = \sum_i \sigma_i(-1)^i \mu_i / i!, \tag{4.22}$$

where μ_i is the ith moment about the origin. For many distributions the sum on the right side will not converge, but in those cases where it does, two obvious inferences about selection for higher-order moments can be made. The first is that selection will favor larger odd moments about the origin and smaller even moments. The second is that the strength of selection on higher moments will diminish as the reciprocal of the factorial of the moment. Whether these inferences hold when the sum, (4.22), does not converge will be interesting to see.

References

Ewens, W. J. 1964. The pseudo-transient distribution and its uses in population genetic. *J. Appl. Prob.* 1: 141–156.

Ewens, W. J. 1965. The adequacy of the diffusion approximation to certain distributions in genetics. *Biometrics* 21: 386–394.

Felsenstein, J. 1976. The theoretical population genetics of variable selection and migration. *Ann. Rev. Genet.* 10: 253–280.

Gillespie, J. H. 1973. Polymorphism in a random environment. *Theor. Pop. Biol.* 4: 193–195.

Gillespie, J. H. 1977a. Sampling theory for alleles in a random environment. *Nature* 266: 443–445.

Gillespie, J. H. 1977b. Natural selection for variance in offspring number: A new evolutionary principle. *Amer. Natur.* 111: 1010–1014.

Gillespie, J. H. 1983a. A simple stochastic gene substitution model. *Theor. Pop. Biol.* 23: 202–215.

Gillespie, J. H. 1983b. Some properties of finite populations experiencing strong selection and weak mutation. *Amer. Natur.* 121: 691–708.

Gillespie, J. H. 1984. Molecular evolution over the molecular landscape. *Evolution* 38: 1116–1129.

Gillespie, J. H. 1986. Natural selection and the molecular clock. *Mol. Biol. Evol.* 3: 138–155.

Harshberger, J. W. 1970. *The Vegetation of the New Jersey Pine Barrens.* Dover, New York.

Karlin, S., and Levikson, B. 1974. Temporal fluctuations in selection intensities: Case of small population size. *Theor. Pop. Biol.* 6: 383–412.

Karlin, S., and Liberman, U. 1974. Random temporal variation in selection intensities: Case of large population size. *Theor. Pop. Biol.* 6: 355–382.

Karlin, S., and McGregor, J. 1964. On some stochastic models in population genetics. In *Stochastic Models in Medicine and Biology*, pp. 245–278, Ed. J. Gurland. University of Wisconsin Press, Madison.

Kauffman, S., and Levin, S. 1987. Towards a general theory of adaptive walks on rugged landscapes. Manuscript.

Kimura, M. 1955. Stochastic processes and distribution of gene frequencies under natural selection. *Cold Spring Harbor Symp. on Quant. Biology* 20: 33–53.

Kimura, M. 1964. Diffusion models in population genetics. *J. Appl. Prob.* 1: 177–232.

Moran, P.A.P. 1962. *The Statistical Processes of Evolutionary Theory.* Clarendon Press, Oxford.

Nagylaki, T. 1980. The strong-migration limit in geographically structured populations. *J. Math. Biology* 9: 101–114.

Norman, M. F. 1975. An ergodic theorem for evolution in a random environment. *J. Appl. Prob.* 12: 661–672.

Ohta, T. 1976. Role of very slightly deleterious mutations in molecular evolution and polymorphism. *Theor. Pop. Biol.* 10: 254–275.

Seno, S., and Shiga, T. 1984. Diffusion models of temporally varying selection in population genetics. *Adv. Appl. Prob.* 16: 260–280.

Slatkin, M. 1981. Fixation probabilities and fixation times in a subdivided population. *Evolution* 35: 477–488.

Watterson, G. A. 1962. Some theoretical aspects of diffusion theory in population genetics. *Annals Math. Stat.* 33: 939–957.

Wright, S. 1931. Evolution in Mendelian populations. *Genetics* 16: 97–159.

Wright, S. 1945. The differential equation of the distribution of gene frequencies. *Proc. Nat. Acad. Sci. USA* 31: 382–389.

CHAPTER FIVE

The Effect of Population Subdivision on Multiple Loci without Selection

Freddy Bugge Christiansen

5.1. Introduction

When previously isolated populations are mixed, the genotypic proportions at a locus with variable gene frequencies show an excess of homozygotes as compared to the Hardy-Weinberg proportions corresponding to the mean gene frequency in the mixed population (Wahlund 1928). This well-known signature of population mixing disappears after just a single breeding by random mating in a population with nonoverlapping generations. A more profound and long-lasting effect of population mixing is the creation of distributional interactions between alleles of a pair of loci with variable gene frequencies (Sinnock and Singh 1972; Prout 1973; Nei and Li 1973; Feldman and Christiansen 1975). The effect is initially the Wahlund effect on the two-locus Hardy-Weinberg proportions giving an excess of homogametic types that disappears after one round of random mating, but at the same time an amount of linkage disequilibrium equal to the covariance in gene frequencies between the two loci is created. This may, in structured populations, for example the steppingstone model (Kimura 1953), lead to a lasting level of linkage disequilibrium proportional to the covariance in gene frequencies between neighboring populations and inversely proportional to the frequency of recombination between the two loci (Feldman and Christiansen 1975; Christiansen 1987).

The two-locus result for the creation of distributional interaction between alleles is expressed easily in terms of the usual multiplicative measure of interaction (Bennet 1954; Slatkin 1972), the linkage disequilibrium. However, in theoretical work, linear measures of interaction have proven very natural and hence productive (Karlin and Feldman 1970; Feldman, Franklin, and Thomson 1974; Christiansen and Feldman 1975), and this parametrization is closely related to the genetic algebra formulation of multiple-locus models (Holgate 1968, 1976; Wörz-Busekros 1980). It turns out that the two-locus result has a very simple generalization to multiple loci, if linear rather than multiplicative measures of interaction are used. In two-locus formulations of population mixing the linear interaction measure is proportional to the multiplicative interaction measure.

Multilocus interactions will be investigated assuming that each locus has only two alleles. This limitation is immaterial, as a proper description of the distributional interaction between particular alleles at different loci with multiple alleles is reached by considering the alternative alleles as one collective allele.

5.2. Multilocus Interaction Measures

Consider n loci, each with two alleles designated allele numbers 0 and 1. The 2^n gametes are represented by n-tuples, $\mathbf{G} = (g_1, g_2, \ldots, g_n) \in E_n = \{0, 1\}^n$, where g_a names the allele carried by the ath locus in the gamete $\mathbf{G}$. The gametes are indexed $1, 2, \ldots, 2^n$ by the definition that the index of $\mathbf{G}$ is $i = I(\mathbf{G})$, where the function I is given by

$$I(g_1, g_2, \ldots, g_n) = g_1 2^{n-1} + g_2 2^{n-2} + \cdots + g_n + 1. \tag{5.1}$$

This corresponds to the commonly used indexing of gametes in considerations of two and three loci (Bodmer and Felsenstein 1967; Feldman, Franklin, and Thomson 1974), and the n-tuples correspond to a representation of the indexing (5.1) in terms of binary numbers. Genotypes with respect to the n loci are represented by an ordered pair of gametes $\mathbf{G}_i \mathbf{G}_j \in E_n^2$.

The segregation of gametes is described by Geiringer's (1944) linkage (or recombination) distribution $R(\mathbf{e})$, $\mathbf{e} \in E_n$. The probability $R(e_1, e_2, \ldots, e_n)$ describes the frequency in which the gamete $[(1 - e_1)g_{i1} + e_1 g_{j1}, (1 - e_2)g_{i2} + e_2 g_{j2}, \ldots, (1 - e_n)g_{in} + e_n g_{jn}]$ is produced by an individual of genotype $\mathbf{G}_i \mathbf{G}_j$ due to the recombination event that picks an allele of locus a from gamete $\mathbf{G}_i$ if $e_a = 0$ and from gamete $\mathbf{G}_j$ if $e_a = 1$ (Geiringer 1944; Schnell 1961; Karlin and Liberman 1978, 1979a,b,c; Østergaard and Christiansen 1981). The recombination events are unique, so the sum of $R(\mathbf{e})$ over all $\mathbf{e} \in E_n$ is one, and as we will presume Mendelian segregation, $R(\mathbf{1} - \mathbf{e}) = R(\mathbf{e})$ for all $\mathbf{e} \in E_n$.

Considerable simplification of the formulation is reached by describing the gametes of this two-allele model in terms of index sets of the n-tuples rather than the n-tuples themselves. An n-tuple in E_n is uniquely described by the subset of indices in $N = N_n = \{1, 2, \ldots, n\}$, where the tuple contains the number 1. The set of these subsets of N is designated $S(N) = S_n$, and for any subset $A \in S(N)$ we use $S(A)$ for the set of subsets of A. Thus the gamete represented by the set $A \in S(N)$ has allele number 1 at the loci $a \in A$ and allele number 0 at the loci $a \in N \backslash A$. The complementary gamete to A is therefore $N\backslash A$ ($\backslash$ designates set difference), and the sets N and $\emptyset$ correspond to the gametes $(1, 1, \ldots, 1)$ and $(0, 0, \ldots, 0)$, respectively. If we define $A * B$, $A, B \in S(N)$, as the disjoint union of A and B ($A * B = [A\backslash B] \cup [B\backslash A]$, where $\cup$ designates set union), the genotype AB is a heterozygote 01 in all the loci $a \in A * B$, a homozygote 00 in all the loci $a \in (N\backslash A) \cap (N\backslash B)$, and a homozygote 11 in all the loci $a \in A \cap B$ (where $\cap$ designates set intersection).

Similarly, the recombination distribution R is viewed as a probability distribution on $S(N)$ (Schnell 1961). Thus, $R(K)$, $K \in S(N)$, is the probability that the genotype AB, say, segregates a gamete with alleles from A at the loci $a \in N\backslash K$ and alleles from B at the loci $a \in K$. The recombination event complementary to K is $N\backslash K$, so the assumption of Mendelian segregation now reads $R(N\backslash K) = R(K)$. This formulation of the recombination distribution eases the definition of the marginal recombination distribution R_A, $A \in S(N)$, describing the segregration of the loci $a \in A$:

$$R_A(B) = \sum_{C \in S(N\backslash A)} R(B \cup C), \quad B \in S(A). \tag{5.2}$$

Now let $\pi(A)$ be the frequency of gamete A in the population in a given generation in a panmictic population. Corresponding to these frequencies the marginal gametic frequencies for the loci in A are addressed as $\pi_A(B)$, $B \in S(A)$. The frequency $\pi'(A)$ of gamete A in the offspring generation is given by

$$\pi'(A) = \sum_{B \in S(N)} \sum_{C \in S(N\backslash(B * A))} R_{(A \cup B) * C}(A * B)\pi(B)\pi(C \cup A\backslash B), \tag{5.3}$$

which may seem a rather complicated formulation, but it nevertheless leads to a very simple and closed form of the recurrence equations for linear interaction measures.

The linear interaction measures are usually defined in terms of tensor powers of the transformation

$$\begin{pmatrix} 1 & 1 \\ 1 & -1 \end{pmatrix}$$

applied to the vector of gametic frequencies with the usual indexing defined by (5.1). The equivalent transformations with the set indexing is given by

$$\theta_A = \sum_{B \in S(N)} \gamma(A \cap B)\pi(B), \tag{5.4}$$

where $\gamma(C) = (-1)^{\#C}$, $C \in S(N)$, with $\#C$ designating the number of elements in the set C. Thus $\gamma(C)$ is 1 if C contains an even number of elements, and -1 if C contains an odd number of elements. Note that by this definition $\theta_\emptyset = 1$, $\theta_{\{a\}}$ is the difference between the frequencies of the two alleles at locus a, that is, $\theta_{\{a\}} = \pi_{\{a\}}(\emptyset) - \pi_{\{a\}}(\{a\})$, and in general

$$\theta_A = \sum_{B \in S(A)} \gamma(B)\pi_A(B). \tag{5.4'}$$

The reverse transformation is given by

$$\pi(A) = 2^{-n} \sum_{B \in S(N)} \gamma(A \cap B)\theta_B, \tag{5.5}$$

which is conserved for the marginal gametic frequencies, where for $B \in S(A)$ we have

$$\pi_A(B) = 2^{-\#A} \sum_{C \in S(A)} \gamma(B \cap C)\theta_C. \tag{5.5'}$$

So the gene frequency of allele 0 at locus a is $p_a = \pi_{\{a\}}(\emptyset) = (1 + \theta_{\{a\}})/2$ and the gene frequency of allele 1 in that locus is $q_{\{a\}} = \pi_{\{a\}}(a) = (1 - \theta_{\{a\}})/2$.

The use of transformation of (5.4) makes the recurrence equations given by (5.3) equivalent to

$$\theta'_A = \sum_{B \in S(A)} R_A(B)\theta_B\theta_{A * B}. \tag{5.6}$$

This is shown by repeatedly using $\gamma(A * B) = \gamma(A)\gamma(B)$ for all $A, B \in S(N)$ and by using

$$\sum_{C \in S(A)} \gamma(B \cap C) = \begin{cases} 2^{\#A} & \text{if} \quad A \cap B = \emptyset, \\ 0 & \text{otherwise,} \end{cases} \tag{5.7}$$

for all $A, B \in S(N)$. Note the symmetry of the two θ-factors in (5.6); because of this symmetry, the sum contains only $2^{\#A-1}$ different terms and not the apparent $2^{\#A}$ terms.

The gametic frequencies converge with time to linkage equilibrium, or to Robbins' equilibrium, where the alleles at the different loci are distributed independently in the gametes, if recombination can occur between any pair of loci (Robbins 1918; Geiringer 1944). These equilibrium gametic proportions, the Robbins proportions, are given in the transformed system as

$$\rho_A = \prod_{a \in A} \theta_{\{a\}}.$$

The deviation from linkage equilibrium, the linkage disequilibrium, may be measured by the deviation $\delta_A = \theta_A - \rho_A$ of the transformed gametic frequencies from the transformed Robbins' proportions. Note that if $\#A < 2$, then $\delta_A = 0$, and that $4\delta_{\{ab\}}$ is the usual linkage disequilibrium measure $D_{\{ab\}}$ for the two loci a and b. The recurrence equations in these deviation measures are given by

$$\delta'_A = \sum_{B \in S(A)} R_A(B)\delta_B(\delta_{A * B} + 2\rho_{A * B}). \tag{5.6'}$$

We will address the measure δ_A as the deviation from linkage equilibrium on the level A.

The convergence to linkage equilibrium is clearly described by the hierarchical nature of the recurrence equations (5.6′). The eigenvalues for the convergence is simply given as $R_A(A)$, $A \in S(N)$, which are nicely ordered as $R_A(A) \leq R_B(B)$ for $B \in S(A)$, so the higher-order deviations show the fastest convergence in the limit. However, only the two-locus linkage disequilibria decrease monotonically to zero, and the higher-order deviations may initially increase because of their dependence on the lower-order deviations.

The usual measures of linkage disequilibrium introduced by Bennett (1954) are given in a modified version by

$$D_A = \sum_{B \in S(A)} \gamma(A * B)\pi_B(\emptyset) \prod_{a \in A * B} p_a \qquad (5.8)$$

for $A \in S(N)$. Note that Bennett's measures are extended to include $D_\emptyset = 1$ and $D_{\{a\}} = 0$ for all $a \in N$. The reverse relation may be given in terms of the marginal gamete frequencies as

$$\pi_A(\emptyset) = \sum_{B \in S(A)} D_B \prod_{a \in A * B} p_a, \qquad (5.9)$$

which may also be used for a recursive definition of the linkage disequilibrium measures (Slatkin 1972), or it may be given as

$$\pi(A) = \sum_{B \in S(N)} \gamma(A \cap B)D_B \prod_{a \in N\backslash(A \cup B)} p_a \prod_{b \in A\backslash B} q_b. \qquad (5.9')$$

In general, the correspondence between Bennett's linkage disequilibrium measures and the present parameters are given by

$$D_A = 2^{-\#A} \sum_{B \in S(A)} \gamma(A * B)\delta_B \rho_{A * B},$$

$$\delta_A = \sum_{B \in S(A)} \sum_{C \in S(B)\backslash\{\phi\}} 2^{\#B}\gamma(A * B)D_C \prod_{c \in B * C} p_c \qquad (5.10)$$

for all $A \in S(N)\backslash\{\emptyset\}$. (In view of Eq. 5.7, the θ's may be substituted for the δ's in the first equation, and the second equation provides θ_A if $C = \emptyset$ is included in the second sum.)

The recurrence equations in Bennett's measures are given by

$$D'_A = \sum_{B \in S(A)} R_A(B)D_B D_{A * B}, \qquad (5.11)$$

which is exactly the same form as (5.6). However, from the definition of the D's the right side of (5.11) depends linearly on D_A and contains products of D's two orders lower than D_A. Thus (5.11) is the very simple, well-known equation for $\#A = 2$ and 3.

The two ways of expressing the deviation from linkage equilibrium measure different aspects of the nature of the deviation. If the alleles of the loci in $A \in S(N)$ are distributed in the gametes independently of the alleles of the loci in $N \backslash A$, then for all $B \in S(N)$: $\pi(B) = \pi_A(A \cap B)\pi_{N \backslash A}(B \backslash A)$, which is equivalent to $\theta_B = \theta_{A \cap B}\theta_{B \backslash A}$, $D_B = D_{A \cap B}D_{B \backslash A}$, and $\delta_B = \delta_{A \cap B}\delta_{B \backslash A} + \delta_{A \cap B}\rho_{B \backslash A} + \rho_{A \cap B}\delta_{B \backslash A}$. If, in addition, the alleles of the loci in A are distributed independently in the gametes, that is, the π_A's are in Robbins' proportions, then $\theta_B = \rho_{A \cap B}\theta_{B \backslash A}$ and $\delta_B = \rho_{A \cap B}\delta_{B \backslash A}$ for all $B \in S(N)$ in the present parametrization, but in Bennett's parametrization $D_B = 0$ if and only if $B \cap A \neq \emptyset$. Thus, even though the simplicity of the linear measures of linkage disequilibrium for theoretical arguments has been emphasized, the interpretation of results are most intuitive in terms of Bennett's multiplicative measures.

5.3. Gametic Proportions in a Mixed Population

Now suppose that the population is divided into k subunits of relative size $c_i, i = 1, 2, \ldots, k$, with $c_1 + c_2 + \cdots + c_k = 1$. Further, let θ_{Ai}, $A \in S(N)$ describe the gametic proportions in population i in a given generation, where all the individuals are mixed and allowed to breed at random.

The interbreeding in this mixed population does not influence the recombination process as it occurs within the individuals, so the proportions among the gametes forming the offspring generation are given by

$$\theta'_A = \sum_{i=1}^{k} c_i \sum_{B \in S(A)} R_A(B)\theta_{Bi}\theta_{A * Bi}, \qquad (5.12)$$

using the recurrence equation (5.6) within each subpopulation.

To produce (5.12) in terms of the measures of deviations from linkage equilibrium, let δ^*_{Ai} be the deviation from linkage equilibrium in the ith population in the offspring generation assuming that the mixing did not occur (δ^*_{Ai} is for each subpopulation produced from Eq. 5.6'). With this definition we get the deviation from Robbins' proportions in the offspring from the mixed population as

$$\delta'_A = \sum_{i=1}^{k} c_i \delta^*_{Ai} + \sum_{i=1}^{k} c_i(\rho_{Ai} - \rho'_A). \qquad (5.13)$$

Thus the deviation from Robbins' proportions in the mixed population is given in the same form as by Feldman and Christiansen (1975), in that it is the sum of the average deviation from Robbins' proportions in the subpopulations without mixing and a term measuring the variation in gene frequencies among the subpopulations. For two-locus interaction measures, which are proportional to the usual two-locus linkage disequilibria, the second term in (5.13) is indeed a correlation in gene frequencies. For measures of higher order interactions the second term becomes the mean of the deviation between the local Robbins' proportions and the Robbins' proportions in the mixed populations. The second term in (5.13) suggests an even more straightforward interpretation in the multilocus version. The production of deviations from linkage equilibrium by mixing equals the difference between the average of the Robbins' proportions in the population before mixing and the Robbins' proportions in the average population, that is, the population after mixing:

$$\delta'_A = \sum_{i=1}^{k} c_i \delta^*_{Ai} + \left(\sum_{i=1}^{k} c_i \rho_{Ai} - \rho'_A \right). \tag{5.13'}$$

An immediate consequence of this result is that the deviation from linkage equilibrium within a subset of loci only depends on the variation in gene frequencies of this subset of loci. If a locus, say, a, does not vary among the subpopulations, then the second term in (5.13) is not necessarily zero for all A with $a \in A$, but the term has a factor of $\theta_{\{a\}}$. Therefore, if all the subpopulations are in linkage equilibrium, then $\delta_{A \cup \{a\}} = \theta_{\{a\}} \delta_A$ for all $A \in S(N \backslash \{a\})$. From (5.10) we then get that mixing does not produce deviation from linkage equilibrium involving locus a, that is, $D_{A \cup \{a\}} = 0$, $A \in S(N \backslash \{a\})$.

The translation of (5.13) to an equation in Bennett's multiplicative measures of linkage disequilibrium does not produce a similar closed form of the result. However, it is illustrative to consider the result to be able to argue that the linear interaction measures are the natural measures for the current problem. The gametic proportions in the offspring of the mixed population are found as the average gametic proportions after breeding in the unmixed populations. So, using (5.8) and (5.9), we get

$$D'_A = \sum_{i=1}^{k} c_i D^*_{Ai} + \sum_{i=1}^{k} c_i \sum_{B \in S(A) \backslash \{A\}} D^*_{Bi} \prod_{a \in A * B} (p_{ai} - p'_a), \tag{5.14}$$

where D^*_{Bi} is given by (5.11) in terms of the linkage disequilibria in the ith population before mixing. If all subpopulations have reached linkage equilibrium before mixing, (5.14) simplifies to

$$D'_A = \sum_{i=1}^{k} c_i \prod_{a \in A} (p_{ai} - p'_a). \tag{5.15}$$

In general, for two loci the rather complicated expression for the second term in (5.14) reduces to (5.15), which in the two-locus case is a simple covariance in gene frequencies, but in the multilocus case it involves a product moment in the gene frequencies and the lower order linkage disequilibria in the population before mixing. However, this term may convey too simple an impression of the creation of multiplicative linkage disequilibria, in that the added linkage disequilibrium depends on the linkage disequilibria before mixing and on the recombination distribution through (5.11). Thus, for five or more loci the second term in (5.14) involves products of lower-order linkage disequilibria. For three and four loci the second term simplifies because D^* for two and three loci involves only a linear term. Thus, as for two loci, covariation in the gene frequencies produces interaction among alleles in their distribution in gametes. For three or more loci, covariation between the disequilibria and the gene frequencies produces interaction among alleles; and for five or more loci, covariation in linkage disequilibria produces a higher-order linkage disequilibrium.

The second term in (5.14) provides a more immediate basis for the evaluation of the contribution of loci that do not vary among the subpopulations than does the second term of (5.13). If a locus does not vary among the subpopulations, then the second term in (5.14) is not necessarily zero for all A with $a \in A$, where a is the constant locus. Rather, it depends on the covariation between the linkage disequilibria involving locus a and the gene frequencies at the other loci. Again, if all the subpopulations are in linkage equilibrium, then mixing does not produce deviations from linkage equilibrium involving a constant locus a.

5.4. The Effect of Recurrent Immigration

Recurrent immigration into a large population changes the gene frequencies of the population toward the value among the immigrants, so eventually the residents and the immigrants will be genetically equal. Thus a classical island model (Wright 1931, 1943) is not suited for the study of the buildup of linkage disequilibrium because of recurrent immigration since the production of distributional interactions among loci requires mixing of genetically different populations. Instead, consider an island population that receives immigrants from two source populations. The gene frequencies of this population will eventually settle on a value determined by the gene frequencies in the source populations and the relative amount of immigration from these two populations. The result of this is that immigration into the population at each generation creates a mixture with the two source populations that differs from one another and form the resident population. Thus, at equilibrium we can study the creation of deviations from Robbins' proportions due to immigration and the erosion of this deviations due to recombination.

The simplest assumption is that, at the time of breeding, a fraction $1 - 2m$ of the individuals in the island population was born in that population, and fractions m originate as immigrants from the two source populations. Let the island population be population number 1, and let the two source populations have numbers 0 and 2 (the model is an extremely short version of the steppingstone cline model of Feldman and Christiansen 1975). Assume further that the two source populations are at linkage equilibrium, that is, $\theta_{A0} = \rho_{A0}$ and $\theta_{A2} = \rho_{A2}$ for all $A \in S(N)$. The deviations from linkage equilibrium in the source populations are therefore zero, that is, $\delta_{A0} = \delta_{A2} = 0$ for all $A \in S(N)$, so we may in the following drop the population index from δ_{A1} and simply write δ_A. Then at equilibrium in the island population we have

$$[1 - 2R_A(A)(1 - 2m)]\delta_A = m(\rho_{A0} + \rho_{A2} - 2\rho_{A1})$$
$$+ (1 - 2m) \sum_{B \in S(A) \setminus \{A, \emptyset\}} R_A(B)\delta_B(\delta_{A*B} + 2\rho_{A*B1}). \tag{5.16}$$

As $\delta_{\{a\}} = 0$ for all $a \in N$, we have at equilibrium that

$$\theta_{\{a\}1} = (\theta_{\{a\}0} + \theta_{\{a\}2})/2, \tag{5.17}$$

and $\theta_{\{a\}1}$ converges to this equilibrium with rate $1 - 2m$. This rate of convergence will dominate the convergence to equilibrium in the island population, because the convergence of the δ's will be at rates $(1 - 2m)R_A(A)$, $A \in S(N)$. Thus, the convergence of this multilocus model is like the convergence of the two-locus steppingstone cline model (Holgate 1976).

With the gene frequencies at equilibrium given by (5.17), (5.16) constitutes an iterative solution for the equilibrium in the island population as a function of the gene frequencies in the source populations. For the special case of no recombination, the second term in (5.16) vanishes, so the solution becomes

$$\delta_A = (\rho_{A0} + \rho_{A2} - 2\rho_{A1})/2. \tag{5.18}$$

For absolute linkage, the Robbins' proportions are not a natural point of reference, so it is natural that (5.18) expresses the same linear relationship among θ_{Ai}, $i = 0, 1, 2$ as in (5.17). With this result, the extension of the model to include multiple alleles becomes trivial: each locus in a multiple allele model is just represented as a subset of absolutely linked loci in the two-allele model. Below, the deviations from linkage equilibrium in the island population are given more explicitly for a low number of loci in the more interesting cases where recombination is allowed.

Corresponding to any number of loci, the first term in (5.16) is given by

$$\rho_{A0} + \rho_{A2} - 2\rho_{A1} = \sum_{B \in S(A) \setminus \{\emptyset\}} \rho_{A*B1}[1 + \gamma(B)] \prod_{b \in B} \Delta_b, \tag{5.19}$$

where $\Delta_a = (\theta_{\{a\}0} - \theta_{\{a\}2})/2$, $a \in N$. Note that (5.19) shows that every part of the first term in (5.16) contains an even number of Δ-factors. For two loci at equilibrium we have for all $a, b \in N$ that

$$\delta_{\{ab\}} = 2m\Delta_a\Delta_b/[1 - 2R_{\{ab\}}(\{ab\})(1 - 2m)], \tag{5.20}$$

which is the result of Feldman and Christiansen (1975) for the simpler model used here. For three loci at equilibrium, we get the following simple expression:

$$\delta_{\{abc\}} = \delta_{\{ab\}}\theta_{\{c\}1} + \delta_{\{ac\}}\theta_{\{b\}1} + \delta_{\{bc\}}\theta_{\{a\}1}. \tag{5.21}$$

For more loci, the deviation from Robbins' proportions, δ_A, at equilibrium has a term similar to the three-locus interaction (5.21) which depends on the even lower order deviations, viz,

$$\sum_{B \in S(A)\backslash\{\emptyset, A\}} \delta_B \rho_{A \ast B1} \Psi(A \ast B)[1 + \gamma(B)]/2, \tag{5.22}$$

where the function Ψ depends on the number of loci in the argument, that is, $\Psi(A) = \Psi(\#A)$ for all $A \in S(N)$. The values of this function are defined recursively for sets $A \in S(N)$ with an even number of loci by $\Psi(\emptyset) = -1$ and

$$\sum_{B \in S(A)} \Psi(B)[1 + \gamma(B)] = -[1 - \gamma(A)]\Psi(A) \tag{5.23}$$

(the right side is zero for even $\#A$) and at the same time this equation defines the values of Ψ for sets $A \in S(N)$ with an odd number of loci. For sets with one or two elements the value is 1, and for sets with 3, 4, 5, 6, 7, 8, and 9 elements the value is $-2, -5, 16, 61, -272, -1385,$ and -2081, respectively.

The four-locus deviation from Robbins' proportions, δ_A with $A = \{abcd\}$, at equilibrium has in addition to the term (5.22) a term that looks more like the two-locus interaction:

$$2m\left(\prod_{x \in A} \Delta_x\right)\bigg/[1 - 2R_A(A)(1 - 2m)] \tag{5.24}$$

(this originates from the term corresponding to $B = A$ in Eq. 5.19) and a term which contains products of the two-locus interactions:

$$(1 - 2m)[\, 2R_{\{abcd\}}(\{ab\})\delta_{\{ab\}}\delta_{\{cd\}}$$
$$+ 2R_{\{abcd\}}(\{ac\})\delta_{\{ac\}}\delta_{\{bd\}}$$
$$+ 2R_{\{abcd\}}(\{ad\})\delta_{\{ad\}}\delta_{\{bc\}}]/[1 - 2R_{\{abcd\}}(\{abcd\})(1 - 2m)], \tag{5.25}$$

which, from (5.19), has a factor of $m\Delta_a\Delta_b\Delta_c\Delta_d$. The value of $\delta_{\{abcd\}}$ at equilibrium is then the sum of the expressions in (5.22), (5.24), and (5.25).

The five-locus deviation is again comparatively simple, in that it is given by (5.22) without a term like (5.24) and without product terms like (5.25). This situation seems general for deviations from Robbins' proportions corresponding to an odd number of loci. I have no general proof of this statement, but it is also true for seven and nine loci. For an even number of loci the deviation again contains (5.22), (5.24), and a product term which is parallel to (5.25). This third term is for six loci, $A = \{abcdef\}$, given by

$$(1 - 2m)\{2R_{\{abcdef\}}(\{ab\})\delta_{\{ab\}}\delta_{\{cdef\}} + 2R_{\{abcdef\}}(\{ac\})\delta_{\{ac\}}\delta_{\{bdef\}} + \cdots$$
$$+ 2R_{\{abcdef\}}(\{ef\})\delta_{\{ef\}}\delta_{\{abcd\}} - [2R_{\{abcdef\}}(\{ab\}) + 2R_{\{abcdef\}}(\{cd\})]$$
$$\times \delta_{\{ab\}}\delta_{\{cd\}}\rho_{\{ef\}1} - \cdots - [2R_{\{abcdef\}}(\{cd\}) + 2R_{\{abcdef\}}(\{ef\})]$$
$$\times \delta_{\{cd\}}\delta_{\{ef\}}\rho_{\{ab\}1}\}/[1 - 2R_{\{abcdef\}}(\{abcdef\})(1 - 2m)]. \qquad (5.26)$$

For eight loci, four different kinds of product terms exist, all with two δ-factors and two kinds with one additional ρ-factor. Therefore, the expression is long compared to (5.26), which contains only two different kinds of product terms. The general form of the product terms for an arbitrary number of loci has not been found.

5.5. Examples and Discussion

As an example, consider the simple situation of symmetry, where $\theta_{\{a\}0} = -\theta_{\{a\}2}$. Then at equilibrium $\theta_{\{a\}1} = 0$ and $\rho_{A1} = 0$ for all $A \in S(N)$ (all gene frequencies in the island population are equal to $1/2$). Therefore, the correspondence between Bennett's multiplicative and the current linear measures of linkage disequilibrium is in this example very simple; from Eq. (5.10) we have $\delta_A = 2^{\#A}D_A$, that is, δ_A is the ratio of the equilibrium value of D_A and the value obtained at equilibrium if the loci in A were absolutely linked. Further, $\delta_{\{abc\}} = \delta_{\{abcde\}} = 0$, and indeed $\delta_A = 0$ whenever $\#A$ is odd, as the term (5.22) is always zero. For $\#A = 2$ the equilibrium linkage disequilibrium becomes

$$\delta_{\{ab\}} = 2m\Delta_a\Delta_b/[2m + 2R_{\{ab\}}(\{a\})(1 - 2m)], \qquad (5.27)$$

where the denominator in (5.20) is now given in terms of the usual recombination fraction for two loci. For m small, the linkage disequilibrium is approximately inversely proportional to the recombination fraction, just as it is in the middle of a long steppingstone cline model (Feldman and Christiansen 1975). For tightly linked loci, that is, $2R_{\{ab\}}(\{a\})$ small, $\delta_{\{ab\}}$ approaches its maximum value of $\Delta_a\Delta_b$.

For four loci the deviation parameter is given by the sum of (5.24) and (5.25). A further simplification of the results can be obtained by assuming $\theta_{\{a\}0} = 1$, so all Δ's become equal to one (this corresponds to considering δ divided by the appropriate product of Δ's). Again, for m small the four-locus linkage disequilibrium is approximately inversely proportional to the probability, $1 - 2R_{\{abcd\}}(\{abcd\})$, that recombination occurs among the four loci. Thus we may expect that in a steppingstone cline model, this result will again hold in the middle of the cline.

The value of the four-locus linkage disequilibrium will depend strongly on the recombination distribution. So assume the linkage map

$$
\begin{array}{cccc}
a & b & c & d
\end{array}
$$

$$- + - - - - - - + - - - - - - - + - - - - - - + - \qquad (5.28)$$

$$
\begin{array}{ccc}
r_1 & r_2 & r_3
\end{array}
$$

and consider two cases: complete interference, where $R_{\{abcd\}}(\emptyset) = 1 - r_1 - r_2 - r_3$, and no interference, where $R_{\{abcd\}}(\emptyset) = (1 - r_1)(1 - r_2)(1 - r_3)$.

The two-locus disequilibria are given as

$$\delta_{\{ab\}} = (1 + Mr_1)^{-1}, \, \delta_{\{bc\}} = (1 + Mr_2)^{-1}, \, \delta_{\{cd\}} = (1 + Mr_3)^{-1}, \quad (5.29)$$

where $M = (1 - 2m)/(2m)$, in both recombination models, as

$$\delta_{\{ac\}} = [1 + M(r_1 + r_2)]^{-1}, \, \delta_{\{bd\}} = [1 + M(r_2 + r_3)]^{-1},$$
$$\delta_{\{ad\}} = [1 + M(r_1 + r_2 + r_3)]^{-1}, \qquad (5.30)$$

in the complete interference model, and as

$$\delta_{\{ac\}} = [1 + M(r_1 + r_2 - 2r_1r_2)]^{-1}, \, \delta_{\{bd\}} = [1 + M(r_2 + r_3 - 2r_2r_3)]^{-1},$$
$$\delta_{\{ad\}} = [1 + M(r_1 + r_2 + r_3 - 2r_1r_2 - 2r_1r_3 - 2r_2r_3 + 4r_1r_2r_3)]^{-1},$$

$$(5.31)$$

in the no-interference model. The four-locus interaction is from (5.24) and (5.25) in the complete-interference model, simplified to

$$\delta_{\{abcd\}} = [1 + M(r_1 + r_2 + r_3)]^{-1} + (1 - 2m)(1 + Mr_1)^{-1}r_2(1 + Mr_3)^{-1}, \qquad (5.32)$$

whereas all terms of (5.24) and (5.25) are present in the no-interference model:

$$
\begin{aligned}
\delta_{\{abcd\}} = {} & [1 + M(r_1 + r_2 + r_3 - r_1r_2 - r_1r_3 - r_2r_3 + r_1r_2r_3)]^{-1} \\
& + (1 - 2m)(1 - r_1)r_2(1 - r_3)(1 + Mr_1)^{-1}(1 + Mr_3)^{-1} \\
& + (1 - 2m)r_1r_2(1 - r_3) \\
& \times [1 + M(r_1 + r_2 - 2r_1r_2)]^{-1}[1 + M(r_2 + r_3 - 2r_2r_3)]^{-1} \\
& + (1 - 2m)r_1(1 - r_2)r_3(1 + Mr_2)^{-1} \\
& \times [1 + M(r_1 + r_2 + r_3 - 2r_1r_2 - 2r_1r_3 - 2r_2r_3 + 4r_1r_2r_3)]^{-1}.
\end{aligned}
$$

$$(5.33)$$

Thus, for comparable linkage disequilibria between neighboring loci, no-interference recombination produces a higher level of interaction than complete-interference recombination.

For six loci a, b, c, d, e, and f in that order on the map and with complete interference recombination, we have from (5.24) and (5.26) that

$$
\begin{aligned}
\delta_{\{abcdef\}} = {} & [1 + M(r_1 + r_2 + r_3 + r_4 + r_5)]^{-1} \\
& + (1 - 2m)(1 + Mr_1)^{-1} r_2 [1 + M(r_3 + r_4 + r_5)]^{-1} \\
& + (1 - 2m)[1 + M(r_1 + r_2 + r_3)]^{-1} r_4 (1 + Mr_5)^{-1} \\
& + 2(1 - 2m)^2 (1 + Mr_1)^{-1} r_2 (1 + Mr_3)^{-1} r_4 (1 + Mr_5)^{-1},
\end{aligned}
$$

(5.34)

whereas again the term (5.26) becomes considerably more complex in the no-interference model.

For a low immigration rate, the linkage disequilibrium measures in the island population are approximately given as the ratio of the immigration rate and the relevant recombination fraction. For instance, $\delta_{\{ab\}} = 2m/r_1$ in both recombination models, $\delta_{\{abcd\}} = 2m/(r_1 + r_2 + r_3)$ in the complete interference model, and $\delta_{\{abcd\}} = 2m/(r_1 + r_2 + r_3 - r_1 r_2 - r_1 r_3 - r_2 r_3 + r_1 r_2 r_3)$ in the no-interference model. In general, for low migration rates, and m small, we have

$$
\delta_A = \frac{m[1 + \gamma(A)]}{1 - 2R_A(A)} \prod_{a \in A} \Delta_a.
$$

(5.35)

For tightly linked loci, the difference between the two models vanishes, and at equilibrium we get approximately

$$
\begin{aligned}
& \delta_{\{ab\}} = 1 - Mr_1, \; \delta_{\{bc\}} = 1 - Mr_2, \; \delta_{\{cd\}} = 1 - Mr_3, \\
& \delta_{\{ac\}} = 1 - M(r_1 + r_2), \; \delta_{\{bd\}} = 1 - M(r_2 + r_3), \\
& \delta_{\{ad\}} = 1 - M(r_1 + r_2 + r_3), \; \delta_{\{abcd\}} = 1 - M(r_1 + r_2 + r_3).
\end{aligned}
$$

(5.36)

This is a particular case of a more general result for tightly linked loci. If the loci are numbered according to their position on a linkage map, then we have approximately that

$$
\delta_{\{i, i+1, \dots, i+2j-1\}} = \left[1 - M \sum_{k=0}^{2j-2} r_{i+k} \right] \sum_{l=0}^{2j-1} \Delta_{i+l},
$$

(5.37)

where r_i is the recombination distance between the neighbor loci i and $i + 1$.

The higher-order linkage disequilibria may therefore build up as a result of recurrent immigration. The pairwise correspondence between alleles at different loci in a geographically structured population (Feldman and Christiansen 1975) is extended to an expectation of stretches of correspondence among alleles at the loci along a chromosome. This result is, of course, expected, considering the two-locus results, but here the qualitative expectation has been amplified by a quantitative designation of the effects. In addition, the interaction among the loci has a simple structure which may be recognizable in data. The deviations from Robbins' proportions considered here depend only on the recombination parameters and the even products of the gene frequency differences between the source populations. This produces the simple relation between the deviations corresponding to an odd number of loci and the lower-order even deviations as expressed by (5.22). The important property of this relation is its independence from the recombination parameters, so it can potentially be applied in the analysis of population data from organisms where a detailed description of the linkage relationship among the considered loci is lacking.

References

Bennett, J. H. 1954. On the theory of random mating. *Ann. Eugenics* 18: 311–317.

Bodmer, W. F., and Felsenstein, J. 1967. Linkage and selection: Theoretical analysis of the deterministic two locus random mating model. *Genetics* 57: 237–265.

Christiansen, F. B. 1987. The deviation from linkage equilibrium with multiple loci varying in a steppingstone cline. *J. Genet.* 66: 45–67

Christiansen, F. B., and Feldman, M. W. 1975. Subdivided populations: A review of the one- and two-locus theory. *Theor. Pop. Biol.* 7: 13–38.

Feldman, M. W., and Christiansen, F. B. 1975. The effect of population subdivision on two loci without selection. *Genet. Res. (Camb.)* 24: 151–162.

Feldman, M. W.; Franklin, I. R.; and Thomson, G. J. 1974. Selection in complex genetic systems, I. The symmetric equilibria of the three-locus symmetric viability model. *Genetics* 76: 135–162.

Geiringer, H. 1944. On the probability theory of linkage in Medelian heredity. *Ann. Math. Statist.* 15: 25–57.

Holgate, P. 1968. The genetic algebra of k-linked loci. *Proc. London Math. Soc.* 18: 315–327.

Holgate, P. 1976. Direct products of genetic algebras and Markov chains. *J. Math. Biol.* 6: 289–295.

Karlin, S., and Feldman, M. W. 1970. Linkage and selection: Two-locus symmetric viability model. *Theor. Pop. Biol.* 1: 39–71.

Karlin, S., and Liberman, U. 1978. Classifications and comparisons of multilocus recombination distributions. *Proc. Natl. Acad. Sci. USA* 75: 6332–6336.

Karlin, S., and Liberman, U. 1979a. Central equilibria in multilocus systems, I. Generalized nonepistatic selection regimes. *Genetics* 91: 777–798.

Karlin, S., and Liberman, U. 1979b. Central equilibria in multilocus systems, II. Bisexual generalized nonepistatic selection models. *Genetics* 91: 799–816.

Karlin, S., and Liberman, U. 1979c. A natural class of multilocus recombination processes and related measures of crossover interference. *Adv. Appl. Prob.* 11: 479–501.

Kimura, M. 1953. "Steppingstone" model of population. *Ann Rept. Nat. Inst. Genet. Japan* 3: 62–63.

Nei, M., and Li, W. 1973. Linkage disequilibrium in subdivided populations. *Genetics* 75: 213–219.

Østergaard, H., and Christiansen, F. B. 1981. Selection component analysis of natural polymorphisms using population samples including mother-offspring combinations II. *Theor. Pop. Biol.* 19: 378–419.

Prout, T. 1973. *Appendix to*: Population genetics of marine pelecypods, III. Epistasis between functionally related isoenzymes in *Mytilus* edulis (by J. B. Mitten and R. C. Koehn). *Genetics* 73: 487–496.

Robbins, R. B. 1918. Some applications of mathematics to breeding problems, III. *Genetics* 3: 375–389.

Schnell, F. W. 1961. Some general formulations of linkage effects in inbreeding. *Genetics* 46: 947–957.

Sinnock, P., and Sing, C. F. 1972. Analysis of multilocus genetic systems in Tecumseh, Michigan, II. Consideration of the correlation between non-alleles in gametes. *Amer. J. Hum. Genet.* 24: 393–415.

Slatkin, M. 1972. On treating the chromosome as the unit of selection. *Genetics* 72: 157–168.

Wahlund, S. 1928. Zusammensetzung von Populationen und Korrelationserscheinungen vom Standpunkt der Vererbungslehre aus Betrachte. *Hereditas* 11: 65–106.

Wörz-Busekros, A. 1980. "Algebras in Genetics." Lecture Notes in Biomathematics, vol. 36. Springer-Verlag, Berlin, Heidelberg, New York.

Wright, S. 1931. Evolution in Mendelian populations. *Genetics* 16: 97–159.

Wright, S. 1943. Isolation by distance. *Genetics* 16: 114–138.

Complete Characterization of Disequilibrium at Two Loci

Bruce S. Weir and C. Clark Cockerham

Summary

All possible disequilibrium coefficients involving one or two genes at each of two loci are defined. The behavior over time of the various classes of coefficients—four digenic, two trigenic, and one quadrigenic—is expressed in terms of two-locus descent measures. Within populations, maximum likelihood estimates of the disequilibrium coefficients are presented, and so are expected values of their sampling variances. Simulations confirm that hypothesis tests can be based on the asymptotic normality of the estimates. It is recommended that all disequilibrium coefficients be considered in the statistical analysis of genotypic data.

Introduction

The transmission of genetic material between generations in gametes, and the pairing of gametes to package genes into zygotes, can lead to an association of sets of genes within and between loci. The associations may simply be a consequence of the sampling of gametes in finite populations, or they may be caused by some groups of genes that confer greater advantages to the individuals carrying them than do other groups. Other forces, such as migration and mutation, may also cause the population frequencies of sets of genes to differ from the products of the frequencies of the individual genes or their subsets.

In theoretical population genetics, we predict the consequences of various evolutionary forces on the frequencies of combinations of genes at one or more loci. To reflect the effects of population size, mating system, and recombination, we have found various measures of identity by descent to be useful (Cockerham and Weir 1973). We showed then how frequencies could be expressed as linear combinations of the descent measures, and we now consider an alternative parametrization in terms of measures of disequilibrium. The formulations apply over the set of replicate populations that can arise under the same set of conditions.

In experimental studies we are generally concerned with making inferences about evolutionary forces from observed genotypic frequencies within a single population. It is desirable to use such frequencies to estimate various measures of association for sets of genes, and here we consider the procedures for making inferences about these measures.

Most published analyses of multilocus genetic data have been of two types. The first type compares multilocus genotypic frequencies to products of either corresponding single-locus frequencies, or simply to products of gene frequencies (e.g., Allard et al. 1972). In either case, some types of disequilibrium are ignored. The other principal type of analysis is concerned with linkage disequilibrium, where gametic frequencies are compared to the products of corresponding gene frequencies (e.g., Clegg, Allard, and Kahler 1972). Apart from the difficulties sometimes found in determining gametic frequencies, this approach ignores the genotypic structure of the data.

In an earlier paper (Weir 1979) we defined a set of two-locus disequilibrium coefficients that are not limited in either of these two ways. Questions of inference in that paper were limited to linkage disequilibrium coefficients, with a variety of methods for estimation and testing being given. Simulation studies confirmed that a satisfactory procedure could be based on unbiased estimators and chi-square goodness-of-fit tests, with a composite coefficient being appropriate when coupling and repulsion double heterozygotes cannot be distinguished. Tests for the composite coefficient were best performed by dividing the squared estimate by an estimate of its variance and appealing to asymptotic normality. This procedure has the advantage of taking account of any departures from Hardy-Weinberg equilibrium at either of the two loci.

In this chapter we provide details for estimating and testing the higher-order disequilibria at two loci. Two-locus genotypic frequencies are expressed as linear combinations of disequilibria for two, three, and four genes. The methodology is again illustrated with simulated samplings, and the approach is related to others such as the use of log-linear models. The theory presented here supports our previous applications of these disequilibrium measures to data on Drosophila isozymes (Barker, East, and Weir 1986) and human restriction fragment polymorphisms (Weir and Brooks 1986).

For convenience, we limit attention to one of the alleles at a locus, and so consider that allele and a composite of all other alleles. Multiple-allele definitions were given by Weir (1979), and there is no conceptual difficulty in extending the present two-allele treatment to many alleles. The more general situation does tend to obscure the methodology, however.

General Theory

One Locus

For a single locus A we consider alleles A and a (a represents not-A) with expected frequencies p_A and p_a. Expectation here refers to an average over all the replicate populations maintained under the same conditions. Genotypic frequencies are written as P_A^A for AA homozygotes and $2P_a^A$ for Aa heterozygotes, so that

$$p_A = P_A^A + P_a^A.$$

Departures from Hardy-Weinberg equilibrium are measured additively as

$$D_A = P_A^A - p_A^2 = P_a^a - p_a^2 = -(P_a^A - p_A p_a).$$

Such additive coefficients are easier to handle for statistical inference than are ratios such as $f = 1 - P_a^A/p_A p_a$, where $f (= F_{IS})$ is the within-population inbreeding coefficient (Cockerham 1969), and we prefer them to the multiplicative coefficients employed by log-linear models, as we explain below.

Two Loci

For loci A, with alleles A and a, and B, with alleles B and b, we again incorporate a "2" into heterozygote frequencies. Gametic genes are written adjacent to each other as subscripts or superscripts, so that the frequencies of the two types of double heterozygote are written as $2P_{ab}^{AB}$ and $2P_{aB}^{Ab}$. Appropriate summations give frequencies of subsets of the four genes at two loci within individuals:

single gene:

$$p_A = P_{AB}^{AB} + 2P_{Ab}^{AB} + 2P_{Ab}^{Ab} + P_{aB}^{AB} + P_{ab}^{AB} + P_{aB}^{Ab} + P_{ab}^{Ab}$$

gametic pair of genes:

$$P_{AB} = P_{AB}^{AB} + P_{Ab}^{AB} + P_{aB}^{AB} + P_{ab}^{AB}$$

nongametic pair of genes:

$$P_{A/B} = P_{AB}^{AB} + P_{Ab}^{AB} + P_{aB}^{AB} + P_{aB}^{Ab}$$

three genes:

$$P_A^{AB} = P_{AB}^{AB} + P_{Ab}^{AB}, \quad P_B^{AB} = P_{AB}^{AB} + P_{aB}^{AB}.$$

For a pair of nonallelic genes, disequilibrium measures can be defined in the same way as for allelic genes at a locus:

$$D_{AB} = P_{AB} - p_A p_B \qquad \text{for the gametic pair AB}$$

$$D_{A/B} = P_{A/B} - p_A p_B \qquad \text{for the nongametic pair A/B.}$$

When the two types of double heterozygote cannot be distinguished, we work with a composite measure of linkage disequilibrium:

$$\Delta_{AB} = P_{AB} + P_{A/B} - 2p_A p_B$$
$$= D_{AB} + D_{A/B}.$$

It is customary to restrict attention to digenic disequilibria, whether at one or two loci, but this is to ignore the possibility of other associations among the genes. Our definitions of trigenic disequilibria measure the associations between triples of genes after the associations between all appropriate pairs of genes have been removed:

$$D_{ABB} = P_B^{AB} - p_B(D_{AB} + D_{A/B}) - p_A D_B - p_A p_B^2$$
$$D_{AAB} = P_A^{AB} - p_A(D_{AB} + D_{A/B}) - p_B D_A - p_A^2 p_B.$$

To complete the characterization of disequilibrium at two loci, we need the following quadrigenic measure, which has been corrected for digenic and trigenic disequilibria:

$$D_{AB}^{AB} = P_{AB}^{AB} - 2p_A D_{ABB} - 2p_B D_{AAB} - 2p_A p_B \Delta_{AB}$$
$$- p_A^2 D_B - p_B^2 D_A - D_{AB}^2 - D_{A/B}^2 - D_A D_B - p_A^2 p_B^2.$$

In the two-locus, two-allele situation we have now defined seven disequilibrium coefficients. Together with the two independent gene frequencies, these allow the expression of all ten genotypic frequencies, as shown in Table 6.1. When there are just nine distinguishable frequencies, we use the composite measure of linkage disequilibrium, Δ_{AB}, and define a composite quadrigenic component as

$$\Delta_{AABB} = D_{AB}^{AB} - 2D_{AB} D_{A/B}.$$

This composite measure removes the product $D_{AB} D_{A/B}$ from consideration. The expression of genotypic frequencies in this situation is shown in Table 6.2. Both tables make use of the notation,

$$\pi_X = p_X(1 - p_X)$$
$$\tau_X = 1 - 2p_X$$

for an allele X.

Table 6.1

Genotypic frequencies in terms of disequilibria (when double heterozygotes can be distinguished).

Freq.	1	$2D_{ABB}$	$2D_{AAB}$	$2D_{AB}$	$2D_{A/B}$	D^2_{AB}	$D^2_{A/B}$	D_B	D_A	$D_A D_B$	D^{AB}_{AB}
P^{AB}_{AB}	$p^2_A p^2_B$	p_A	p_B	$p_A p_B$	$p_A p_B$	1	1	p^2_A	p^2_B	1	1
$2P^{AB}_{Ab}$	$2p^2_A \pi_B$	$-2p_A$	τ_B	$p_A \tau_B$	$p_A \tau_B$	-2	-2	$-2p^2_A$	$2\pi_B$	-2	-2
P^{Ab}_{Ab}	$p^2_A p^2_b$	p_A	$-p_b$	$-p_A p_b$	$-p_A p_b$	1	1	p^2_A	p^2_b	1	1
$2P^{AB}_{aB}$	$2\pi_A p^2_B$	τ_A	$-2p_B$	$\tau_A p_B$	$\tau_A p_B$	-2	-2	$2\pi_A$	$-2p^2_B$	-2	-2
$2P^{AB}_{ab}$	$2\pi_A \pi_B$	$-\tau_A$	$-\tau_B$	$p_A p_B + p_a p_b$	$p_A p_B + p_a p_b - 1$	2	2	$-2\pi_A$	$-2\pi_B$	2	2
$2P^{Ab}_{aB}$	$2\pi_A \pi_B$	$-\tau_A$	$-\tau_B$	$p_A p_B + p_a p_b - 1$	$p_A p_B + p_a p_b$	2	2	$-2\pi_A$	$-2\pi_B$	2	2
$2P^{Ab}_{ab}$	$2\pi_A p^2_b$	τ_A	$2p_b$	$-\tau_A p_b$	$-\tau_A p_b$	-2	-2	$2\pi_A$	$-2p^2_b$	-2	-2
P^{aB}_{aB}	$p^2_a p^2_B$	$-p_a$	p_B	$-p_a p_B$	$-p_a p_B$	1	1	p^2_a	p^2_B	1	1
$2P^{aB}_{ab}$	$2p^2_a \pi_B$	$2p_a$	τ_B	$-p_a \tau_B$	$-p_a \tau_B$	-2	-2	$-2p^2_a$	$2\pi_B$	-2	-2
P^{ab}_{ab}	$p^2_a p^2_b$	$-p_a$	$-p_b$	$p_a p_b$	$p_a p_b$	1	1	p^2_a	p^2_b	1	1

Table 6.2
Genotypic frequencies in terms of disequilibria
(when double heterozygotes cannot be distinguished).

Freq.	1	$2D_{ABB}$	$2D_{AAB}$	$2\Delta_{AB}$	Δ^2_{AB}	D_B	D_A	$D_A D_B$	Δ_{AABB}
P^{AB}_{AB}	$p^2_A p^2_B$	p_A	p_B	$p_A p_B$	1	p^2_A	p^2_B	1	1
$2P^{AB}_{Ab}$	$2p^2_A \pi_B$	$-2p_A$	τ_B	$p_A \tau_B$	-2	$-2p^2_A$	$2\pi_B$	-2	-2
P^{Ab}_{Ab}	$p^2_A p^2_b$	p_A	$-p_b$	$-p_A p_b$	1	p^2_A	p^2_b	1	1
$2P^{AB}_{aB}$	$2\pi_A p^2_B$	τ_A	$-2p_B$	$\tau_A p_B$	-2	$2\pi_A$	$-2p^2_B$	-2	-2
$\left.\begin{array}{l}2P^{AB}_{ab}\\2P^{Ab}_{aB}\end{array}\right\}$	$4\pi_A \pi_B$	$-2\tau_A$	$-2\tau_B$	$\tau_A \tau_B$	4	$-4\pi_A$	$-4\pi_B$	4	4
$2P^{Ab}_{ab}$	$2\pi_A p^2_b$	τ_A	$2p_b$	$-\tau_A p_b$	-2	$2\pi_A$	$-2p^2_b$	-2	-2
P^{aB}_{aB}	$p^2_a p^2_B$	$-p_a$	p_B	$-p_a p_B$	1	p^2_a	p^2_B	1	1
$2P^{aB}_{ab}$	$2p^2_a \pi_B$	$2p_a$	τ_B	$-p_a \tau_B$	-2	$-2p^2_a$	$2\pi_B$	-2	-2
P^{ab}_{ab}	$p^2_a p^2_b$	$-p_a$	$-p_b$	$p_a p_b$	1	p^2_a	p^2_b	1	1

Disequilibria and Descent Measures

The relation of these additive disequilibrium coefficients, which are functions of frequencies of gene combinations, to various descent measures will now be discussed for the drift situation. It is necessary to keep in mind the distinction between quantities defined over replicate populations and those defined for replicate samples from a single population.

Over all replicate populations, it is known that one-locus homozygote frequencies can be written as

$$P^A_A = p^2_A + F\pi_A,$$

where F (or F_{IT}) is the inbreeding coefficient, or probability of identity by descent of genes within individuals, as an expectation over populations. If frequencies are drawn from a single population, however,

$$P^A_A = p^2_A + f\pi_A,$$

where f (or F_{IS}) reflects the difference in identity probabilities for genes within (F or F_{IT}) and between (θ or F_{ST}) individuals. Specifically,

$$f = (F - \theta)/(1 - \theta).$$

The total (or value over populations) one-locus disequilibrium D_A is therefore

$$D_A = F\pi_A,$$

while the within-population disequilibrium is

$$D_A = f\pi_A.$$

A random mating population has $f = 0$, and there is not expected to be a one-locus disequilibrium, while for a monoecious population of size N mating at random, we know that

$$F_t = 1 - (1 - 1/2N)^t$$

after t generations from a noninbred initial population. When fixation is reached, there remains no disequilibrium within populations, but there is an expected disequilibrium of

$$D_A = \pi_A$$

between populations, which reflects the fixation of alternative alleles.

We previously (Cockerham and Weir 1973, 1977) expressed two-locus genotypic frequencies in terms of two-locus descent measures, and the two-locus disequilibria can be similarly expressed. The necessary descent measures are shown in Table 6.3, and all refer to the relations among genes within individuals, as expectations over all replicate populations. Genes shown there as being equivalent by descent are those that have descended from the same initial gamete. Combining Tables 6.1 and 6.2 with the previous results leads to the expressions in Table 6.4.

The arguments leading to Table 6.4 assume an initial population formed by the random union of gametes, taken from an infinite population in which the frequency of AB gametes was

$$P_{AB} = p_A p_B + D_{AB},$$

so that all initial disequilibrium is digenic. It can be seen there that linkage and trigenic disequilibria remain proportional to D_{AB}, while the Hardy-Weinberg and quadrigenic disequilibria have components that do not depend on the initial value D.

Evaluation of the descent measures follows a well-defined algorithm (Cockerham and Weir 1973) and requires the introduction of measures for four genes located on three or four gametes as well as the two-gamete measures shown in Table 6.3. To illustrate the behavior of the measures, we display some values for drift in a monoecious population in Table 6.5. A random mating population of size $N = 100$ was taken, and a linkage parameter of $\lambda = 0.5$ (recombination coefficient $r = 0.25$) was used. All

Table 6.3

Two-locus descent measures (for individual formed by union of gametes ab and a'b').

Measure[a]	Genes that are equivalent[b]
F_1	(a, a') or (b, b')[c]
F^1	(a, b) or (a', b')
$_1F$	(a, b') or (a', b)
$_1F_1^1$	(a, a', b) or (a, a', b') or (a, b, b') or (a', b, b')
F_{11}	(a, a') and (b, b')
F^{11}	(a, b) and (a', b')
$_{11}F$	(a, b') and (a', b)
F_{11}^{11}	(a, a', b, b')

[a] Measure is probability of corresponding equivalence relations.
[b] Equivalent genes are descended from genes on the same initial gamete.
[c] Alternative events are considered to be equally probable.

other forces, such as selection, are assumed to be absent. The necessary transition equations were given by Cockerham and Weir (1973). Evidently the higher-order measures can take a considerable period of time to reach equilibrium, and may decrease initially before increasing to a final value.

Table 6.4

Disequilibrium coefficients in terms of descent measures.

$$D_A = F_1\pi_A \qquad D_B = F_1\pi_B$$
$$D_{AB} = F^1 D_{AB} \qquad D_{A/B} = {}_1F D_{AB}$$
$$D_{AAB} = {}_1F_1^1\tau_A D_{AB} \qquad D_{ABB} = {}_1F_1^1\tau_B D_{AB}$$
$$D_{AB}^{AB} = \eta_{11}\pi_A\pi_B + (\eta^{11} + {}_{11}\eta)D_{AB}^2 + F_{11}^{11}(\tau_A\tau_B D_{AB} - 2D_{AB}^2)$$
$$\Delta_{AB} = (F^1 + {}_1F)D_{AB}$$
$$\Delta_{AABB} = (\eta_{11}\pi_A\pi_B + F_{11}^{11}\tau_A\tau_B D_{AB}) + (\eta^{11} + {}_{11}\eta - 2F_{11}^{11} - 2_1FF^1)D_{AB}^2,$$

where

$$\eta_{11} = F_{11} - (F_1)^2, \qquad \eta^{11} = F^{11} - (F^1)^2, \qquad {}_{11}\eta = {}_{11}F - ({}_1F)^2$$

Table 6.5

Values of two-locus descent measures for monoecious populations.
(Population size $N = 100$, linkage parameter $\lambda = 0.5$.)

Generation	F_1	F^1	$_1F$	$_1F^1_1$	F_{11}	F^{11}	$_{11}F$	F^{11}_{11}
0	0.0000	1.0000	0.0000	0.0000	0.0000	1.0000	0.0000	0.0000
1	0.0050	0.7500	0.0050	0.0037	0.0031	0.5625	0.0031	0.0028
5	0.0248	0.2446	0.0151	0.0061	0.0071	0.0612	0.0031	0.0029
10	0.0489	0.0712	0.0186	0.0036	0.0090	0.0059	0.0013	0.0013
50	0.2217	0.0196	0.0196	0.0046	0.0536	0.0014	0.0014	0.0012
100	0.3942	0.0196	0.0196	0.0079	0.1582	0.0034	0.0034	0.0032
∞*	1.0000	0.0196	0.0196	0.0196	1.0000	0.0196	0.0196	0.0196

* $F_1 = F_{11} = 1$, $F^1 = {}_1F = {}_1F^1_1 = F^{11} = {}_{11}F = F^{11}_{11} = 1/[1 + N(1 - \lambda)]$.

A comparison of Tables 6.4 and 6.5 shows that D_{AB} and $D_{A/B}$ can reach their final values very quickly, even before the single-locus disequilibria. For intermediate gene frequencies, the trigenic disequilibria will always be very small, but the quadrigenic coefficient can behave in a nonmonotonic way. We have previously stressed (Weir, Allard, and Kahler 1974) that the higher-order disequilibria can take a long time to reach final values. This is of some importance because these coefficients measure the departure of two-locus genotypic frequencies from the products of one-locus frequencies. We see that, even for populations initially in linkage equilibrium, there will be some quadrigenic disequilibrium, in an amount proportional to the identity coefficient η_{11}. We have discussed the behavior of this quantity previously (Cockerham and Weir 1973).

These theoretical expressions for total disequilibria take into account the variation between replicate populations maintained under the same conditions, and so they are appropriate for predicting future values when the particular replicate population is not known.

For the study of a single population, however, we need alternative expressions and, for simplicity, consider here only the case of finite random-mating populations. With gametes uniting at random, we do not expect there to be any disequilibrium between gametes, and the only disequilibrium will be between loci within gametes. We have previously (Weir and Cockerham 1974) derived the expression for gametic linkage disequilibrium within populations:

$$D_{AB} = (F^1 - {}_1F)D_{AB}$$

so that, in generation t,

$$D_{AB(t)} = (1 - 1/2N - r)^t D_{AB}.$$

A more complete discussion of the differences between within-population and total linkage disequilibrium is given by Cockerham and Weir (1977).

Within-Population Inference

We now take up a consideration of the methods for making inferences about disequilibria from observations taken from a single population. For populations mating at random, we expect there to be no disequilibrium involving genes on different gametes, and the gametic linkage disequilibrium is expected to dissipate. Any evidence for disequilibrium provided by the data therefore suggests the actions of some disturbing forces in the population.

ESTIMATION

The procedures of inference are based on the assumption of multinomial sampling of *individuals* from a population. We denote observed frequencies with tildes, and recognize that expectations over all samples of these observations provide the population frequencies. For example,

$$E(\tilde{P}_{AB}^{AB}) = P_{AB}^{AB}$$

$$E(\tilde{P}_A^A) = P_A^A$$

$$E(\tilde{p}_A) = p_A.$$

Because our additive model has been constructed with the same number of parameters as there are degrees of freedom, maximum likelihood estimates of disequilibria follow by replacing all frequencies used in the definitions with corresponding observed values.

At a single locus, the estimate is

$$\tilde{D}_A = \tilde{P}_A^A - \tilde{p}_A^2.$$

If gametic frequencies were available, we could estimate gametic disequilibria

$$\tilde{D}_{AB} = \tilde{P}_{AB} - \tilde{p}_A \tilde{p}_B,$$

but with genotypic frequencies we work with composite frequencies,

$$\tilde{P}_{AB} + \tilde{P}_{A/B} = 2P_{AB}^{AB} + (2P_{Ab}^{AB} + 2P_{aB}^{AB}) + (2P_{ab}^{AB} + 2P_{aB}^{Ab})/2,$$

which are directly observable, and with composite disequilibria

$$\tilde{\Delta}_{AB} = \tilde{P}_{AB} + \tilde{P}_{A/B} - 2\tilde{p}_A \tilde{p}_B.$$

For the trigenic measures, we have

$$\tilde{D}_{AAB} = \tilde{P}_A^{AB} - \tilde{p}_A \tilde{\Delta}_{AB} - \tilde{p}_B \tilde{D}_A - \tilde{p}_A^2 \tilde{p}_B,$$

and for four genes, the use of genotypic data requires us to work with the composite disequilibrium

$$\tilde{\Delta}_{AABB} = \tilde{P}_{AB}^{AB} - 2\tilde{p}_A \tilde{D}_{ABB} - 2\tilde{p}_B \tilde{D}_{AAB} - 2\tilde{p}_A \tilde{p}_B \tilde{\Delta}_{AB} - \tilde{\Delta}_{AB}^2$$
$$- \tilde{p}_A^2 \tilde{D}_B - \tilde{p}_B^2 \tilde{D}_A - \tilde{D}_A \tilde{D}_B - \tilde{p}_A^2 \tilde{p}_B^2.$$

These maximum likelihood estimates are biased, since they involve products of multinomial variables. The expectation of the squared gene frequency over replicate samples of size n is

$$E(\tilde{p}_A^2) = p_A^2 + [p_A(1 - p_A) + D_A]/2n,$$

leading to

$$E(\tilde{D}_A) = D_A - [p_A(1 - p_A) + D_A]/2n.$$

We are going to invoke large-sample theory for testing hypotheses about disequilibria, and any bias corrections we made would not affect the test statistics. We choose, therefore, to ignore the biases of order $1/n$ and stay with the maximum likelihood estimates.

VARIANCES

Variances of linear combinations of multinomial variables are known exactly, so that for samples of n individuals, for example,

$$\text{var}(\tilde{P}_{AB}^{AB}) = P_{AB}^{AB}(1 - P_{AB}^{AB})/n$$

$$\text{var}(\tilde{P}_A^A) = P_A^A(1 - P_A^A)/n$$

$$\text{var}(\tilde{p}_A) = (\pi_A + D_A)/2n.$$

The variances of disequilibrium estimates, which involve quadratic functions of observations, also follow from multinomial theory, and with the assumption of large samples we can use Fisher's expression for the approximate variance of a function T of multinomial observations n_i, $\sum_i n_i = n$, whose expectations are $E(n_i) = n\phi_i$:

$$\text{var}(T)/n = \sum_i (\partial T/\partial n_i)^2 \phi_i - (\partial T/\partial n)^2$$

At a single locus, if the counts for genotypes AA, Aa, aa are written as n_1, n_2, n_3, so that

$$\phi_1 = P_A^A, \qquad \phi_2 = 2P_a^A, \qquad \phi_3 = P_a^a;$$

then writing

$$T = \tilde{D}_A = n_1/n - (n_1 + n_2/2)^2/n^2,$$

and applying Fisher's result, gives

$$\text{var}(\tilde{D}_A) = [\pi_A^2 + \tau_A^2 D_A - D_A^2]/n.$$

The large-sample variances for the digenic measures that gametic information allows are

$$\text{var}(\tilde{D}_{AB}) = [\pi_A \pi_B + \tau_A \tau_B D_{AB} + D_A D_B - D_{AB}^2 + D_{A/B}^2 + D_{AB}^{AB}]/2n$$

$$\text{var}(\tilde{D}_{A/B}) = [\pi_A \pi_B + \tau_A \tau_B D_{A/B} + D_A D_B + D_{AB}^2 - D_{A/B}^2 + D_{AB}^{AB}]/2n,$$

while for the composite measure used for genotypic data,

$$\begin{aligned}
\text{var}(\tilde{\Delta}_{AB}) = [&(\pi_A + D_A)(\pi_B + D_B) + \tau_A \tau_B \Delta_{AB}/2 \\
&+ \tau_A D_{ABB} + \tau_B D_{AAB} + \Delta_{AABB}]/n.
\end{aligned}$$

Although this variance follows from Fisher's result, it can also be found by dropping terms of order $1/n^2$ in the exact expression given earlier (Weir 1979).

Variances for trigenic and quadrigenic disequilibria are obtained similarly, but there is always the possibility of making algebraic errors. We are grateful to Dr. Lisa D. Brooks for obtaining them for us with the computer algebra package MACSYMA. The results are

$$\begin{aligned}
2n\,\text{var}(\tilde{D}_{AAB}) = &(\pi_A^2 + \tau_A^2 D_A - D_A^2)(\pi_B + D_B) + \pi_A \tau_A \tau_B \Delta_{AB} \\
&+ (1 - 5\pi_A + D_A)\Delta_{AB}^2 + 2\pi_A \tau_A D_{ABB} \\
&+ (\tau_A^2 \tau_B - 2D_A \tau_B - 4\tau_A \Delta_{AB})D_{AAB} - 2D_{AAB}^2 \\
&+ (\tau_A^2 - 2D_A)(D_{AB}^{AB} - 2D_{AB}D_{A/B})
\end{aligned}$$

$$\begin{aligned}
n\,\text{var}(\tilde{\Delta}_{AABB}) = &(\pi_A^2 + \tau_A^2 D_A - D_A^2)(\pi_B^2 + \tau_B^2 D_B - D_B^2) \\
&+ 2\tau_A \tau_B(\pi_A \pi_B - 4D_A D_B)\Delta_{AB} \\
&+ (\tau_A^2 \tau_B^2 - 4\tau_B^2 D_A - 4\tau_A^2 D_B + 4D_A D_B + 2D_A + 2D_B)\Delta_{AB}^2 \\
&- 6\tau_A \tau_B \Delta_{AB}^3 + 3\Delta_{AB}^4 \\
&+ [2\pi_B \tau_B(\tau_A^2 - 2D_A) + 4\Delta_{AB}(2\tau_A D_B - 2\tau_A \pi_B + \tau_B \Delta_{AB})]D_{AAB} \\
&+ [2\pi_A \tau_A(\tau_B^2 - 2D_B) + 4\Delta_{AB}(2\tau_B D_A - 2\tau_B \pi_A + \tau_A \Delta_{AB})]D_{ABB} \\
&+ 2(3D_B - \pi_B)D_{AAB}^2 + 2(3D_A - \pi_A)D_{ABB}^2 + 20\Delta_{AB}D_{AAB}D_{ABB} \\
&+ [(\tau_A^2 - 2D_A)(\tau_B^2 - 2D_B) - 8\tau_A \tau_B \Delta_{AB} + 6\Delta_{AB}^2 \\
&- 4\tau_A D_{ABB} - 4\tau_B D_{AAB}]\Delta_{AABB} - \Delta_{AABB}^2.
\end{aligned}$$

When all ten genotypes are available, we need the variance of D_{AB}^{AB}.

$$n \, \text{var}(\tilde{D}_{AB}^{AB}) = (\pi_A^2 + \tau_A^2 D_A - D_A^2)(\pi_B^2 + \tau_B^2 D_B - D_B^2)$$

$$+ (D_{AB}^2 + D_{A/B}^2)^2$$

$$+ (\tau_A^2 \tau_B^2 - 2\pi_A \pi_B - 2\pi_A^2 D_B - 2\pi_B^2 D_A + 2D_A D_B)(D_{AB}^2 + D_{A/B}^2)$$

$$- 4\tau_A \tau_B D_{AB} D_{A/B}(D_{AB} + D_{A/B}) + 12(\pi_A D_B + \pi_B D_A)D_{AB} D_{A/B}$$

$$- 2\tau_A \tau_B (D_{AB}^3 + D_{A/B}^3) + 12 D_{AAB} D_{ABB}(D_{AB} + D_{A/B})$$

$$+ 2\tau_A \tau_B (\pi_A \pi_B - 2D_A D_B)(D_{AB} + D_{A/B})$$

$$+ 2(3D_B - 2\pi_B)D_{AAB}^2 + (3D_A - \pi_A)D_{ABB}^2$$

$$+ [2\pi_B \tau_B (\tau_A^2 - 2D_A) + 4\tau_B D_{AB} D_{A/B} + 4\tau_A (D_B - \pi_B)(D_{AB} + D_{A/B})]D_{AAB}$$

$$+ [2\pi_A \tau_A (\tau_B^2 - 2D_B) + 4\tau_A D_{AB} D_{A/B} + 4\tau_B (D_A - \pi_A)(D_{AB} + D_{A/B})]D_{ABB}$$

$$+ [(\tau_A^2 - 2D_A)(\tau_B^2 - 2D_B) - 4\tau_A \tau_B (D_{AB} + D_{A/B}) + 4(D_{AB}^2 + D_{A/B}^2)$$

$$- 4\tau_A D_{ABB} - 4\tau_B D_{AAB}]D_{AB}^{AB} - (D_{AB}^{AB})^2.$$

We have checked these two quadrigenic variances by verifying that they differ by appropriate terms involving the variance of $D_{AB}D_{A/B}$ and the covariance of D_{AB}^{AB} and $D_{AB}D_{A/B}$.

TESTING

With a maximum likelihood estimate $\tilde{D}$ of a disequilibrium D, and an expression $\text{var}(\tilde{D})$ for its variance, we appeal to asymptotic normality and use the statistic

$$X^2 = \tilde{D}^2/\text{var}(\tilde{D})$$

to test the hypothesis H: D = 0. When the hypothesis is true, this statistic is expected to be distributed as chi-square with one degree of freedom. Of course, we do not know the parametric values to substitute into the variance expressions, and so we use observed or estimated values. We also invoke the hypothesis being tested, and set that particular D to zero in the variance expression.

At a single locus, we recover a familiar expression for testing for Hardy-Weinberg equilibrium:

$$X_A^2 = n\tilde{D}_A^2/\tilde{\pi}_A^2.$$

For all other disequilibria, though, the variance formulas involve the other coefficients of disequilibrium as well, and, for example, the test statistic for composite linkage disequilibrium is

$$X_{AB}^2 = n\tilde{\Delta}_{AB}^2/[(\tilde{\pi}_A + \tilde{D}_A)(\tilde{\pi}_B + \tilde{D}_B) + \tilde{\tau}_A \tilde{D}_{ABB} + \tilde{\tau}_B \tilde{D}_{AAB} + \tilde{\Delta}_{AABB}].$$

A procedure that is more in line with what is done in testing for statistical interactions in analyses of variance is to test for the higher-order disequilibria first. If there is found to be evidence for such disequilibria, then these terms must be retained in the variances of lower-order disequilibria. In the absence of significant high-order disequilibria, though, these terms may be dropped in lower-order tests. If there is no evidence for trigenic or quadrigenic disequilibrium in a particular population then, the test for composite linkage disequilibrium may be conducted with the statistic

$$X^2_{AB} = n\tilde{\Delta}^2_{AB}/[(\tilde{\pi}_A + \tilde{D}_A)(\tilde{\pi}_B + \tilde{D}_B)].$$

This is the quantity we have used in the past (Weir 1979; Laurie-Ahlberg and Weir 1979; Barker, East, and Weir 1986; Weir and Brooks 1986). The use of this test does require the prior exclusion of trigenic and quadrigenic disequilibrium, however.

We emphasize that the test statistics use estimated variances in the denominators, but that these estimates incorporate the hypotheses and maybe the results of prior tests. It would not be appropriate to use variances estimated by numerical resampling since these ignore the hypotheses. For example, we found by simulation that Hardy-Weinberg tests based on jackknife variance estimates performed poorly when gene frequencies took values away from 0.5 and the term $2\tilde{\tau}^2_A\tilde{D}_A$ had a large effect.

Numerical Results

The performance of the estimators and test statistics for the six disequilibrium coefficients D_A, D_B, Δ_{AB}, D_{AAB}, D_{ABB}, and Δ_{AABB} was investigated by simulation. For specified values of the six disequilibria and the two gene frequencies p_A, p_B, the eight independent two-locus genotypic frequencies were constructed using the expressions in Table 6.2, and 10,000 replicate samples of size $n = 100$ were drawn from such populations. The disequilibria were estimated and tested with the equations given above. Tests were performed for higher-order disequilibria first, and such terms were used in the lower-order tests only if found to be significantly different from zero.

In Tables 6.6 and 6.7 we present the simulation results for the case where the population had no disequilibrium, and all the D's were zero. Overall, the results are very satisfactory, and several points can be made. Bias is seen to be negligible, as expected, since it results only from factors such as $(2n - 1)/2n$. The variance and bias of each estimate increases as population gene frequencies become more intermediate (π becomes larger), also as expected. Variances of the estimates were less for the higher-order

Table 6.6

Properties of disequilibrium estimates when there is no disequilibrium.
(From 10,000 samples of size 100.)

$p_A = p_B$		$\tilde{D}_A$	$\tilde{D}_B$	$\tilde{\Delta}_{AB}$	$\tilde{D}_{AAB}$	$\tilde{D}_{ABB}$	$\tilde{\Delta}_{AABB}$
0.1	Mean	−0.0005	−0.0004	0.0000	0.0000	0.0000	−0.0001
	Std. Dev.	0.0088	0.0091	0.0089	0.0019	0.0019	0.0008
0.3	Mean	−0.0007	−0.0010	0.0002	0.0000	0.0000	−0.0004
	Std. Dev.	0.0210	0.0209	0.0210	0.0067	0.0067	0.0043
0.5	Mean	−0.0011	−0.0015	0.0000	0.0000	0.0001	−0.0007
	Std. Dev.	0.0250	0.0249	0.0246	0.0087	0.0087	0.0062

Table 6.7

Properties of test statistics when there is no disequilibrium.
(From 10,000 samples of size 100.)

$p_A = p_B$		X_A^2	X_B^2	X_{AB}^2	X_{AAB}^2	X_{ABB}^2	X_{AABB}^2
0.1	Mean	0.93	0.98	0.98	0.72	0.77	0.17
	Std. Dev.	1.30	1.43	1.35	2.12	4.59	0.90
	Power*	0.04	0.04	0.04	0.03	0.03	0.00
0.3	Mean	1.01	0.99	1.02	1.01	1.02	1.08
	Std. Dev.	1.41	1.36	1.44	1.43	1.37	3.48
	Power*	0.05	0.05	0.05	0.05	0.05	0.05
0.5	Mean	1.01	1.01	1.00	1.01	1.00	1.04
	Std. Dev.	1.44	1.43	1.38	1.41	1.42	1.46
	Power*	0.05	0.06	0.05	0.05	0.05	0.05

* Proportion of time that X^2 exceeded 3.84, the 5% critical value of $\chi^2_{(1)}$.

than for the lower-order disequilibria. If the test statistics are judged by power, then their performance is very good, with the possible exception of tests for composite quadrigenic disequilibrium at extreme gene frequencies—although, even here, the power is less rather than greater the nominal value. The power is expected to be 0.05 since we are using a 5% significance level and rejecting hypotheses when the test statistics exceed 3.84. (Power and significance levels are equal when the hypothesis being tested is true.) The means and standard deviations of the test statistics are also close to the chi-square values of 1.00 and 1.41, again with better agreement for intermediate gene frequencies.

One aspect of the tests featured in Table 6.7 needs some comment. When observed values are substituted for parameters in variance formulas for the various disequilibria, it can happen that the resulting expression is negative. We have taken a conservative attitude in saying that such cases give no basis for testing the hypothesis of that particular disequilibrium being zero, and we have excluded those simulation runs from Table 6.7. The problem arises only for trigenic and quadrigenic disequilibria, and only for extreme gene frequencies. The numbers of tests performed for the coefficients D_{AAB}, D_{ABB}, Δ_{AABB} when $p_A = p_B = 0.1$ were 7997, 8305, and 5074, respectively, out of 10,000 simulated data sets. Similar figures in the same situation were found in the tests shown in later tables.

We undertook a study of the effects of nonzero values for each of the various disequilibria on estimates and tests for the remaining coefficients. The results are now discussed, but we should say that the nonzero values were chosen to represent fairly extreme situations rather than those likely to be met in practice. In Tables 6.8 and 6.9 we illustrate the effect of departures from Hardy-Weinberg equilibrium on the estimates and test statistics. We took disequilibrium to be the same at each of the two loci, and equal to half the maximum or half the minimum possible values [i.e., for

Table 6.8

Properties of estimates when there is Hardy-Weinberg disequilibrium. (From 10,000 samples of size 100.)

$p_A = p_B$	$D_A = D_B$		$\tilde{D}_A$	$\tilde{D}_B$	$\tilde{\Delta}_{AB}$	$\tilde{D}_{AAB}$	$\tilde{D}_{ABB}$	$\tilde{\Delta}_{AABB}$
0.1	−0.0050	Mean	−0.0053	−0.0054	0.0000	0.0000	0.0000	−0.0001
		Std. Dev.	0.0070	0.0070	0.0085	0.0014	0.0014	0.0005
0.1	0.0450	Mean	0.0442	0.0447	−0.0002	−0.0001	0.0000	−0.0002
		Std. Dev.	0.0185	0.0185	0.0133	0.0047	0.0047	0.0033
0.3	−0.0450	Mean	−0.0459	−0.0458	0.0000	0.0000	0.0000	−0.0003
		Std. Dev.	0.0186	0.0186	0.0162	0.0053	0.0053	0.0034
0.3	0.1050	Mean	0.1036	0.1035	−0.0003	0.0000	−0.0001	−0.0011
		Std. Dev.	0.0222	0.0221	0.0315	0.0087	0.0087	0.0050
0.5	−0.1250	Mean	−0.1255	−0.1257	−0.0001	0.0000	0.0001	−0.0002
		Std. Dev.	0.0215	0.0216	0.0126	0.0054	0.0053	0.0045
0.5	0.1250	Mean	0.1226	0.1233	−0.0002	0.0002	0.0002	−0.0014
		Std. Dev.	0.0218	0.0217	0.0368	0.0093	0.0095	0.0051

Table 6.9

Properties of test statistics when there is Hardy-Weinberg disequilibrium.
(From 10,000 samples of size 100.)

$p_A = p_B$	$D_A = D_B$		X^2_A	X^2_B	X^2_{AB}	X^2_{AAB}	X^2_{ABB}	X^2_{AABB}
0.1	−0.0050	Mean	0.89	0.87	1.01	0.61	0.60	0.09
		Std. Dev.	0.80	0.80	1.41	4.61	2.14	0.20
		Power*	0.01	0.01	0.04	0.03	0.03	0.00
0.1	0.0450	Mean	25.82	26.14	0.90	0.83	0.85	0.65
		Std. Dev.	14.36	14.38	1.12	1.04	1.04	0.65
		Power*	0.96	0.96	0.03	0.02	0.02	0.00
0.3	−0.0450	Mean	5.49	5.48	0.98	1.02	1.03	1.40
		Std. Dev.	3.71	3.73	1.39	1.53	1.50	13.79
		Power*	0.61	0.61	0.05	0.05	0.05	0.06
0.3	0.1050	Mean	25.52	25.46	1.03	1.00	1.00	0.97
		Std. Dev.	9.36	9.31	1.42	1.37	1.38	1.31
		Power*	1.00	1.00	0.05	0.05	0.05	0.05
0.5	−0.1255	Mean	26.07	26.17	1.02	1.04	1.01	1.06
		Std. Dev.	8.64	8.69	1.39	1.44	1.41	1.47
		Power*	1.00	1.00	0.05	0.05	0.05	0.05
0.5	0.1250	Mean	25.20	25.44	0.98	0.99	1.05	0.92
		Std. Dev.	8.66	8.64	1.38	1.38	1.48	1.23
		Power*	1.00	1.00	0.05	0.05	0.06	0.04

* Proportion of time that X^2 exceeded 3.84, the 5% critical value of $\chi^2_{(1)}$.

locus A, $p_A(1 - p_A)/2$ or $-p_A^2/2$, respectively]. The estimated disequilibria remained close to the theoretical values, but with variances reflecting the values of D_A and D_B, while the test statistics for the higher-order disequilibria continued to be satisfactory and were essentially unchanged by the single-locus values as required. Problems with negative variances were more likely to be found for negative disequilibria. We note a very high standard deviation for the quadrigenic coefficient at one point in Table 6.9.

The effects of linkage disequilibria on the test statistics are shown in Tables 6.10 and 6.11. In Table 6.10 just the gametic disequilibrium, D_{AB}, was set to one-fourth of its minimum value $(-p_A p_B/4)$ or one-fourth its maximum value $[p_A(1 - p_B)/4]$ (these extrema assume all other disequilibria are zero), and the tests performed quite well. Nonzero values for

Table 6.10

Properties of test statistics when there is gametic linkage disequilibrium. (From 10,000 samples of size 100.)

$p_A = p_B$	D_{AB}		X_A^2	X_B^2	X_{AB}^2	X_{AAB}^2	X_{ABB}^2	X_{AABB}^2
0.1	-0.0025	Mean	0.98	0.95	0.97	0.70	0.71	0.11
		Std. Dev.	1.50	1.40	1.22	2.17	2.93	0.25
		Power*	0.04	0.04	0.03	0.04	0.03	0.00
0.1	0.0225	Mean	0.98	0.99	7.48	0.75	0.77	0.47
		Std. Dev.	1.45	1.47	6.25	0.96	1.01	0.62
		Power*	0.04	0.04	0.65	0.02	0.02	0.00
0.3	-0.0225	Mean	0.98	1.02	2.08	1.02	1.01	1.10
		Std. Dev.	1.37	1.43	2.37	1.46	1.44	4.11
		Power*	0.05	0.05	0.18	0.05	0.05	0.05
0.3	0.0525	Mean	1.00	0.97	7.29	1.02	1.03	1.03
		Std. Dev.	1.39	1.36	5.19	1.39	1.41	1.44
		Power*	0.05	0.04	0.71	0.05	0.05	0.05
0.5	-0.0625	Mean	1.00	1.02	7.15	1.03	1.03	0.99
		Std. Dev.	1.41	1.45	4.89	1.44	1.42	1.42
		Power*	0.05	0.06	0.72	0.05	0.05	0.05
0.5	0.0625	Mean	1.03	1.00	7.20	1.00	1.02	1.01
		Std. Dev.	1.45	1.42	4.95	1.38	1.42	1.38
		Power*	0.06	0.06	0.71	0.05	0.05	0.05

* Proportion of time that X^2 exceeded 3.84, the 5% critical value of $\chi_{(1)}^2$.

the composite coefficient Δ_{AB} could be achieved with other arrangements however, and Table 6.11 shows what happens when the gametic and non-gametic coefficients are both allowed to be nonzero. This leads to nonzero values for the composite quadrigenic coefficient Δ_{AABB}, as is reflected in the increased power for the test statistic Δ_{AABB}. When D_{AB} and $D_{A/B}$ are equal but of opposite signs, Δ_{AB} is zero, and the test statistic X_{AB}^2 reflects this, but there are high values for X_{AABB}^2.

The effects of trigenic and quadrigenic disequilibria are shown in Tables 6.12 and 6.13. Again it can be seen that the presence of neither order of disequilibrium has any effect on the tests performed on Hardy-Weinberg or linkage disequilibrium. We used the values $-p_A^2 p_B/4$ and $p_A^2 p_b/4$ for the trigenic case and $-p_A^2 p_B^2/4$ and $p_A^2(1 - p_b^2)/4$ in the quadrigenic case.

Table 6.11
Properties of test statistics when there is linkage disequilibrium.
(From 10,000 samples of size 100.)

$p_A = p_B$	D_{AB}	$D_{A/B}$		X_A^2	X_B^2	X_{AB}^2	X_{AAB}^2	X_{ABB}^2	X_{AABB}^2
0.1	−0.0025	−0.0025	Mean	0.96	0.98	1.10	0.69	0.86	0.08
			Std. Dev.	1.37	1.47	1.26	3.31	9.88	0.16
			Power*	0.04	0.04	0.04	0.03	0.03	0.00
0.1	−0.0025	0.0225	Mean	0.97	0.97	6.23	0.77	0.76	0.46
			Std. Dev.	1.41	1.43	5.65	1.01	1.02	0.63
			Power*	0.04	0.04	0.57	0.02	0.02	0.00
0.1	0.0225	0.0225	Mean	0.96	0.98	26.00	0.73	0.74	0.78
			Std. Dev.	1.43	1.44	15.42	0.87	0.89	0.81
			Power*	0.04	0.04	0.99	0.01	0.02	0.01
0.3	−0.0225	−0.0225	Mean	1.02	0.98	5.35	1.04	1.02	1.20
			Std. Dev.	1.40	1.34	3.84	1.62	1.51	6.95
			Power*	0.05	0.05	0.59	0.05	0.05	0.04
0.3	−0.0225	0.0525	Mean	1.01	1.03	3.02	1.05	1.05	1.19
			Std. Dev.	1.43	1.41	3.24	1.46	1.45	1.67
			Power*	0.05	0.05	0.30	0.06	0.06	0.07
0.3	0.0525	0.0525	Mean	1.02	1.00	27.86	1.02	1.03	1.97
			Std. Dev.	1.41	1.38	11.42	1.55	1.53	2.16
			Power*	0.05	0.05	1.00	0.05	0.05	0.16
0.5	−0.0625	−0.0625	Mean	1.00	1.02	26.60	1.00	1.01	1.90
			Std. Dev.	1.40	1.45	8.12	1.41	1.39	1.76
			Power*	0.05	0.06	1.00	0.05	0.05	0.14
0.5	−0.0625	0.0625	Mean	1.01	0.99	1.09	1.02	1.00	2.34
			Std. Dev.	1.40	1.42	1.57	1.44	1.44	2.75
			Power*	0.05	0.05	0.06	0.05	0.05	0.21
0.5	0.0625	0.0625	Mean	1.01	1.02	26.49	1.02	1.03	1.87
			Std. Dev.	1.41	1.44	8.17	1.46	1.43	1.74
			Power*	0.05	0.06	1.00	0.05	0.05	0.13

* Proportion of time that X^2 exceeded 3.84, the 5% critical value of $\chi_{(1)}^2$.

Recall that the higher-order terms were included, via the variance formulas, in the tests for lower-order tests only if they were judged to be significantly nonzero by the tests discussed here.

The only feature of Tables 6.9 to 6.13 that causes us some concern is the large standard deviation for X_{AABB}^2, especially when gene frequencies

Table 6.12
Properties of test statistics when there is trigenic disequilibrium.
(From 10,000 samples of size 100.)

$p_A = p_B$	D_{AAB}	D_{ABB}		X_A^2	X_B^2	X_{AB}^2	X_{AAB}^2	X_{ABB}^2	X_{AABB}^2
0.1	−0.0002	zero	Mean	0.96	0.98	0.98	0.66	0.66	0.14
			Std. Dev.	1.41	1.46	1.39	2.28	1.65	0.31
			Power*	0.04	0.04	0.04	0.02	0.03	0.00
0.1	−0.0002	0.0022	Mean	0.96	0.96	1.06	0.61	2.91	0.28
			Std. Dev.	1.38	1.40	1.43	4.25	10.05	0.54
			Power*	0.04	0.04	0.05	0.02	0.25	0.00
0.1	0.0022	zero	Mean	0.98	0.97	1.06	2.73	0.74	0.29
			Std. Dev.	1.45	1.50	1.41	4.81	2.56	0.56
			Power*	0.04	0.04	0.05	0.25	0.04	0.00
0.1	0.0022	0.0022	Mean	0.98	0.99	1.21	2.90	3.08	0.36
			Std. Dev.	1.47	1.51	1.55	10.12	17.93	0.59
			Power*	0.04	0.04	0.07	0.23	0.23	0.00
0.3	−0.0067	zero	Mean	1.04	1.00	0.97	1.89	0.92	1.14
			Std. Dev.	1.44	1.40	1.39	2.13	1.28	2.76
			Power*	0.05	0.05	0.04	0.16	0.04	0.06
0.3	−0.0067	0.0157	Mean	1.01	0.98	1.02	1.84	6.47	1.05
			Std. Dev.	1.40	1.35	1.45	2.20	4.50	1.42
			Power*	0.05	0.05	0.05	0.16	0.67	0.05
0.3	0.0157	zero	Mean	1.02	1.00	1.05	6.26	1.06	1.04
			Std. Dev.	1.40	1.39	1.47	4.79	1.46	1.41
			Power*	0.05	0.05	0.05	0.63	0.06	0.05
0.3	0.0157	0.0157	Mean	1.00	1.02	1.10	5.44	5.44	1.07
			Std. Dev.	1.38	1.42	1.54	4.62	4.27	1.44
			Power*	0.05	0.05	0.06	0.56	0.57	0.06
0.5	−0.0313	zero	Mean	1.02	1.05	1.01	13.35	1.02	1.03
			Std. Dev.	1.43	1.47	1.43	6.30	1.46	1.45
			Power*	0.06	0.06	0.05	0.96	0.05	0.06
0.5	0.0313	zero	Mean	1.02	1.00	1.00	13.24	1.04	1.02
			Std. Dev.	1.45	1.38	1.41	6.29	1.46	1.42
			Power*	0.06	0.05	0.05	0.96	0.06	0.05

* Proportion of time that X^2 exceeded 3.84, the 5% critical value of $\chi_{(1)}^2$.

Table 6.13

Properties of test statistics when there is quadrigenic disequilibrium.
(From 10,000 samples of size 100.)

$p_A = p_B$	D_{AB}^{AB}		X_A^2	X_B^2	X_{AB}^2	X_{AAB}^2	X_{ABB}^2	X_{AABB}^2
0.1	-0.0000	Mean	1.01	0.95	1.00	0.80	0.75	0.14
		Std. Dev.	1.59	1.38	1.43	5.05	2.39	0.27
		Power*	0.04	0.04	0.04	0.03	0.04	0.00
0.1	0.0005	Mean	0.99	0.97	1.00	0.71	0.73	0.40
		Std. Dev.	1.53	1.43	1.38	1.89	2.55	0.86
		Power*	0.04	0.04	0.04	0.03	0.04	0.00
0.3	-0.0020	Mean	1.01	0.99	0.99	0.99	1.03	1.26
		Std. Dev.	1.40	1.37	1.39	1.42	1.44	1.85
		Power*	0.05	0.05	0.05	0.05	0.05	0.07
0.3	0.0115	Mean	1.00	1.01	1.08	1.07	1.05	9.27
		Std. Dev.	1.39	1.40	1.55	1.49	1.45	34.35
		Power*	0.05	0.05	0.06	0.06	0.05	0.73
0.5	-0.0156	Mean	1.02	1.03	1.01	1.02	1.01	7.60
		Std. Dev.	1.44	1.45	1.43	1.43	1.37	5.26
		Power*	0.06	0.06	0.05	0.05	0.05	0.73
0.5	0.0469	Mean	1.01	1.01	1.07	1.06	1.02	54.34
		Std. Dev.	1.43	1.44	1.55	1.50	1.47	12.05
		Power*	0.05	0.05	0.06	0.06	0.05	1.00

* Proportion of time that X^2 exceeded 3.84, the 5% critical value of $\chi_{(1)}^2$.

are 0.3. We can offer no explanation for this, but have noticed that very large values of the test statistic are obtained for data sets with large numbers of double heterozygotes.

Bounds on Disequilibria

Recently (Weir and Brooks 1986) we presented expressions that showed how disequilibria of one order may impose bounds on the range of values that may be taken by those of higher order. This work continued that for gametic disequilibria given by Thomson and Baur (1984).

For linkage disequilibria, the same bounds apply for D_{AB} and $D_{A/B}$ but are shown here for the gametic term:

$$D_{min} \leq D_{AB} \leq D_{max},$$

where

$$D_{max} = p_A p_b \quad \text{if} \quad D_{AB} > 0, P_A^A < P_B^B, P_a^a > P_b^b$$
$$= X - Y \quad \text{if} \quad D_{AB} > 0, P_A^A < P_B^B, P_a^a < P_b^b$$
$$= p_a p_B \quad \text{if} \quad D_{AB} > 0, P_A^A > P_B^B, P_a^a < P_b^b$$
$$= X + Y \quad \text{if} \quad D_{AB} > 0, P_A^A > P_B^B, P_a^a > P_b^b$$
$$D_{min} = -p_a p_B \quad \text{if} \quad D_{AB} < 0, P_a^a > P_B^B, P_A^A < P_b^b$$
$$= -Z - Y \quad \text{if} \quad D_{AB} < 0, P_a^a > P_B^B, P_A^A > P_b^b$$
$$= -p_a p_b \quad \text{if} \quad D_{AB} < 0, P_a^a < P_B^B, P_A^A > P_b^b$$
$$= -Z + Y \quad \text{if} \quad D_{AB} < 0, P_a^a < P_B^B, P_A^A < P_b^b,$$

and $X = (p_A p_b + p_a p_B)/2$, $Y = (P_a^A - P_b^B)/4$, $Z = (p_A p_B + p_a p_b)/2$.
For three genes, the bounds are exemplified by

$$D_{min} \le D_{AAB} \le D_{max},$$

where

$$D_{max} = \min[(P_A^A p_b - p_A \Delta_{AB}), (P_a^a p_b + p_a \Delta_{AB}), (P_a^A p_B + \tau_A \Delta_{AB})/2]$$
$$D_{min} = \max[-(P_A^A p_B + p_A \Delta_{AB}), -(P_a^a p_B - p_a \Delta_{AB}), -(P_a^A p_b - \tau_A \Delta_{AB})/2].$$

These bounds are such that the three-gene disequilibria may sometimes be bounded away from zero. The bounds are all sample phenomena, and we recommend that the usual testing procedures be conducted. Further work is needed for tests conditioned on constraints, however.

Log-Linear Models

Our approach is to express disequilibrium in a series of additive components, with genotypic frequencies being expressed as a sum of terms involving the disequilibria and gene frequencies. This method allows the relationship between disequilibria and descent measures, and has the advantage that estimates can be given explicitly. Besides allowing for a discussion of the sampling properties of these estimates, the explicit expressions make it clear that the disequilibria are unaffected by terms of different order. We will always estimate p_A, for example, by the sample frequency of allele A regardless of whether or not we make inferences about any of the disequilibria. Similarly, D_A is estimated as $\tilde{P}_A^A - \tilde{p}_A^2$ whether or not we are making inferences about other disequilibria. The disadvantage of our approach is that it does not allow an easy generalization to more loci, and even the extension to more alleles at two loci entails some algebraic care.

A different approach is often advocated under the general heading of log-linear models. As they are generally applied, however, the models are not fitted on a logarithmic scale, but departures from the model are considered to be added to a multiplicative model in the original data scale. We have shown (Weir and Wilson 1986) that the full model for two loci can be written as

$$P_{AB}^{AB} = MM_A^2M_B^2M_{AA}M_{BB}M_{AB}^2M_{A/B}^2M_{AAB}^2M_{ABB}^2M_{AB}^{AB},$$

where each of the multiplicative terms M_X corresponds to one of the disequilibria D_X in our additive model. The first term M is a normalizing factor. There is still overparametrization when the two double heterozygotes cannot be distinguished, and we pointed out that one approach would be to set

$$M_{AB}^{AB}/(M_{AB}M_{A/B})^2 = 1,$$

which is equivalent to setting

$$\Delta_{AABB} = 0$$

(i.e., ignoring these terms).

Testing for disequilibria in this framework requires the fitting of the model with and without the terms of interest, and comparing the resulting multinomial likelihoods with a likelihood ratio test statistic.

The multiplicative model has the disadvantages of not allowing explicit expressions for the estimates of the component terms, and of not therefore having an easily interpretable meaning for the various terms. Estimates of the gene effects M_A and M_B, for example, will change according to which other terms are included in the model. The use of numerical methods to fit and test the models, without deriving estimates and their variances, does give the advantage of easy extension to more complex situations.

Discussion

This chapter has been concerned with extracting as much information as possible about the associations between pairs of loci from genotypic data. In the common two-allele situation, the nine distinguishable genotypic classes allow the estimation of two gene frequencies, and six additional quantities. We suggest that these all be estimated as a routine step in the analysis of such data.

A common objection to this procedure is that the three- and four-gene measures are not biologically meaningful, a viewpoint that is consistent with the traditional equating of interlocus interaction to the gametic linkage disequilibrium coefficient D_{AB}. Apart from the drawback that such a gametic measure is not generally estimable from genotypic data, this viewpoint suggests that it is more meaningful biologically to consider the interaction of genes on gametes than in genotypes. While the higher-order terms are generally smaller than digenic measures, as statistical measures of frequency association they appear to us to all have equal biological relevance.

It is true that intergametic disequilibria are expected to be very small when there is random mating, but there are situations, such as in mixed self- and random-mating populations (Weir, Allard, and Kahler 1974) where the quadrigenic disequilibrium can exceed the gametic linkage disequilibrium. This phenomenon was first discussed by Bennett and Binet (1956).

The six single degree-of-freedom chi-square test statistics presented here do not necessarily correspond to other partitionings. It is possible, for example, to arrange the nine distinguishable two-locus frequencies as a three-by-three contingency table with single-locus genotypic frequencies as row and column totals. The four degree-of-freedom contingency table chi-square that can be calculated, which is conditional on the one-locus frequencies and disequilibria, does not correspond to the sum of X^2_{AB}, X^2_{AAB}, X^2_{ABB}, and X^2_{AABB}.

The increase from one disequilibrium measure at a single locus to six measures for two loci does not augur well for multilocus data analysis, and there remains the problem of how to summarize associations within groups of genes (Karlin and Piazza 1981). There are good reasons for providing a comprehensive treatment of the two-locus situation, however. Principally, the two-locus measures are those needed to express the variances of single-locus statistics averaged over loci. One of the most common statistics used in the presentation of population genetic data is the heterozygosity averaged over all the loci scored. The variance of such an average statistic necessarily involves two-locus measures, as shown by Brown, Feldman, and Nevo (1980), although these authors did not go beyond the digenic coefficients. In another direction, we have shown here that the variance of digenic measures, such as D_{AB}, involve the quadrigenic measures.

Finally, in the face of disequilibrium of any order in their data, population geneticists need to consider why the loci they have studied cannot be regarded as having been sampled from a population free from disturbing forces.

Acknowledgments

This is Paper No. 10662 of the Journal Series of the North Carolina Agricultural Research Service, Raleigh, NC 27695-7601. This investigation was supported in part by NIH Research Grant GM 11546 from the National Institute of General Medical Sciences.

References

Allard, R. W.; Babbel, G. R.; Clegg, M. T.; and Kahler, A. L. 1972. Evidence for coadaptation in *Avena barbata*. *Proc. Natl. Acad. Sci. USA* 69: 3043–3048.

Barker, J.S.F.; East, P. D.; and Weir, B. S. 1986. Temporal and microgeographic variation in allozyme frequencies in a natural population of *Drosophila buzzatii*. *Genetics* 112: 577–611.

Bennett, J. H., and Binet, F. E. 1956. Association between Mendelian factors with mixed selfing and random mating. *Heredity* 10: 51–55.

Brown, A.H.D.; Feldman, M. W.; and Nevo, E. 1980. Multilocus structure of natural populations of *Hordeum spontaneum*. *Genetics* 96: 523–536.

Clegg, M. T.; Allard, R. W.; and Kahler, A. L. 1972. Is the gene the unit of selection? Evidence from two experimental plant populations. *Proc. Natl. Acad. Sci. USA* 69: 2474–2478.

Cockerham, C. Clark. 1969. Variance of gene frequencies. *Evolution* 23: 72–84.

Cockerham, C. Clark, and Weir, B. S. 1973. Descent measures for two loci with some applications. *Theor. Pop. Biol.* 4: 300–330.

Cockerham, C. Clark, and Weir, B. S. 1977. Digenic descent measures for finite populations. *Genet. Res.* 30: 121–147.

Karlin, S., and Piazza, A. 1981. Statistical methods for assessing linkage disequilibrium at the HLA, -A, B, C loci. *Ann. Hum. Genet.* 45: 79–94.

Laurie-Ahlberg, C. C., and Weir, B. S. 1979. Allozymic variation and linkage disequilibrium in some laboratory populations of *Drosophila melanogaster*. *Genetics* 92: 1295–1314.

Thomson, G., and Baur, M. 1984. Third-order linkage disequilibrium. *Tissue Antigens* 24: 250–255.

Weir, B. S. 1979. Inferences about linkage disequilibrium. *Biometrics* 35: 235–254.

Weir, B. S.; Allard, R. W.; and Kahler, A. L. 1974. Further analysis of complex allozyme polymorphisms in a barley population. *Genetics* 78: 911–919.

Weir, B. S., and Brooks, L. D. 1986. Disequilibrium on human chromosome 11p. *Genetic Epidemiology*, Supplement 1: 177–183.

Weir, B. S., and Cockerham, C. Clark. 1974. Behavior of pairs of genes in finite monoecious populations. *Theor. Pop. Biol.* 6: 323–354.

Weir, B. S., and Wilson, S. R. 1986. Log-linear models for linked loci. *Biometrics* 42: 665–670.

CHAPTER SEVEN

The Reduction Principle for Genetic Modifiers of the Migration Rate

Uri Liberman and Marcus W. Feldman

With an Appendix by

Kent E. Holsinger, Marcus W. Feldman, and Uri Liberman

Introduction

The extent to which the demes in a system of subpopulation are separated is one of the key parameters in the genetic evolution of each deme, and of the system considered as a whole. The interaction between migration and selection in such systems has recently been surveyed by Karlin (1982), especially with regard to conditions for the maintenance of genetic polymorphism. In the present paper we shall take the equilibrium structure engendered by the interaction of migration and selection as the point of departure from which to examine the evolution of a gene that controls the rate of migration in the system.

Recent theoretical studies on the evolution of dispersal can be partitioned into three approaches. The first is ecological and allows the population sizes of a single type of individual in different patches to obey growth laws characteristic of the patch. Passive dispersal is introduced via linear changes in the numbers in the different niches. Hastings (1983) showed that a new type of individual can invade only if its dispersal coefficient is less than that of the resident majority. Levin, Cohen, and Hastings (1984) generalized this situation to include temporal and spatial variation in the growth functions for the populations. In this case increased dispersal may succeed.

The second approach to the evolution of dispersal is that of Hamilton and May (1977) and Motro (1982a,b, 1983). Here a genetically determined fraction of the infinite population disperses uniformly over the whole range. Dispersal itself entails a fitness loss which is independent of the genotype. In this framework an *optimal* rate of dispersal exists, that is, a value of the dispersal rate such that any new genotype with a different dispersal rate cannot succeed. The "range" in these models consists of a set of sites that are reoccupied at every generation by a *single* individual. It is the competition for sites that induces a type of frequency-dependent selection and gives rise to the optimum.

The third class of models for the evolution of dispersal includes a gene that controls the rate of exchange between two populations. This migration-modifying locus is linked to a locus whose genotypes are under natural selection that is different in the two niches. Balkau and Feldman (1973) considered such a two-locus haploid system where the selection favored different alleles in the different niches. They showed that near an equilibrium of the system with an allele M_1 fixed at the modifying locus, a new allele, M_2, could succeed only if it produced less migration than M_1. Karlin and McGregor (1974) showed the same result to be true for a special class of diploid selection schemes where M_1M_1 was fixed and M_1M_2 increased only if it lowered the migration rate. This finding for an initial state of fixation at the modifier locus was generalized to arbitrary selection regimes by Teague (1977).

Asmussen (1976, 1983) combined the ecological and genetic modifier approach by taking the fitness (determined at the major gene) of the haploid genotypes in the two niches to be density dependent but of opposite direction in their response to density. Again a new migration-controlling allele succeeded only if it reduced migration below that of the resident modifier homozygote.

In all three approaches, if the selection acting on the system is constant in time, reduction of dispersal is favored unless those genetic differences that affect dispersal also affect fitness. In the latter case optimal rates of migration may exist. The present paper is a generalization of the modifier approach to the evolution of migration between two populations of diploids under different but constant selection. We remove the restriction that the populations be initially fixed at the modifier locus. This leads us naturally to examine polymorphic equilibria of the modifying gene and hence to questions of long-term evolution and evolutionary genetic stability.

7.1. The Model

Consider a population of diploid individuals divided into two demes. Our attention is focused on two gene loci. The first, with two alleles, A and a, controls the viability selection on each individual and is referred to as the "major" locus. The second locus has alleles $M_1, M_2, \ldots, M_n$ and its only function is to determine the rate of migration between the demes. Since the second locus has no effect on genotypic viabilities, it is referred to as the "modifier" locus. With both loci considered, the gametes are of the form

$$AM_i, aM_i, \qquad i = 1, 2, \ldots, n, \qquad (7.1.1)$$

while the diploid genotypes of zygotes are

$$\frac{AM_i}{AM_j}, \frac{AM_i}{aM_j}, \frac{aM_i}{aM_j}, \qquad i, j = 1, 2, \ldots, n. \qquad (7.1.2)$$

The two-population system is censused at the gametic stage, and each generation is defined by the following ordered sequence of events between two successive census points:

1. Random union of gametes in each deme is separate to produce diploid zygotes.
2. Differential survival among the zygotes in each deme with viabilities of the genotypes is determined by the major locus and the deme in which the zygotes reside.
3. Migration between the demes of the mature survivors of selection is at rates determined by the genotype at the modifier locus.
4. Recombination and Mendelian segregation occur in each deme to produce the next generation of gametes.

The state of the system is determined by the gamete frequencies in each deme. Denote by x_1 and $x_2 = 1 - x_1$ (y_1 and $y_2 = 1 - y_1$) the frequencies of A and a in deme I (deme II) at the present generation. Following the notation used by Lessard (1985), we represent the gamete frequencies as follows:

Gamete	AM_i	aM_i	
Deme I frequency	$x_1 p_i$	$x_2 q_i$	
Deme II frequency	$y_1 r_i$	$y_2 s_i$	(7.1.3)

where $0 \leq p_i, q_i, r_i, s_i \leq 1$ and $\sum_i p_i = \sum_i q_i = \sum_i r_i = \sum_i s_i = 1$. The population state is then fully specified by the two sets of vectors:

Deme I	Deme II	
$\mathbf{x} = (x_1, x_2)$	$\mathbf{y} = (y_1, y_2)$	
$\mathbf{p} = (p_1, p_2, \ldots, p_n)$	$\mathbf{r} = (r_1, r_2, \ldots, r_n)$	
$\mathbf{q} = (q_1, q_2, \ldots, q_n)$	$\mathbf{s} = (s_1, s_2, \ldots, s_n).$	(7.1.4)

The viabilities are determined by the genotypes at the major locus as follows:

Genotype	AA	Aa	aa
Deme I viability	w_{11}	w_{12}	w_{22}
Deme II viability	v_{11}	v_{12}	$v_{22}.$

$$(7.1.5)$$

The rates of migration are determined by the genotypes at the modifier locus independently of the deme in which the individual resides. Hence in either deme any individual of genotype $M_i M_j$ migrates to the other deme with probability m_{ij}, or remains in the same deme with probability $1 - m_{ij}$. The modifier locus therefore determines a (forward) migration matrix $\mathbf{M} = \|m_{ij}\|$, $i, j = 1, 2, \ldots, n$, which has nonnegative entries. We further assume that $\mathbf{M}$ is symmetric, that is,

$$m_{ij} = m_{ji} \quad \text{and} \quad 0 \le m_{ij} \le 1 \quad \text{for} \quad i, j = 1, 2, \ldots, n. \quad (7.1.6)$$

Transformation of Frequencies

Let $\mathbf{x}'$, $\mathbf{y}'$, $\mathbf{p}'$, $\mathbf{q}'$, $\mathbf{r}'$, $\mathbf{s}'$ be the frequency vectors of the population at the next generation. After random union of gametes and viability selection in each deme, migration between the demes, and recombination with Mendelian segregation, it is easily verified that the gametic frequencies at the next generation are given by

$$
Px_1' p_i' = \frac{w_{11} x_1^2}{w} p_i \sum_j (1 - m_{ij}) p_j + \frac{w_{12} x_1 x_2}{w} p_i \sum_j (1 - m_{ij}) q_j (1 - R)
$$

$$
+ \frac{w_{12} x_1 x_2}{w} q_i \sum_j (1 - m_{ij}) p_j R + \frac{v_{11} y_1^2}{v} r_i \sum_j m_{ij} r_j
$$

$$
+ \frac{v_{12} y_1 y_2}{v} r_i \sum_j m_{ij} s_j (1 - R) + \frac{v_{12} y_1 y_2}{v} s_i \sum_j m_{ij} r_j R
$$

$$
Px_2' q_i' = \frac{w_{22} x_2^2}{w} q_i \sum_j (1 - m_{ij}) q_j + \frac{w_{12} x_1 x_2}{w} q_i \sum_j (1 - m_{ij}) p_j (1 - R)
$$

$$
+ \frac{w_{12} x_1 x_2}{w} p_i \sum_j (1 - m_{ij}) q_j R + \frac{v_{22} y_2^2}{v} s_i \sum_j m_{ij} s_j
$$

$$
+ \frac{v_{12} y_1 y_2}{v} s_i \sum_j m_{ij} r_j (1 - R) + \frac{v_{12} y_1 y_2}{v} r_i \sum_j m_{ij} s_j R \quad (7.1.7)
$$

$$Qy_1'r_i' = \frac{v_{11}y_1^2}{v} r_i \sum_j (1 - m_{ij})r_j + \frac{v_{12}y_1y_2}{v} r_i \sum_j (1 - m_{ij})s_j(1 - R)$$

$$+ \frac{v_{12}y_1y_2}{v} s_i \sum_j (1 - m_{ij})r_j R + \frac{w_{11}x_1^2}{w} p_i \sum_j m_{ij}p_j$$

$$+ \frac{w_{12}x_1x_2}{w} p_i \sum_j m_{ij}q_j(1 - R) + \frac{w_{12}x_1x_2}{w} q_i \sum_j m_{ij}p_j R$$

$$Qy_2's_i' = \frac{v_{22}y_2^2}{v} s_i \sum_j (1 - m_{ij})s_j + \frac{v_{12}y_1y_2}{v} s_i \sum_j (1 - m_{ij})r_j(1 - R)$$

$$+ \frac{v_{12}y_1y_2}{v} r_i \sum_j (1 - m_{ij})s_j R + \frac{w_{22}x_2^2}{w} q_i \sum_j m_{ij}q_j$$

$$+ \frac{w_{12}x_1x_2}{w} q_i \sum_j m_{ij}p_j(1 - R) + \frac{w_{12}x_1x_2}{w} p_i \sum_j m_{ij}q_j R,$$

where $w = w_{11}x_1^2 + 2w_{12}x_1x_2 + w_{22}x_2^2$, $v = v_{11}y_1^2 + 2v_{12}y_1y_2 + v_{22}y_2^2$ are the mean fitnesses in each deme. P and Q are normalization factors given by

$$P = \frac{w_{11}x_1^2}{w} \langle \mathbf{p}, (\mathbf{E} - \mathbf{M})\mathbf{p} \rangle + \frac{2w_{12}x_1x_2}{w} \langle \mathbf{p}, (\mathbf{E} - \mathbf{M})\mathbf{q} \rangle$$

$$+ \frac{w_{22}x_2^2}{w} \langle \mathbf{q}, (\mathbf{E} - \mathbf{M})\mathbf{q} \rangle + \frac{v_{11}y_1^2}{v} \langle \mathbf{r}, \mathbf{Mr} \rangle + \frac{2v_{12}y_1y_2}{v} \langle \mathbf{r}, \mathbf{Ms} \rangle$$

$$+ \frac{v_{22}y_2^2}{v} \langle \mathbf{s}, \mathbf{Ms} \rangle, \tag{7.1.8}$$

$$Q = \frac{w_{11}x_1^2}{w} \langle \mathbf{p}, \mathbf{Mp} \rangle + \frac{2w_{12}x_1x_2}{w} \langle \mathbf{p}, \mathbf{Mq} \rangle + \frac{w_{22}x_2^2}{w} \langle \mathbf{q}, \mathbf{Mq} \rangle$$

$$+ \frac{v_{11}y_1^2}{v} \langle \mathbf{r}, (\mathbf{E} - \mathbf{M})\mathbf{r} \rangle + \frac{2v_{12}y_1y_2}{v} \langle \mathbf{r}, (\mathbf{E} - \mathbf{M})\mathbf{s} \rangle$$

$$+ \frac{v_{22}y_2^2}{v} \langle \mathbf{s}, (\mathbf{E} - \mathbf{M})\mathbf{s} \rangle, \tag{7.1.9}$$

where $\mathbf{E}$ is the $n \times n$ matrix, all of whose entries are unity and $\langle \mathbf{a}, \mathbf{b} \rangle$ is the scalar product of the vectors $\mathbf{a}$ and $\mathbf{b}$.

The new allelic frequencies of A and a in the two demes are easily found from (7.1.7) to be

$$Px_1' = \sum_{i=1}^{n} Px_1' p_i', \qquad Px_2' = \sum_{i=1}^{n} Px_2' q_i'$$

$$Qy_1' = \sum_{i=1}^{n} Qy_1' r_i', \qquad Qy_2' = \sum_{i=1}^{n} Qy_2' s_i'. \tag{7.1.10}$$

The frequency transformations (7.1.7) can be rewritten in vector notation as follows using $\circ$ to denote the Schur product of vectors:

$$Px_1'\mathbf{p}' = \frac{w_{11}x_1^2}{w} \mathbf{p} \circ (\mathbf{E} - \mathbf{M})\mathbf{p} + \frac{w_{12}x_1x_2}{w} [(1 - R)\mathbf{p} \circ (\mathbf{E} - \mathbf{M})\mathbf{q}$$

$$+ R\mathbf{q} \circ (\mathbf{E} - \mathbf{M})\mathbf{p}] + \frac{v_{11}y_1^2}{v} \mathbf{r} \circ \mathbf{M}\mathbf{r}$$

$$+ \frac{v_{12}y_1y_2}{v} [(1 - R)\mathbf{r} \circ \mathbf{M}\mathbf{s} + R\mathbf{s} \circ \mathbf{M}\mathbf{r}].$$

$$Px_2'\mathbf{q}' = \frac{w_{22}x_2^2}{w} \mathbf{q} \circ (\mathbf{E} - \mathbf{M})\mathbf{q} + \frac{w_{12}x_1x_2}{w} [(1 - R)\mathbf{q} \circ (\mathbf{E} - \mathbf{M})\mathbf{p}$$

$$+ R\mathbf{p} \circ (\mathbf{E} - \mathbf{M})\mathbf{q}] + \frac{v_{22}y_2^2}{v} \mathbf{s} \circ \mathbf{M}\mathbf{s}$$

$$+ \frac{v_{12}y_1y_2}{v} [(1 - R)\mathbf{s} \circ \mathbf{M}\mathbf{r} + R\mathbf{r} \circ \mathbf{M}\mathbf{s}].$$

$$Qy_1'\mathbf{r}' = \frac{v_{11}y_1^2}{v} \mathbf{r} \circ (\mathbf{E} - \mathbf{M})\mathbf{r} + \frac{v_{12}y_1y_2}{v} [(1 - R)\mathbf{r} \circ (\mathbf{E} - \mathbf{M})\mathbf{s} \tag{7.1.11}$$

$$+ R\mathbf{s} \circ (\mathbf{E} - \mathbf{M})\mathbf{r}] + \frac{w_{11}x_1^2}{w} \mathbf{p} \circ \mathbf{M}\mathbf{p}$$

$$+ \frac{w_{12}x_1x_2}{w} [(1 - R)\mathbf{p} \circ \mathbf{M}\mathbf{q} + R\mathbf{q} \circ \mathbf{M}\mathbf{p}].$$

$$Qy_2'\mathbf{s}' = \frac{v_{22}y_2^2}{v} \mathbf{s} \circ (\mathbf{E} - \mathbf{M})\mathbf{s} + \frac{v_{12}y_1y_2}{v} [(1 - R)\mathbf{s} \circ (\mathbf{E} - \mathbf{M})\mathbf{r}$$

$$+ R\mathbf{r} \circ (\mathbf{E} - \mathbf{M})\mathbf{s}] + \frac{w_{22}x_2^2}{w} \mathbf{q} \circ \mathbf{M}\mathbf{q}$$

$$+ \frac{w_{12}x_1x_2}{w} [(1 - R)\mathbf{q} \circ \mathbf{M}\mathbf{p} + R\mathbf{p} \circ \mathbf{M}\mathbf{q}].$$

7.2. Existence of Hardy Weinberg Equilibria

A special class of equilibria is the class of two-locus Hardy-Weinberg (HW) equilibria such that the two loci are in linkage equilibrium in each of the two demes. In this case at equilibrium $\mathbf{p} = \mathbf{q}$, $\mathbf{r} = \mathbf{s}$ and the gametic frequencies in the two demes are given by the Kronecker product vectors $\mathbf{x} \otimes \mathbf{p}$, $\mathbf{y} \otimes \mathbf{r}$, respectively. We prove now:

Result 1

If a HW polymorphic equilibrium exists, then it is either a viability analogous (VAHW) equilibrium with $\mathbf{p} = \mathbf{q} = \mathbf{r} = \mathbf{s}$ where $\mathbf{p} \circ \mathbf{Mp} = m\mathbf{p}$ and $m = \langle \mathbf{p}, \mathbf{Mp} \rangle$, or the equilibrium frequencies of A and a do not depend on the migration rates.

Proof

Suppose that a HW equilibrium exists so that $\mathbf{p} = \mathbf{q}$ and $\mathbf{r} = \mathbf{s}$. In this case, from (7.1.8) and (7.1.9) and using $w = w_{11}x_1^2 + 2w_{12}x_1x_2 + w_{22}x_2^2$ and $v = v_{11}y_1^2 + 2v_{12}y_1y_2 + v_{22}y_2^2$ we find that

$$P = 1 - m_{\mathbf{p}} + m_{\mathbf{r}}, \qquad Q = 1 - m_{\mathbf{r}} + m_{\mathbf{p}}, \qquad (7.2.1)$$

where $m_{\mathbf{p}} = \langle \mathbf{p}, \mathbf{Mp} \rangle$, $m_{\mathbf{r}} = \langle \mathbf{r}, \mathbf{Mr} \rangle$. Using (7.1.11) at equilibrium we have

$$Px_1\mathbf{p} = \frac{w_1 x_1}{w}[\mathbf{p} - \mathbf{p} \circ \mathbf{Mp}] + \frac{v_1 y_1}{v}[\mathbf{r} \circ \mathbf{Mr}]$$

$$Px_2\mathbf{p} = \frac{w_2 x_2}{w}[\mathbf{p} - \mathbf{p} \circ \mathbf{Mp}] + \frac{v_2 y_2}{v}[\mathbf{r} \circ \mathbf{Mr}]$$

$$\qquad\qquad (7.2.2)$$

$$Qy_1\mathbf{r} = \frac{v_1 y_1}{v}[\mathbf{r} - \mathbf{r} \circ \mathbf{Mr}] + \frac{w_1 x_1}{w}[\mathbf{p} \circ \mathbf{Mp}]$$

$$Qy_2\mathbf{r} = \frac{v_2 y_2}{v}[\mathbf{r} - \mathbf{r} \circ \mathbf{Mr}] + \frac{w_2 x_2}{w}[\mathbf{p} \circ \mathbf{Mp}],$$

where

$$w_1 = w_{11}x_1 + w_{12}x_2, \qquad v_1 = v_{11}y_1 + v_{12}y_2,$$

$$\qquad\qquad (7.2.3)$$

$$w_2 = w_{12}x_1 + w_{22}x_2, \qquad v_2 = v_{12}y_1 + v_{22}y_2$$

are the marginal fitnesses of A and a at each of the two demes. Since $w_1 x_1 + w_2 x_2 = w$ and $v_1 y_1 + v_2 y_2 = v$ it follows that

$$(m_{\mathbf{r}} - m_{\mathbf{p}})\mathbf{p} = -\mathbf{p} \circ \mathbf{Mp} + \mathbf{r} \circ \mathbf{Mr},$$

$$\qquad\qquad (7.2.4)$$

$$(m_{\mathbf{p}} - m_{\mathbf{r}})\mathbf{r} = -\mathbf{r} \circ \mathbf{Mr} + \mathbf{p} \circ \mathbf{Mp}.$$

Thus $(m_r - m_p)(\mathbf{p} - \mathbf{r}) = 0$. Hence either $\mathbf{p} = \mathbf{r}$ or $m_p = m_r$. If $m_p = m_r$, then $P = Q = 1$, and from (7.2.4) $\mathbf{p} \circ \mathbf{Mp} = \mathbf{r} \circ \mathbf{Mr}$. From (7.2.2) this produces

$$\left(\frac{w_1}{w} - 1\right)x_1\mathbf{p} = k\mathbf{p} \circ \mathbf{Mp}$$

$$\left(\frac{w_2}{w} - 1\right)x_2\mathbf{p} = -k\mathbf{p} \circ \mathbf{Mp}$$

$$\left(\frac{v_1}{v} - 1\right)y_1\mathbf{r} = -k\mathbf{p} \circ \mathbf{Mp} \qquad (7.2.5)$$

$$\left(\frac{v_2}{v} - 1\right)y_2\mathbf{r} = k\mathbf{p} \circ \mathbf{Mp},$$

where $k = \dfrac{w_1 x_1}{w} - \dfrac{v_1 y_1}{v} = \dfrac{v_2 y_2}{v} - \dfrac{w_2 x_2}{w}$. If $k = 0$, then at a polymorphic equilibrium $w_1 = w_2 = w$, $v_1 = v_2 = v$ and also $x_1 = y_1$, $x_2 = y_2$, and in fact the equilibrium frequencies of A and a in each of the two demes are independent of the migration rates. If $k \neq 0$, then from (7.2.5)

$$C\mathbf{p} = \mathbf{p} \circ \mathbf{Mp} \qquad D\mathbf{r} = \mathbf{p} \circ \mathbf{Mp}, \qquad (7.2.6)$$

where, in fact, $C = D = m = \langle \mathbf{p}, \mathbf{Mp} \rangle$. Hence, if $k \neq 0$, then $\mathbf{p} = \mathbf{r}$.

Thus at a HW polymorphic equilibrium, either the frequencies of A and a are independent of the migration parameters or $\mathbf{p} = \mathbf{q} = \mathbf{r} = \mathbf{s}$ with $m\mathbf{p} = \mathbf{p} \circ \mathbf{Mp}$.

Remarks
1. If it is assumed that at equilibrium $\mathbf{p} = \mathbf{r}$ and $\mathbf{q} = \mathbf{s}$, then for $R > 0$ it is easy to see that necessarily $\mathbf{p} = \mathbf{q} = \mathbf{r} = \mathbf{s}$. For $R = 0$, however, a curve of equilibria with $\mathbf{p} = \mathbf{r}$ and $\mathbf{q} = \mathbf{s}$ but $\mathbf{p} \neq \mathbf{q}$ may exist, with the initial conditions determining the equilibrium attained. At $R = 0$ these equilibria may have one or more chromosomes absent from the system. Two cases in which both A and a are present can be seen when there are two modifying alleles, M_1 and M_2. In one case, AM_2, aM_1, and aM_2 are present and AM_1 is absent, and in the other just AM_1 and aM_2 are present. These are cases of high linkage disequilibrium between the major and modifying loci, commonly referred to as "high complementarity" equilibria (Franklin and Lewontin 1970).

2. The solution $\mathbf{p}$ of $m\mathbf{p} = \mathbf{p} \circ \mathbf{Mp}$ is the same as that obtained as the polymorphic equilibrium in the one-locus multiple-allele viability model with viability matrix $\mathbf{M}$ (see, e.g., Karlin 1984). In other words, the solution $\mathbf{p}$ is *viability analogous*.

3. If $\mathbf{M}$ is a nonsingular and positive matrix, then there is at most one normalized positive vector $\mathbf{p}$ for which $\mathbf{p} \circ \mathbf{Mp} = m\mathbf{p}$, and $\mathbf{p}$ is given by

$$p_i = z_i \bigg/ \sum_{j=1}^{n} z_j \qquad i = 1, 2, \ldots, n, \tag{7.2.7}$$

where $\mathbf{z}$ is the unique positive solution (if it exists) of the set of linear equations

$$\mathbf{Mz} = \mathbf{e} \qquad \mathbf{e} = (1, 1, \ldots, 1). \tag{7.2.8}$$

4. The equilibrium vectors $\mathbf{x} = (x_1, x_2)$ and $\mathbf{y} = (y_1, y_2)$ associated with $\mathbf{p}$ satisfy (see 7.2.5) the set of equations

$$
\begin{aligned}
\left(\frac{w_1}{w} - 1\right) x_1 &= mk \\[6pt]
\left(\frac{w_2}{w} - 1\right) x_2 &= -mk \\[6pt]
\left(\frac{v_1}{v} - 1\right) y_1 &= -mk \\[6pt]
\left(\frac{v_2}{v} - 1\right) y_2 &= mk,
\end{aligned}
\tag{7.2.9}
$$

where $m = \langle \mathbf{p}, \mathbf{Mp} \rangle$ and $k = \dfrac{w_1 x_1}{w} - \dfrac{v_1 y_1}{v} = \dfrac{v_2 y_2}{v} - \dfrac{w_2 x_2}{w}$.

Our interest is in those equilibrium frequencies of A and a that depend on the migration parameters, so we assume from now on that $k \neq 0$.

7.3. "External" Stability of VAHW Polymorphic Equilibria

Suppose that a stable VAHW equilibrium $\{\mathbf{x}^*, \mathbf{y}^*, \mathbf{p}^*\}$ exists where the frequency vectors at the two demes are $\mathbf{x}^* \otimes \mathbf{p}^*$ and $\mathbf{y}^* \otimes \mathbf{p}^*$ with average migration rate m^* given by

$$m^* = \langle \mathbf{p}^*, \mathbf{Mp}^* \rangle, \tag{7.3.1}$$

where there are n alleles $M_1, M_2, \ldots, M_n$ at the modifier locus. We investigate the stability of this equilibrium toward the invasion by a new allele, M_{n+1}, at the modifier locus. This is termed "external" stability.

Let $\tilde{m}$ be the "new" mean migration rate averaged over all gametes carrying the new allele M_{n+1}, that is,

$$\tilde{m} = \sum_{j=1}^{n} m_{n+1,j} p_j^*. \tag{7.3.2}$$

Near the VAHW equilibrium the linear approximations for p_{n+1}', q_{n+1}', r_{n+1}', s_{n+1}' from (7.1.7) are given by

$$x_1^* p_{n+1}' = \frac{w_{11} x_1^{*2}(1-\tilde{m})}{w^*} p_{n+1} + \frac{w_{12} x_1^* x_2^*(1-\tilde{m})}{w^*}\left[(1-R)p_{n+1} + Rq_{n+1}\right]$$

$$+ \frac{v_{11} y_1^{*2}\tilde{m}}{v^*} r_{n+1} + \frac{v_{12} y_1^* y_2^* \tilde{m}}{v^*}\left[(1-R)r_{n+1} + Rs_{n+1}\right].$$

$$x_2^* q_{n+1}' = \frac{w_{22} x_2^{*2}(1-\tilde{m})}{w^*} q_{n+1} + \frac{w_{12} x_1^* x_2^*(1-\tilde{m})}{w^*}\left[(1-R)q_{n+1} + Rp_{n+1}\right]$$

$$+ \frac{v_{22} y_2^{*2}\tilde{m}}{v^*} s_{n+1} + \frac{v_{12} y_1^* y_2^* \tilde{m}}{v^*}\left[(1-R)s_{n+1} + Rr_{n+1}\right].$$

$$\tag{7.3.3}$$

$$y_1^* r_{n+1}' = \frac{v_{11} y_1^{*2}(1-\tilde{m})}{v^*} r_{n+1} + \frac{v_{12} y_1^* y_2^*(1-\tilde{m})}{v^*}\left[(1-R)r_{n+1} + Rs_{n+1}\right]$$

$$+ \frac{w_{11} x_1^{*2}}{w^*} p_{n+1} + \frac{w_{12} x_1^* x_2^* \tilde{m}}{w^*}\left[(1-R)p_{n+1} + Rq_{n+1}\right].$$

$$y_2^* s_{n+1}' = \frac{v_{22} y_2^{*2}(1-\tilde{m})}{v^*} s_{n+1} + \frac{v_{12} y_1^* y_2^*(1-\tilde{m})}{v^*}\left[(1-R)s_{n+1} + Rr_{n+1}\right]$$

$$+ \frac{w_{22} x_2^{*2}\tilde{m}}{w^*} q_{n+1} + \frac{w_{12} x_1^* x_2^* \tilde{m}}{w^*}\left[(1-R)q_{n+1} + Rp_{n+1}\right],$$

where we have used the fact that $P = Q = 1$ at the VAHW equilibrium. Let

$$\alpha = x_1^* p_{n+1}, \qquad \beta = x_2^* q_{n+1}, \qquad \gamma = y_1^* r_{n+1}, \qquad \delta = y_2^* s_{n+1} \tag{7.3.4}$$

then

$$\begin{bmatrix} \alpha' \\ \beta' \\ \gamma' \\ \delta' \end{bmatrix} = \mathbf{L} \begin{bmatrix} \alpha \\ \beta \\ \gamma \\ \delta \end{bmatrix} \tag{7.3.5}$$

where $\mathbf{L} = [l_{ij}]_{i,j=1}^4$ is the 4×4 matrix whose entries are given by

$$l_{11} = (1 - \tilde{m}) \frac{w_{11}x_1^* + w_{12}x_2^*(1 - R)}{v^*} \qquad l_{12} = \frac{w_{12}x_1^*(1 - \tilde{m})R}{w^*}$$

$$l_{13} = \tilde{m} \frac{v_{11}y_1^* + v_{12}y_2^*(1 - R)}{v^*} \qquad l_{14} = \frac{v_{12}y_1^* \tilde{m} R}{v^*}$$

$$l_{21} = \frac{w_{12}x_2^*(1 - \tilde{m})R}{w^*} \qquad l_{22} = (1 - \tilde{m}) \frac{w_{22}x_2^* + w_{12}x_1^*(1 - R)}{w^*}$$

$$l_{23} = \frac{v_{12}y_1^* \tilde{m} R}{v^*} \qquad l_{24} = \tilde{m} \frac{v_{22}y_2^* + v_{12}y_1^*(1 - R)}{v^*}$$

$$l_{31} = \tilde{m} \frac{w_{11}x_1^* + w_{12}x_2^*(1 - R)}{w^*} \qquad l_{32} = \frac{w_{12}x_1^* \tilde{m} R}{w^*}$$

$$l_{33} = (1 - \tilde{m}) \frac{v_{11}y_1^* + v_{12}y_2^*(1 - R)}{v^*} \qquad l_{34} = \frac{v_{12}y_1^*(1 - \tilde{m})R}{v^*}$$

$$l_{41} = \frac{w_{12}x_2^* \tilde{m} R}{w^*} \qquad l_{42} = \tilde{m} \frac{w_{22}x_2^* + w_{12}x_1^*(1 - R)}{w^*}$$

$$l_{43} = \frac{v_{12}y_2^*(1 - \tilde{m})R}{v^*} \qquad l_{44} = (1 - \tilde{m}) \frac{v_{22}y_2^* + v_{12}y_1^*(1 - R)}{v^*}.$$

$$(7.3.6)$$

Here,

$$w^* = x_1^{*2}w_{11} + 2x_1^*x_2^*w_{12} + x_2^{*2}w_{22} \text{ and } v^* = y_1^{*2}v_{11} + 2y_1^*y_2^*v_{12} + y_2^{*2}v_{22}.$$

The "external" stability of the VAHW polymorphism is determined by the absolute value of the largest eigenvalue of $\mathbf{L}$. Since $\mathbf{L}$ is a positive matrix, the Perron-Frobenius theory of matrices entails that this largest eigenvalue is simple and positive. Let $Q(z)$ be the characteristic polynomial of $\mathbf{L}$. The analysis is based on the following two observations.

OBSERVATION 1

When $R = 0$ the 4×4 matrix $\mathbf{L}$ factors into two 2×2 positive matrices $\tilde{\mathbf{L}}_1$ and $\tilde{\mathbf{L}}_2$, where

$$\tilde{\mathbf{L}}_i = \begin{bmatrix} \dfrac{w_i^*(1 - \tilde{m})}{w^*} & \dfrac{v_i^* \tilde{m}}{v^*} \\[2mm] \dfrac{w_i^* \tilde{m}}{w^*} & \dfrac{v_i^*(1 - \tilde{m})}{v^*} \end{bmatrix}, \qquad (7.3.7)$$

with $w_i^* = w_{i1}x_1^* + w_{i2}x_2^*$, $v_i^* = v_{i1}y_1^* + v_{i2}y_2^*$ for $i = 1. 2$, *and*

(i) *When $\tilde{m} > m^*$ each of the matrices $\tilde{\mathbf{L}}_1$ and $\tilde{\mathbf{L}}_2$ have two real eigenvalues smaller than* 1.
(ii) *When $\tilde{m} < m^*$ each of the matrices $\tilde{\mathbf{L}}_1$ and $\tilde{\mathbf{L}}_2$ have two real eigenvalues, one larger than* 1 *and one smaller than* 1.

Proof

When $R = 0$ $l_{12} = l_{14} = l_{21} = l_{23} = l_{32} = l_{34} = l_{41} = l_{43} = 0$ and $\mathbf{L}$ clearly factors into the two 2×2 positive matrices $\tilde{\mathbf{L}}_1$ and $\tilde{\mathbf{L}}_2$. At the VAHW equilibrium $P = Q = 1$, $\mathbf{p} = \mathbf{q} = \mathbf{r} = \mathbf{s} = \mathbf{p}^*$ and the equilibrium equations (7.2.2) imply that

$$x_i^* = \frac{w_i^*(1 - m^*)}{w^*} x_i^* + \frac{v_i^* m^*}{v^*} y_i^*$$
$$y_i^* = \frac{v_i^*}{v^*}(1 - m^*)y_i^* + \frac{w_i^* m^*}{w^*} x_i^*$$

$i = 1, 2,$ (7.3.8)

or, in matrix notation,

$$\begin{bmatrix} \dfrac{w_i^*(1 - m^*)}{w^*} & \dfrac{v_i^* m^*}{v^*} \\ \dfrac{w_i^* m^*}{w^*} & \dfrac{v_i^*(1 - m^*)}{v^*} \end{bmatrix} \begin{bmatrix} x_i^* \\ y_i^* \end{bmatrix} = \begin{bmatrix} x_i^* \\ y_i^* \end{bmatrix} \qquad i = 1, 2.$$ (7.3.9)

Therefore the positive 2×2 matrix $\mathbf{L}_i^*$ where

$$\mathbf{L}_i^* = \begin{bmatrix} \dfrac{w_i^*(1 - m^*)}{w^*} & \dfrac{v_i^* m^*}{v^*} \\ \dfrac{w_i^* m_i^*}{w^*} & \dfrac{v_i^*(1 - m^*)}{v^*} \end{bmatrix} \qquad i = 1, 2$$ (7.3.10)

has one eigenvalue of 1, with positive eigenvector (x_i^*, y_i^*) and another eigenvalue less than 1. Thus its characteristic polynomial $Q_i^*(z)$ satisfies

$$Q_i^*(1) = 0, \qquad Q_i^{*\prime}(1) > 0 \qquad i = 1, 2.$$ (7.3.11)

Observe that if $\tilde{Q}_i(z)$ is the characteristic polynomial of $\tilde{\mathbf{L}}_i$, for $i = 1, 2,$

then

$$Q_i^*(1) = 1 - \left(\frac{w_i^*}{w^*} + \frac{v_i^*}{v^*}\right)(1 - m^*) + \frac{w_i^* v_i^*}{w^* v^*}(1 - 2m^*)$$

$$\tilde{Q}_i(1) = 1 - \left(\frac{w_i^*}{w^*} + \frac{v_i^*}{v^*}\right)(1 - \tilde{m}) + \frac{w_i^* v_i^*}{w^* v^*}(1 - 2\tilde{m})$$

$$Q_i^{*\prime}(1) = 2 - \left(\frac{w_i^*}{w^*} + \frac{v_i^*}{v^*}\right)(1 - m^*) \qquad (7.3.12)$$

$$\tilde{Q}_i'(1) = 2 - \left(\frac{w_i^*}{w^*} + \frac{v_i^*}{v^*}\right)(1 - \tilde{m})$$

Thus using (7.3.11) we get that

$$\tilde{Q}_i(1) = \left(\frac{w_i^*}{w^*} + \frac{v_i^*}{v^*} - 2\frac{w_i^* v_i^*}{w^* v^*}\right)(\tilde{m} - m^*)$$

$$\tilde{Q}_i'(1) - Q_i^{*\prime}(1) = \left(\frac{w_i^*}{w^*} + \frac{v_i^*}{v^*}\right)(\tilde{m} - m^*).$$

We will show that at equilibrium $\frac{w_i^*}{w^*} + \frac{v_i^*}{v^*} > 2\frac{w_i^* v_i^*}{w^* v^*}$. In this case, if $\tilde{m} < m^*$, then $\tilde{Q}_i(1) < 0$ for $i = 1, 2$ and so each of the two 2×2 positive matrices $\tilde{L}_i$ has one eigenvalue less than 1 and one greater than 1. On the other hand, if $\tilde{m} > m^*$, then $\tilde{Q}_i(1) > 0$ for $i = 1, 2$. Since $Q_i^{*\prime}(1) > 0$ it then follows that $\tilde{Q}_i'(1) > 0$ for $i = 1, 2$. Thus in this case each of the positive matrices $\tilde{L}_i$ has both eigenvalues less than unity. We now prove the desired inequality.

Using the equilibrium equation (7.3.8) to solve for $\frac{w_i^*}{w^*}$ and $\frac{v_i^*}{v^*}$, $i = 1, 2$, in terms of x_i^*, y_i^*, m^*, we obtain

$$\frac{w_i^*}{w^*} = \frac{y_i^*[x_i^*(1 - m^*) - y_i^* m^*]}{x_i^* y_i^*(1 - 2m^*)}, \qquad \frac{v_i^*}{v^*} = \frac{x_i^*[y_i^*(1 - m^*) - x_i^* m^*]}{x_i^* y_i^*(1 - 2m^*)},$$

$$(7.3.13)$$

provided $m^* \neq 1/2$. These two equalities produce

$$\frac{w_i^*}{w^*} + \frac{v_i^*}{v^*} - 2\frac{w_i^* v_i^*}{w^* v^*} = \frac{m^*(x_i^* - y_i^*)^2}{x_i^* y_i^*(1 - 2m^*)^2}, \qquad (7.3.14)$$

so that $\frac{w_i^*}{w^*} + \frac{v_i^*}{v^*} - 2\frac{w_i^*}{w^*}\frac{v_i^*}{v^*} > 0$ unless either $x_i^* = y_i^*$ or $m^* = 1/2$.

If $m^* = 1/2$, then from (7.3.8) $x_i^* = y_i^*$ and $\dfrac{w_i^*}{w^*} + \dfrac{v_i^*}{v^*} = 2$. If $x_i^* = y_i^*$ it is

clear from (7.3.8) that $\dfrac{w_i^*}{w^*} + \dfrac{v_i^*}{v^*} = 2$ whatever the value of m^*. Suppose

that $\dfrac{w_i^*}{w^*} + \dfrac{v_i^*}{v^*} = 2$ and let $\dfrac{w_i^*}{w^*} = \xi$. Consider $\dfrac{w_i^* v_i^*}{w^* v^*} = \xi(2 - \xi) = H(\xi)$, say.

Obviously $H(\cdot)$ attains its maximum value of 1 at $\xi = 1$. Thus $\dfrac{2w_i^* v_i^*}{w^* v^*} \leq 2 =$

$\dfrac{w_i^*}{w^*} + \dfrac{v_i^*}{v^*}$ with equality if and only if $\dfrac{w_i^*}{w^*} = \dfrac{v_i^*}{v^*} = 1$. Equality is impossible

since from (7.2.9) it would entail $k^* = 0$, contrary to our assumption.

OBSERVATION 2

Let $Q(z)$ be the characteristic polynomial of **L**; *then its value $Q(1)$ at $z = 1$ is given by*

$$Q(1) = \frac{m^* k^{*2}(\tilde{m} - m^*)}{x_1^* x_2^* y_1^* y_2^*} \left[m^* k^{*2}(\tilde{m} - m^*) + \tilde{m} R \left(\frac{w_{12} x_1^* x_2^*}{w^*} + \frac{v_{12} y_1^* y_2^*}{v^*} \right) \right]$$

(7.3.15)

The proof, based on the expansion of the 4th degree determinant associated with $Q(1)$, is very similar to that given in Appendix A of Liberman and Feldman (1986b).

With the two observations above the main result may now be proved.

Result 2

    A VAHW polymorphic equilibrium $\{\mathbf{x}^*, \mathbf{y}^*, \mathbf{p}^*\}$ with $k^* \neq 0$ is externally stable if $\tilde{m} > m^*$ and externally unstable if $\tilde{m} < m^*$.

Proof

    Fix all other parameters and examine the behavior of the positive matrix **L** as a function of R. To this end let $\lambda_0 = \lambda_0(R)$ be the largest positive eigenvalue of **L**. The Perron-Frobenius theory ensures that $\lambda_0(R)$ is simple and it is clearly a continuous function of R.

    When $\tilde{m} > m^*$, then by Observation 1 if $R = 0$ all four eigenvalues of **L** are less than 1 and in particular $\lambda_0(0) < 1$. From (7.3.15), since $k^* \neq 0$ $Q(1) > 0$ for all R if $\tilde{m} > m^*$. Thus $\lambda_0(R) \neq 1$ for all R, and from the continuity of $\lambda_0(R)$ it follows that $\lambda_0(R) < 1$ for all R and the VAHW polymorphism is externally stable.

When $\tilde{m} < m^*$, then by Observation 1, for $R = 0$ L has two eigenvalues larger than 1. But (7.3.15) reveals that $Q(1) = Q(1; R)$ is a linear function of R with $Q(1; 0) > 0$. Thus $Q(1; R)$ can vanish at most once as a function of R. Since $\lambda_0(R)$ is a simple root and a continuous function of R, it follows that $\lambda_0(R) > 1$ for all $R \geq 0$ and the VAHW polymorphism is externally unstable.

7.4. The "Internal" Stability of the VAHW Polymorphism

Suppose that a VAHW polymorphic equilibrium $\{\mathbf{x}^*, \mathbf{y}^*, \mathbf{p}^*\}$ exists. We discuss here some of the local stability properties of this equilibrium in the two-locus–two-deme system with respect to the closed system of n alleles $M_1, M_2, \ldots, M_n$ at the modifier locus, and two alleles A, a at the selected locus. This is termed "internal" stability.

Let $\{\mathbf{x}, \mathbf{y}, \mathbf{p}, \mathbf{q}, \mathbf{r}, \mathbf{s}\}$ be a point "near" the VAHW equilibrium such that

$$\mathbf{x} = \mathbf{x}^* + \boldsymbol{\varepsilon}, \qquad \mathbf{p} = \mathbf{p}^* + \mathbf{a}, \qquad \mathbf{r} = \mathbf{p}^* + \mathbf{c},$$
$$\mathbf{y} = \mathbf{y}^* + \boldsymbol{\eta}, \qquad \mathbf{q} = \mathbf{p}^* + \mathbf{b}, \qquad \mathbf{s} = \mathbf{p}^* + \mathbf{d}, \tag{7.4.1}$$

where the components of $\boldsymbol{\varepsilon}, \boldsymbol{\eta}, \mathbf{a}, \mathbf{b}, \mathbf{c}, \mathbf{d}$ are small and $\langle \mathbf{e}, \mathbf{a} \rangle = \langle \mathbf{e}, \mathbf{b} \rangle = \langle \mathbf{e}, \mathbf{c} \rangle = \langle \mathbf{e}, \mathbf{d} \rangle = 0$ ($\mathbf{e}$ is of order n here), $\langle \mathbf{e}, \boldsymbol{\varepsilon} \rangle = \langle \mathbf{e}, \boldsymbol{\eta} \rangle = 0$ ($\mathbf{e}$ is of order 2 here), where $\mathbf{e}$ is a vector of the indicated size all of whose components are units.

Since $\mathbf{E}\mathbf{p}^* = \mathbf{e}$, $\mathbf{M}\mathbf{p}^* = m^*\mathbf{e}$ with $m^* = \langle \mathbf{p}^*, \mathbf{M}\mathbf{p}^* \rangle$ it is clear from the transformation equations (7.1.8)–(7.1.11) that the linear approximations for $\mathbf{x}, \mathbf{y}, P, Q$ do not involve $\mathbf{a}, \mathbf{b}, \mathbf{c}$, or $\mathbf{d}$. Thus for internal stability of the VAHW polymorphism, it is necessary that $\{\mathbf{x}^*, \mathbf{y}^*\}$ is stable as a two-deme–one-locus polymorphism with migration rate m^*, and that the linear approximation $\mathbf{F}$ to the system (7.1.11) with $\boldsymbol{\varepsilon} = \boldsymbol{\eta} = \mathbf{0}$ and such that $\mathbf{x} = \mathbf{x}^*$, $\mathbf{y} = \mathbf{y}^*$, $P = Q = 1$, namely,

$$\begin{bmatrix} \mathbf{a}' \\ \mathbf{b}' \\ \mathbf{c}' \\ \mathbf{d}' \end{bmatrix} = \mathbf{F} \begin{bmatrix} \mathbf{a} \\ \mathbf{b} \\ \mathbf{c} \\ \mathbf{d} \end{bmatrix}, \tag{7.4.2}$$

has all of its eigenvalues less than one in magnitude. $\mathbf{F}$ is specifically given by

$$x_1^* \mathbf{a}' = \frac{w_{11} x_1^{*2}}{w^*} \left[(1 - m^*)\mathbf{a} - \mathbf{p}^* \circ \mathbf{Ma} \right] + \frac{w_{12} x_1^* x_2^*}{w^*}$$

$$\times \left[(1-R)(1-m^*)\mathbf{a} - (1-R)\mathbf{p}^* \circ \mathbf{Mb} + R(1-m^*)\mathbf{b} - R\mathbf{p}^* \circ \mathbf{Ma} \right]$$

$$+ \frac{v_{11} y_1^{*2}}{v^*} \left[m^*\mathbf{c} + \mathbf{p}^* \circ \mathbf{Mc} \right]$$

$$+ \frac{v_{12} y_1^* y_2^*}{v^*} \left[(1-R)m^*\mathbf{c} + (1-R)\mathbf{p}^* \circ \mathbf{Md} + Rm^*\mathbf{d} + R\mathbf{p}^* \circ \mathbf{Mc} \right].$$

$$x_1^* \mathbf{b}' = \frac{w_{22} x_2^{*2}}{w^*} \left[(1 - m^*)\mathbf{b} - \mathbf{p}^* \circ \mathbf{Mb} \right] + \frac{w_{12} x_1^* x_2^*}{w^*}$$

$$\times \left[(1-R)(1-m^*)\mathbf{b} - (1-R)\mathbf{p}^* \circ \mathbf{Ma} + R(1-m^*)\mathbf{a} - R\mathbf{p}^* \circ \mathbf{Mb} \right]$$

$$+ \frac{v_{22} y_2^{*2}}{v^*} \left[m^*\mathbf{d} + \mathbf{p}^* \circ \mathbf{Md} \right]$$

$$+ \frac{v_{12} y_1^* y_2^*}{v^*} \left[(1-R)m^*\mathbf{d} + (1-R)\mathbf{p}^* \circ \mathbf{Mc} + Rm^*\mathbf{c} + R\mathbf{p}^* \circ \mathbf{Md} \right].$$

$$y_1^* \mathbf{c}' = \frac{v_{11} y_1^{*2}}{v^*} \left[(1 - m^*)\mathbf{c} - \mathbf{p}^* \circ \mathbf{Mc} \right] + \frac{v_{12} y_1^* y_2^*}{v^*}$$

$$\times \left[(1-R)(1-m^*)\mathbf{c} - (1-R)\mathbf{p}^* \circ \mathbf{Md} + R(1-m^*)\mathbf{d} - R\mathbf{p}^* \circ \mathbf{Mc} \right]$$

$$+ \frac{w_{11} x_1^{*2}}{w^*} \left[m^*\mathbf{a} + \mathbf{p}^* \circ \mathbf{Ma} \right]$$

$$+ \frac{w_{12} x_1^* x_2^*}{w^*} \left[(1-R)m^*\mathbf{a} + (1-R)\mathbf{p}^* \circ \mathbf{Mb} + Rm^*\mathbf{b} + R\mathbf{p}^* \circ \mathbf{Ma} \right].$$

$$y_2^* \mathbf{d}' = \frac{v_{22} y_2^{*2}}{v^*} \left[(1 - m^*)\mathbf{d} - \mathbf{p}^* \circ \mathbf{Md} \right] + \frac{v_{12} y_1^* y_2^*}{v^*}$$

$$\times \left[(1-R)(1-m^*)\mathbf{d} - (1-R)\mathbf{p}^* \circ \mathbf{Mc} + R(1-m^*)\mathbf{c} - R\mathbf{p}^* \circ \mathbf{Md} \right]$$

$$+ \frac{w_{22} x_2^{*2}}{w^*} \left[m^*\mathbf{b} + \mathbf{p}^* \circ \mathbf{Mb} \right]$$

$$+ \frac{w_{12} x_1^* x_2^*}{w^*} \left[(1-R)m^*\mathbf{b} + (1-R)\mathbf{p}^* \circ \mathbf{Ma} + Rm^*\mathbf{a} + R\mathbf{p}^* \circ \mathbf{Mb} \right].$$

$$(7.4.3)$$

Let

$$\alpha = x_1\mathbf{a}, \qquad \beta = x_2\mathbf{b}, \qquad \gamma = y_1\mathbf{c}, \qquad \delta = y_2\mathbf{d}. \qquad (7.4.4)$$

Then

$$\begin{bmatrix} \alpha' \\ \beta' \\ \gamma' \\ \delta' \end{bmatrix} = \mathbf{G} \begin{bmatrix} \alpha \\ \beta \\ \gamma \\ \delta \end{bmatrix}, \qquad (7.4.5)$$

where the stability properties of $\mathbf{F}$ and $\mathbf{G}$ are the same and, writing

$$\mathbf{G} = \begin{bmatrix} \mathbf{A}_1 & \mathbf{B}_1 & \mathbf{C}_1 & \mathbf{D}_1 \\ \mathbf{A}_2 & \mathbf{B}_2 & \mathbf{C}_2 & \mathbf{D}_2 \\ \mathbf{A}_3 & \mathbf{B}_3 & \mathbf{C}_3 & \mathbf{D}_3 \\ \mathbf{A}_4 & \mathbf{B}_4 & \mathbf{C}_4 & \mathbf{D}_4 \end{bmatrix}, \qquad (7.4.6)$$

we have

$$\mathbf{A}_1 = \frac{w_{11}x_1^*}{w^*}\left[(1 - m^*)\mathbf{I} - \mathbf{p}^* \circ \mathbf{M}\right]$$

$$+ \frac{w_{12}x_2^*}{w^*}\left[(1 - R)(1 - m^*)\mathbf{I} - R\mathbf{p}^* \circ \mathbf{M}\right]$$

$$\mathbf{A}_2 = \frac{w_{12}x_2^*}{w^*}\left[-(1 - R)\mathbf{p}^* \circ \mathbf{M} + R(1 - m^*)\mathbf{I}\right]$$

$$\mathbf{A}_3 = \frac{w_{11}x_1^*}{w^*}\left[m^*\mathbf{I} + \mathbf{p} \circ \mathbf{M}\right] + \frac{w_{12}x_2^*}{w^*}\left[(1 - R)m^*\mathbf{I} + R\mathbf{p}^* \circ \mathbf{M}\right]$$

$$\mathbf{A}_4 = \frac{w_{12}x_2^*}{w^*}\left[(1 - R)\mathbf{p}^* \circ \mathbf{M} + Rm^*\mathbf{I}\right]$$

$$\mathbf{B}_1 = \frac{w_{12}x_1^*}{w^*}\left[-(1 - R)\mathbf{p}^* \circ \mathbf{M} + R(1 - m^*)\mathbf{I}\right]$$

$$\mathbf{B}_2 = \frac{w_{22}x_2^*}{w^*}\left[(1 - m^*)\mathbf{I} - \mathbf{p}^* \circ \mathbf{M}\right]$$

$$+ \frac{w_{12}x_1^*}{w^*}\left[(1 - R)(1 - m^*)\mathbf{I} - R\mathbf{p}^* \circ \mathbf{M}\right]$$

$$\mathbf{B}_3 = \frac{w_{12}x_1^*}{w^*}\left[(1 - R)\mathbf{p}^* \circ \mathbf{M} + Rm^*\mathbf{I}\right]$$

$$(7.4.7)$$

$$\mathbf{B}_4 = \frac{w_{22}x_2^*}{w^*}\left[m^*\mathbf{I} + \mathbf{p}^* \circ \mathbf{M}\right] + \frac{w_{12}x_1^*}{w^*}\left[(1 - R)m^*\mathbf{I} + R\mathbf{p}^* \circ \mathbf{M}\right]$$

$$\mathbf{C}_1 = \frac{v_{11}y_1^*}{v^*}\left[m^*\mathbf{I} + \mathbf{p}^* \circ \mathbf{M}\right] + \frac{v_{12}y_2^*}{v^*}\left[(1-R)m^*\mathbf{I} + R\mathbf{p}^* \circ \mathbf{M}\right]$$

$$\mathbf{C}_2 = \frac{v_{12}y_2^*}{v^*}\left[(1-R)\mathbf{p}^* \circ \mathbf{M} + Rm^*\mathbf{I}\right]$$

$$\mathbf{C}_3 = \frac{v_{11}y_1^*}{v^*}\left[(1-m^*)\mathbf{I} - \mathbf{p}^* \circ \mathbf{M}\right]$$

$$+ \frac{v_{12}y_2^*}{v^*}\left[(1-R)(1-m^*)\mathbf{I} - R\mathbf{p}^* \circ \mathbf{M}\right]$$

$$\mathbf{C}_4 = \frac{v_{12}y_2^*}{v^*}\left[-(1-R)\mathbf{p}^* \circ \mathbf{M} + R(1-m^*)\mathbf{I}\right]$$

$$\mathbf{D}_1 = \frac{v_{12}y_1^*}{v^*}\left[(1-R)\mathbf{p}^* \circ \mathbf{M} + Rm^*\mathbf{I}\right]$$

$$\mathbf{D}_2 = \frac{v_{22}y_2^*}{v^*}\left[m^*\mathbf{I} + \mathbf{p}^* \circ \mathbf{M}\right] + \frac{v_{12}y_1^*}{v^*}\left[(1-R)m^*\mathbf{I} + R\mathbf{p}^* \circ \mathbf{M}\right]$$

$$\mathbf{D}_3 = \frac{v_{12}y_1^*}{v^*}\left[-(1-R)\mathbf{p}^* \circ \mathbf{M} + R(1-m^*)\mathbf{I}\right]$$

$$\mathbf{D}_4 = \frac{v_{22}y_2^*}{v^*}\left[(1-m^*)\mathbf{I} - \mathbf{p}^* \circ \mathbf{M}\right]$$

$$+ \frac{v_{12}y_1^*}{v^*}\left[(1-R)(1-m^*)\mathbf{I} - R\mathbf{p}^* \circ \mathbf{M}\right].$$

We use the eigenvectors of the positive symmetrizable matrix $\mathbf{p}^* \circ \mathbf{M}$. Since $\mathbf{p}^* \circ \mathbf{M}\mathbf{p}^* = m^*\mathbf{p}^*$, by the Perron-Frobenius theory, m^* is the largest eigenvalue of $\mathbf{p}^* \circ \mathbf{M}$ in absolute value and its eigenvector $\mathbf{p}^*$ is the unique (up to a multiplicative constant) positive eigenvector of $\mathbf{p}^* \circ \mathbf{M}$. Any other eigenvector $\boldsymbol{\xi}$ of $\mathbf{p}^* \circ \mathbf{M}$ satisfies $\mathbf{p}^* \circ \mathbf{M}\boldsymbol{\xi} = \lambda\boldsymbol{\xi}$, $\langle \boldsymbol{\xi}, \mathbf{e} \rangle = 0$ and $|\lambda| < m^*$. Using the same technique as in Liberman and Feldman (1986b) makes it evident that the set of eigenvalues of $\mathbf{G}$ coincides with the eigenvalues of the set of matrices $\mathbf{H} = \mathbf{H}(\lambda)$, where

$$\mathbf{H} = \begin{bmatrix} \tilde{\alpha}_1 & \tilde{\beta}_1 & \tilde{\gamma}_1 & \tilde{\delta}_1 \\ \tilde{\alpha}_2 & \tilde{\beta}_2 & \tilde{\gamma}_2 & \tilde{\delta}_2 \\ \tilde{\alpha}_3 & \tilde{\beta}_3 & \tilde{\gamma}_3 & \tilde{\delta}_3 \\ \tilde{\alpha}_4 & \tilde{\beta}_4 & \tilde{\gamma}_4 & \tilde{\delta}_4 \end{bmatrix}, \tag{7.4.8}$$

$$\tilde{\alpha}_i\boldsymbol{\xi} = \mathbf{A}_i\boldsymbol{\xi}, \qquad \tilde{\beta}_i\boldsymbol{\xi} = \mathbf{B}_i\boldsymbol{\xi}, \qquad \tilde{\gamma}_i\boldsymbol{\xi} = \mathbf{C}_i\boldsymbol{\xi}, \qquad \tilde{\delta}_i\boldsymbol{\xi} = \mathbf{D}_i\boldsymbol{\xi}, \tag{7.4.9}$$

and $\boldsymbol{\xi}$ is any eigenvector of $\mathbf{p}^* \circ \mathbf{M}$ with eigenvalue λ ($\lambda \neq m^*$).

For given ξ and its associated λ, we have

$$\tilde{\alpha}_1 = \frac{w_{11}x_1^*}{w^*}(1 - m^* - \lambda) + \frac{w_{12}x_2^*}{w^*}[(1 - R)(1 - m^*) - R\lambda]$$

$$\tilde{\alpha}_2 = \frac{w_{12}x_2^*}{w^*}[R(1 - m^*) - (1 - R)\lambda]$$

$$\tilde{\alpha}_3 = \frac{w_{11}x_1^*}{w^*}(m^* + \lambda) + \frac{w_{12}x_2^*}{w^*}[(1 - R)m^* + R\lambda]$$

$$\tilde{\alpha}_4 = \frac{w_{12}x_2^*}{w^*}[Rm^* + (1 - R)\lambda]$$

$$\tilde{\beta}_1 = \frac{w_{12}x_1^*}{w^*}[R(1 - m^*) - (1 - R)\lambda]$$

$$\tilde{\beta}_2 = \frac{w_{22}x_2^*}{w^*}(1 - m^* - \lambda) + \frac{w_{12}x_1^*}{w^*}[(1 - R)(1 - m^*) - R\lambda]$$

$$\tilde{\beta}_3 = \frac{w_{12}x_1^*}{w^*}[Rm^* + (1 - R)\lambda]$$

$$\tilde{\beta}_4 = \frac{w_{22}x_2^*}{w^*}(m^* + \lambda) + \frac{w_{12}x_1^*}{w^*}[(1 - R)m^* + R\lambda]$$

$$\tilde{\gamma}_1 = \frac{v_{11}y_1^*}{v^*}(m^* + \lambda) + \frac{v_{12}y_2^*}{v^*}[(1 - R)m^* + R\lambda]$$

(7.4.10)

$$\tilde{\gamma}_2 = \frac{v_{12}y_2^*}{v^*}[Rm^* + (1 - R)\lambda]$$

$$\tilde{\gamma}_3 = \frac{v_{11}y_1^*}{v^*}[1 - m^* - \lambda] + \frac{v_{12}y_2^*}{v^*}[(1 - R)(1 - m^*) - R\lambda]$$

$$\tilde{\gamma}_4 = \frac{v_{12}y_2^*}{v^*}[R(1 - m^*) - (1 - R)\lambda]$$

$$\tilde{\delta}_1 = \frac{v_{12}y_1^*}{v^*}[Rm^* + (1 - R)\lambda]$$

$$\tilde{\delta}_2 = \frac{v_{22}y_2^*}{v^*}(m^* + \lambda) + \frac{v_{12}y_1^*}{v^*}[(1 - R)m^* + R\lambda]$$

$$\tilde{\delta}_3 = \frac{v_{12}y_1^*}{v^*}[R(1 - m^*) - (1 - R)\lambda]$$

$$\tilde{\delta}_4 = \frac{v_{22}y_2^*}{v^*}(1 - m^* - \lambda) + \frac{v_{12}y_1^*}{v^*}[(1 - R)(1 - m^*) - R\lambda].$$

Observe that the matrix **H** of (7.4.8) and (7.4.10) resembles the matrix **L** of (7.3.5) and (7.3.6). Using this similarity we have

Result 3

When the major and modifier loci are loosely linked ($R \approx 1/2$), then the VAHW polymorphism $\{\mathbf{x}^*, \mathbf{y}^*, \mathbf{p}^*\}$ with $k^* \neq 0$ is stable only when all the eigenvalues of $\mathbf{p}^* \circ \mathbf{M}$ are positive. This condition is also sufficient provided $m^* \leq 1/2$ and $\{\mathbf{x}^*, \mathbf{y}^*\}$ is stable as a two-locus two-deme polymorphism with migration rate m^*.

Proof

If $R = 1/2$, then for any eigenvalue λ of $\mathbf{p}^* \circ \mathbf{M}$ different from m^* the matrix $\mathbf{H}(\lambda)$ of (7.4.10) coincides with the matrix **L** of (7.3.5) under the identification $\tilde{m} = m^* + \lambda$. Since $|\lambda| < m^*$, it is clear that $\tilde{m} > 0$. Suppose that $\lambda < 0$, then as $m^* < 1$ we have $\tilde{m} < m^*$, and by Result 2 the matrix $\mathbf{H}(\lambda)$ has a positive eigenvalue larger than 1. Thus for the VAHW polymorphism to be stable at $R = 1/2$, it is necessary that all eigenvalues λ of $\mathbf{p}^* \circ \mathbf{M}$ will be positive.

If all such λ's are positive and $m^* < 1/2$, then in the identification $\tilde{m} = m^* + \lambda$ we have $0 < \tilde{m} < 1$ in addition to $\tilde{m} > m^*$. Thus by Result 2 we know that for all λ's the eigenvalues of their corresponding matrices $\mathbf{H}(\lambda)$ are less than 1 in magnitude, and the VAHW polymorphism is stable under the provision stated above, when $R = 1/2$. By continuity the same result is true if $R \approx 1/2$.

In order to investigate the internal stability of the VAHW polymorphism for other values of R, compute the value of the characteristic polynomial $S(z; \lambda, R)$ of $\mathbf{H}(\lambda)$ at $z = 1$. The computation is very similar to that of $Q(1)$ in (7.3.15) and reduces to

$$S(1; \lambda, R) = \frac{m^* k^{*2} \lambda}{x_1^* x_2^* y_1^* y_2^*} \left[m^* k^{*2} \lambda + (m^* + \lambda)R \left(\frac{w_{12} x_1^* x_2^*}{w^*} + \frac{v_{12} y_1^* y_2^*}{v^*} \right) \right].$$

(7.4.11)

From (7.4.12) we derive the following conclusion.

CONCLUSION 7.4.1

If $m^* \leq 1/2$ and all the eigenvalues $\lambda \neq m^*$ of $\mathbf{p}^* \circ \mathbf{M}$ are positive, then the VAHW polymorphism with $k^* \neq 0$ is internally stable for all $R \geq R_0$ where

$$R_0 = \max_{\lambda \neq m^*} \left\{ \frac{\lambda}{\lambda + (1 - m^*)} \right\} \text{ provided } \{\mathbf{x}^*, \mathbf{y}^*\} \text{ is stable as a one-locus two-}$$

deme polymorphism with migration rate m^*.

Proof

When all the λ's are positive, then $S(1; \lambda, R)$ is positive for all R. If $R \geq R_0$, then clearly $\mathbf{H}(\lambda)$ is a positive matrix, and since its largest eigenvalue at $R = 1/2$ is positive and less than 1, it follows from the continuity of the largest eigenvalue as a function of R that for all R such that $1/2 \geq R \geq R_0$ the largest eigenvalue is less than unity, and the VAHW polymorphism $\{\mathbf{x}^*, \mathbf{y}^*, \mathbf{p}^*\}$ with $k^* \neq 0$ is internally stable provided $\{\mathbf{x}^*, \mathbf{y}^*\}$ is stable.

Remark

Observe that when $\lambda = 0$, $\mathbf{H}(\lambda)$ is a positive matrix whose largest eigenvalue is 1. If λ is small (positive or negative) and R small, then $S(1; \lambda, R) > 0$. Thus the largest eigenvalue of $\mathbf{H}(\lambda)$ is less than 1 under these conditions. Hence it is not clear that all the eigenvalues of $\mathbf{p}^* \circ \mathbf{M}$ must be positive for a VAHW polymorphism to be stable (see the numerical analysis that follows).

7.5. Numerical Analysis of Internal Stability at $R = 0$ (See Appendix)

Even when $R < R_0$ the eigenvalues of $\mathbf{H}$ may be less than unity in absolute value. The precise conditions on the viabilities and migration rates for this to occur are extremely difficult to write down. Thus if $m^* < 1/2$, it is conceivable that the VAHW polymorphism is stable at $R = 0$, and we have not ruled out analytically that this may occur even when the positivity condition on the eigenvalues of $\mathbf{p}^* \circ \mathbf{M}$ (Result 3) is violated. In view of the incompleteness of our analysis of internal stability for tight linkage, we made a numerical investigation of the case $R = 0$ with two modifier alleles M_1 and M_2. The methods and results of this numerical analysis are described in detail in the Appendix to this chapter.

7.6. Discussion

This is the third in our series of multiple-allele modifier studies. The major qualitative features of these analyses have been the characterization of the VAHW polymorphic equilibria and their external instability to alleles that reduce mutation, recombination, and, in this study, migration rates. The external stability properties of other equilibria revealed by the numerical analysis in the present study, and not of the VAHW type, remain to be explored. It would be very surprising if these equilibria were not also externally unstable to parametric reduction. If all equilibria that depend on the parameter under study were externally unstable to reduction, then, in the terminology of Eshel and Feldman (1982), zero would have the property of evolutionary genetic stability (EGS) for mutation, recombination, and migration rates.

Many scholars of evolutionary theory have addressed the issue that is an obvious consequence of the modifier theory developed in our three studies: what prevents mutation, recombination, and migration rates from converging to zero in nature? The two standard answers are (1) As the parameter in question approaches zero, successive mutations that cause further reduction are unlikely to be neutral and, indeed, are likely to be deleterious. Thus, for example, too strong a reduction in migratory ability of an insect may also entail reduced probability of mating; and (2) in fluctuating environments the reduction principle fails. This has been validated for recombination modification by Charlesworth (1976), under conditions that produced cyclic sign changes in the linkage disequilibrium. In cases where the fluctuation in the selection regime is not as orderly, the results are not as clear-cut (Gillespie 1981).

Migration and recombination modification differ from the mutation case in that the former two allow nontrivial equilibria, apart from the VAHW polymorphism, that depend on the parameter of interest. Migration modification differs from recombination modification in that in the latter case at $R = 0$ the VAHW polymorphism is unstable, while as we have seen in the numerical analysis section of this paper the VAHW point is quite likely to be stable at $R = 0$ in the former case. There is a second difference between the internal stability properties of the VAHW point with migration and recombination modification. In the latter case for stability of the VAHW polymorphism, it is necessary for all R that the matrix $\mathbf{p}^* \circ \mathbf{R}$ have all positive eigenvalues (Liberman and Feldman 1986b, Result 3). In the migration case the analogous condition on $\mathbf{p}^* \circ \mathbf{M}$ is necessary at $R = 1/2$, but as described in the Appendix we have numerical examples at $R = 0$, where with modifier alleles M_1 and M_2 such that $m_{12} > m_{11}, m_{22}$ the VAHW point is stable. Clearly, the internal stability properties of the VAHW points depend on what is being modified, in sharp contrast to their external stability properties. We are safe in claiming, however, that for loose linkage between major and modifier genes the internal stability properties of the VAHW points in all three models are similar.

The existence of cycles revealed by the numerical analysis of the Appendix poses something of a mathematical dilemma. Hastings (1981) required hundreds of thousands of parameter samples before finding one that produced cycling in a two-locus selection model. Here we have four examples from roughly one hundred sample sets of parameters. We have suggested that extreme differences in migration rates between heterozygote and homozygote modifiers may be the case, but the viability regime must play some role. In view of the rarity of cycling in standard population genetic models with biologically reasonable parameters, this property of migration modification is worth further study.

The reduction principle for mutation rates fails when the selection is fertility-based (Holsinger, Feldman, and Altenberg 1986). We conjecture that the same will be true in the other modification models and are investigating this. Also under investigation are haploid analogues of the multiallele modification models described in this series of studies.

Acknowledgment

This research was supported in part by NIH grants GM 28016 and GM 10452.

Appendix
Numerical Analysis of Internal Stability at $R = 0$

Kent E. Holsinger, Marcus W. Feldman, and Uri Liberman

In this Appendix we investigate the migration modifier model with two modifying alleles M_1 and M_2 and with absolutely linked modifier locus and major gene. In this case there are four possible gametic types: AM_1, AM_2, aM_1, aM_2. The procedure we adopted was to choose two sets of three viabilities at random from a uniform distribution on $[0, 1]$. One set was assigned to each population. Three migration rates, m_{11}, m_{12}, m_{22}, were chosen at random, uniformly on $[0, 1]$. Ten sets of values for $x_1, y_1, p_1, q_1, r_1, s_1$ were chosen at random, uniformly on $[0, 1]$, to produce the initial gamete frequencies in the two populations. The numerical analysis was then divided according only to assumptions made on the three m_{ij} values. Three arrangements are possible: $m_{12} < m_{11}, m_{22}$, that is, heterozygotes have the lowest migration rate; $m_{12} > m_{11}, m_{22}$, that is, heterozygotes have the highest migration rate; $m_{11} > m_{12} > m_{22}$ or $m_{22} > m_{12} > m_{11}$, that is, directional ordering in favor of one of M_1 or M_2. In each case the system (7.1.7) with $R = 0$ was iterated (in double precision) until the maximum over the eight gametes (in the two populations) of the difference between the frequencies of a gamete in successive generations was less than 10^{-14}.

CASE 1. $m_{12} < m_{11}, m_{22}$

We chose one hundred and six cases of this kind. Recall that at $R = 1/2$ in this case the VAHW equilibrium would be stable if $\{x^*, y^*\}$ were stable as a two-locus, two-deme polymorphism with migration rate m^*. In fifty seven cases the VAHW was stable. In eight of these cases there was an additional equilibrium with one of A or a fixed but where the frequencies of M_1 and M_2 depended on the initial frequencies. In one case, in addition to the VAHW polymorphism there was an equilibrium with three chromosomes present, aM_2, AM_1, AM_2. Thus linkage equilibrium and linkage disequilibrium between the major locus and the modifier can simultaneously be stable here.

In thirty three cases fixation on one of A or a occurred with all starting conditions, with a curve of M_1 and M_2 frequencies, and the precise point achieved depended on the initial frequency of M_1. In seven cases fixation of either A or a occurred depending on the initial condition, with M_1, M_2 curves in all seven. In one case all ten initial vectors resulted in convergence to a three-chromosome equilibrium.

Four cases approached neutral equilibria of the form $\mathbf{p} = \mathbf{r}$; $\mathbf{q} = \mathbf{s}$ (see Remark 1 following Result 1). In these cases convergence was slow with up to 500,000 generations required for equilibrium to be reached. A most surprising finding was that in the remaining four cases each initial condition resulted in a different set of frequencies that were achieved cyclically with periods of many thousands of generations. These cycles were revealed at first by the lack of convergence even to 10^{-4} after 100,000 generations. We then stored the frequency vectors for one hundred consecutive generations and determined that these were again achieved after a number of generations that appeared to be characteristic of the parameter set and appeared to be the same for each starting condition. In view of the extremely large estimated period of the cycles, we made two further numerical checks to ensure that these were in fact cycles. From one of the starting conditions after the cycle was achieved, the population frequency vector was stored for one hundred consecutive generations. We then observed that the other starting conditions produced dynamics that eventually attained this set of one hundred vectors in sequence, and that the period was exactly the same from each starting vector. Our second check was simply to observe that each of the highest and lowest value of each component of the frequency vector (during the fluctuation) was the same for all starting conditions, that is, the amplitude of the cycle did not depend on initial conditions. The four sets of parameters that gave rise to these cycles are recorded in Table 7.1 together with the number of generations between successive recurrences of the same frequency vector. In the fourth case of Table 7.1 some starting conditions appeared to approach the $\mathbf{p} = \mathbf{r}$, $\mathbf{q} = \mathbf{s}$ surface while others produced the cyclic behavior.

The cyclic behavior remains something of a mystery. There may be a clue in the fact that each of the four migration arrays in Table 7.1 show m_{12} to be very small and at least one of m_{11} and m_{22} to be very big. That this may have some significance is suggested by the fact that when we reduced the highest value (m_{11} or m_{22}) by either 0.05 or 0.1 and increased m_{12} by the same amount, the cyclic behavior disappeared and a VAHW polymorphism resulted.

CASE 2. $m_{12} > m_{11}, m_{22}$

Here we tried thirty eight parameter sets with ten starting conditions and $R = 0$. In twenty two cases fixation, both of M_1 and M_2, was stable (as the theory above, for $R = 1/2$, might suggest). In five cases, however, other equilibria were also stable; in one case one high complementarity (AM_1, aM_2) point was stable, in another both high complementarity points were stable, in another a three-chromosome face (aM_1, AM_1, AM_2) was stable, in another a high complementarity point

Table 7.1
Parameter sets producing cycles.

Viability set 1	Viability set 2	Migration rates
$w_{11} = 0.9111$	$v_{11} = 0.3854$	$m_{11} = 0.4051$
$w_{12} = 0.4462$	$v_{12} = 0.6467$	$m_{12} = 0.0534$
$w_{22} = 0.4024$	$v_{22} = 0.0030$	$m_{22} = 0.1962$

Estimated period of cycle: 155,276 generations

Viability set 1	Viability set 2	Migration rates
$w_{11} = 0.0420$	$v_{11} = 0.2226$	$m_{11} = 0.4726$
$w_{12} = 0.3152$	$v_{12} = 0.5401$	$m_{12} = 0.0205$
$w_{22} = 0.1550$	$v_{22} = 0.4834$	$m_{22} = 0.4142$

Estimated period of cycle: 123,600 generations

Viability set 1	Viability set 2	Migration rates
$w_{11} = 0.8635$	$v_{11} = 0.3461$	$m_{11} = 0.4838$
$w_{12} = 0.4121$	$v_{12} = 0.9160$	$m_{12} = 0.0074$
$w_{22} = 0.5112$	$v_{22} = 0.0871$	$m_{22} = 0.1637$

Estimated period of cycle: 14,704 generations

Viability set 1	Viability set 2	Migration rates
$w_{11} = 0.1260$	$v_{11} = 0.6655$	$m_{11} = 0.4499$
$w_{12} = 0.9092$	$v_{12} = 0.7837$	$m_{12} = 0.0494$
$w_{22} = 0.5539$	$v_{22} = 0.8142$	$m_{22} = 0.2621$

Estimated period of cycle: 43,348 generations

(AM_1, aM_2) and the opposite three-chromosome face (aM_1, AM_2, AM_1) were stable, and in another the VAHW point was stable.

In six cases, one of M_1 and M_2 was fixed and in addition to overlaps with high complementarity points, and three chromosome faces, fixation of A or a was possible. In one case we found fixation of M_1, fixation of A and a, each with $M_1 - M_2$ curves, and also a stable high-complementarity edge. In nine cases, either A or a or both became fixed, with $M_1 - M_2$ curves. Finally, in one case all ten initial conditions led to the VAHW polymorphism.

CASE 3. Directionally Ordered Migration: $m_{11} > m_{12} > m_{22}$

We chose thirty-nine cases of this kind, and in twenty-six we saw fixation of M_2. In eight of these, we also saw fixation of A or a (with the $M_1 - M_2$ curve), and in one case a high complementarity edge was also stable. In thirteen cases we saw fixation of either of A or a or both, with $M_1 - M_2$ curves in each case.

From the numerical analysis it is clear that at $R = 0$, the VAHW point can be stable when $\mathbf{p} \circ \mathbf{M}$ has all positive eigenvalues or, as occurs in the case of $m_{12} > m_{11}, m_{22}$, it has one positive and one negative eigenvalue. In addition, points with linkage disequilibrium may be stable at $R = 0$ and may overlap in a complex way with either the VAHW point or with modified fixation states.

References

Asmussen, M. A. 1976. Topics in the theory of ecological genetics. Ph.D. thesis, Stanford University.

Asmussen, M. A. 1983. Evolution of dispersal in density-regulated populations: A haploid model. *Theor. Pop. Biol.* 23: 281–299.

Balkau, B. J., and Feldman, M. W. 1973. Selection for migration modification. *Genetics* 74: 171–174.

Charlesworth, B. 1976. Recombination modification in a fluctuating environment. *Genetics* 83: 181–195.

Eshel, I., and Feldman, M. W. 1982. On evolutionary genetic stability of the sex-ratio. *Theor. Pop. Biol.* 21: 430–439.

Feldman, M. W. 1972. Selection for linkage modification: I. Random mating populations. *Theor. Pop. Biol.* 3: 324–346.

Feldman, M. W., and Krakauer, J. 1976. Genetic modification and modifier polymorphism. In *Population Genetics and Ecology*, pp. 547–582. Ed. S. Karlin and E. Nevo. Academic Press, New York.

Franklin, I., and Lewontin, R. C. 1970. Is the gene the unit of selection? *Genetics* 65: 707–734.

Gillespie, J. H. 1981. Mutation modification in a random environment. *Evolution* 35: 468–476.

Hamilton, W., and May, R. 1977. Dispersal in stable habitats. *Nature* 269: 578–581.

Hastings, A. 1981. Stable cycling in discrete time genetic models. *Proc. Natl. Acad. Sci. USA* 78: 7224–7225.

Hastings, A. 1983. Can spatial variation alone lead to selection for dispersal? *Theor. Pop. Biol.* 24: 244–251.

Holsinger, K. E.; Feldman, M. W.; and Altenberg, L. 1986. Selection for increased mutation rates with fertility differences between matings. *Genetics* 112: 909–922.

Karlin, S. 1982. Classifications of selection-migration structures and conditions for a protected polymorphism. *Evolutionary Biology* 14: 61–203.

Karlin, S. 1984. Mathematical models, problems and controversies of evolutionary biology. *Bull. Am. Math. Soc.* 10: 221–273.

Karlin, S., and McGregor, J. L. 1974. Towards a theory of the evolution of modifier genes. *Theor. Pop. Biol.* 5: 59–103.

Levin, S. A.; Cohen, D.; and Hastings, A. 1984. Dispersal strategies in patchy environments. *Theor. Pop. Biol.* 26: 165–191.

Liberman, U., and Feldman, M. W. 1986a. Modifiers of mutation rate: A general reduction principle. *Theor. Pop. Biol.* 30: 125–142.

Liberman, U., and Feldman, M. W. 1986b. A general reduction principle for genetic modifiers of recombination. *Theor. Pop. Biol.* 30: 341–371.

Motro, U. 1982a. Optimal rates of dispersal. I. Haploid populations. *Theor. Pop. Biol.* 21: 394–411.

Motro, U. 1982b. Optimal rates of dispersal. II. Diploid populations. *Theor. Pop. Biol.* 21: 412–429.

Motro, U. 1983. Optimal rates of dispersal. III. Parent-offspring conflict. *Theor. Pop. Biol.* 23: 159–168.

Teague, R. 1977. A model of migration modification. *Theor. Pop. Biol.* 12: 86–94.

PART II

Behavior, Ecology, and
Evolutionary Genetics

The papers of this section are somewhat more applied in their approach than those of Part I. The first two address the evolution of behavior but from very different points of view, one in terms of cultural evolution, the other in terms of kin selection. In Chapter 8 Feldman and Cavalli-Sforza continue their theoretical studies on the coevolution of genes and culture with their model for the simultaneous evolution of lactose absorption and milk use. Lactose absorption and malabsorption is a genetically based phenotypic distinction manifest in the presence of a cultural distinction, namely, use or nonuse of milk. In their first models, the genetics are haploid and there is vertical transmission from parent to child followed by oblique transmission. The rates of cultural transmission are genetically determined. In a departure from Feldman and Cavalli-Sforza's previous models, the Darwinian fitnesses of the phenotypes depend on the genotypes as well. The diploid model is most relevant to the lactose problem and involves one locus with two alleles and with lactose absorption dominant to malabsorption.

The authors develop conditions for the initial increase of a new phenotype, such as milk use, in the presence of the genetic polymorphism, and conditions for increase of the absorption allele in the presence of both phenotypes. A numerical study of the complete dynamical system of six phenogenotypes is carried out in order to determine the level of Darwinian fitness advantage that must accrue to milk-using lactose absorbers so that current genotype frequencies could have been achieved in three hundred generations. This is the amount of time widely believed to have passed since the advent of dairying. It is concluded that the fitness advantage required for the model to produce this rate of evolution is higher than would generally be regarded as biologically realistic.

The past twelve years have seen rapid development in the evolutionary theory of behavior. Much of this was stimulated by the work of Hamilton (1964; see Chap. 8) that originated the mathematical theory of kin selection. In Chapter 9, Uyenoyama first reviews some of the recent work on the evolution of altruism when the fitness loss to the donors and gain to the recipients of altruism combine additively. In the one-locus multiple-allele model, when parents mate at random, the correct analog to what is usually called the coefficient of relationship is the regression coefficient of the additive genetic value of the recipient on that of the donor. The initial increase of the new altruistic alleles at one locus in the additive case depends on this regression coefficient.

Uyenoyama develops the evolutionary theory of altruism when it depends on the genotype at two sex-linked loci. She examines two cases, one where initially one chromosome is fixed and the other where initially one locus is fixed and the other is segregating. She shows that an additive scale of genotypic values based on allelic contributions at each locus does allow a statistical version of Hamilton's rule to hold. The relevant analogs to the coefficient of relationship are, in general, not the same as the numbers in the classical one-locus case, such as 3/4 in the sister-to-sister case. They depend on such features of the two-locus model as epistasis and the amount of linkage. Nevertheless, by appropriate choice of an additive scale for the effects of genetic substitutions, an analog to Hamilton's rule in terms of the regression of (two-locus) additive genetic values is derived.

For one locus with an arbitrary number of alleles it has been known since the early 1960s that the mean viability of a population that mates at random increases over time, reaching a maximum at equilibrium. We have also known since Moran's work in 1965 that this Fundamental Theorem of Natural Selection is not generally true when the viability acts on two or more loci because recombination (like mutation) constitutes an additional process in the system that can interfere with the dynamics of the selection process. The search for optimality principles in evolution has continued, however, through the game-theoretic approach, with its origins in ecology, and through the theory of modifiers with a more population-genetic context. An attempt to reconcile results from these approaches was made by Eshel and Feldman (1982a,b; see Chap. 10); through their property of Evolutionary Genetic Stability, EGS. A particular parameter π^* of the population has this property if only those mutant alleles that bring the population parameter closer to π^* can initially increase. Karlin and Lessard (1983, 1984; see Chap. 10) took this notion further by examining the sequence of equilibria attained by successive mutant alleles.

In Chapter 10 Lessard generalizes the classical viability selection framework to allow the sexes to have different viabilities. He begins with a single locus with multiple alleles and first develops a general criterion for the initial increase of a mutant allele near a stable equilibrium. The well-known result when the sexes do not have different viabilities is that the marginal fitness of the new allele should exceed the mean fitness of the resident genotypes for the new allele to increase when rare. Lessard shows that the analogous condition with sex-differentiated fitnesses involves the marginal reproductive value of male and female gametes [formula (3.14)].

When the set S of male and female fitness pairs is compact and the maximum value of the product $m \times f$ of the average male and average female fitnesses is taken over the convex hull of S, then this maximum is unique. An equilibrium where this maximum occurs cannot be invaded by any new mutant allele. It is possible, however, that this product decreases between equilibria induced by successive mutant alleles. Extensions to multiple loci and frequency-dependent selection are indicated.

O'Donald and Majerus, in Chapter 11, compare theoretical results derived from mathematical models of sexual selection and experimental observations on the Two-spot Ladybird. In this species there is a series of melanic forms in which the more melanic are genetically dominant to the less melanic phenotypes.

Melanic males are preferred as mates by females. Two of these melanic forms, the four-spot, Q, and the six-spot, X, coexist in populations and appear to be preferred to the same extent. Two classes of mathematical models developed previously by Karlin and O'Donald (1981; see Chap. 11) suggest that these genetically determined preferred male phenotypes should not coexist in a population. O'Donald and Majerus extend and combine these previous models with reference to the four-allele series that determines the melanic phenotypes. This extended model does permit stable coexistence of Q and X. The chapter concludes with experimental data that produce estimates for the levels of female preference among the various morphs.

The last fifteen years have also witnessed the initiation and development of evolutionary ecology, a subject where ecological and evolutionary theory hybridize with introgression. In Chapter 12, Jon Roughgarden develops a theory for the evolution of adult and larval phases of a marine organism. The theory takes as its point of departure a model for the demographic dynamics of adults and larvae. Besides the usual birth and mortality parameters, the model includes the dynamics of settlement appropriate for a benthic habitat.

On the assumption that the nontrivial steady state of this model is stable, Roughgarden demonstrates that the optimal life cycle depends on whether F, the steady-state degree of vacant space in the benthic habitat, is maximized or minimized. If the reproductive success in benthic habitats is greater than in pelagic habitats, then the optimal life cycle includes some benthic component; otherwise, the species should be purely pelagic. A series of simplifying assumptions on the demographic parameters and the settlement pattern allows the author to determine the optimal life history when F is very small, in which case it will include some benthic component. Depending on the mortality and relative growth rates of adults and larvae, five different optimal benthic life cycles are shown to be possible, including direct development, a pelagic larval feeding phase with benthic adults, and budding leading to coloniality. This theory, combined with previous theoretical analyses, gives rise to evolutionary interpretations of observed marine life cycles and their geographic distribution from the equator to the poles.

Families of repetitive DNA sequences have been discovered in many mammalian species, and in some cases these are dispersed throughout the genome. One such family, called Alu, is discussed by Kaplan and Hudson in Chapter 13. They suppose that any copy of the repeated sequence, when inserted into the genome, occupies a site not previously occupied. There is a fixed probability that any copy may be excised during the transmission from one generation to the next. Once again this transmission is accomplished via a Wright-Fisher sampling scheme. A key assumption that differentiates this from previous treatments of the evolution of repeated sequences is that only one parental copy of the sequence generates the new copies, and the number of these that are inserted has a Poisson distribution. The authors focus on the expected time to the most recent common ancestor of a pair of randomly chosen copies.

A statistical analysis of Alu sequences in humans, chimpanzees and galago produces estimators for the parameters in the analytical model. These lead to reasonable estimates for the mutation rate per nucleotide site and per Alu sequence, which may indicate that their model is biologically reasonable.

Walter Bodmer and Samuel Karlin together founded the mathematical biology program at Stanford in the early 1960s. It was Bodmer's theoretical work on linkage and selection that set the stage for much of the subsequent mathematical development of multilocus theory. It also played an important role in the Bodmers' subsequent work on the HLA system in man. Of course this theory informed the interpretation of the patterns of linkage disequilibrium that were found in the system.

In Chapter 14, Bodmer and Bodmer survey the statistical and evolutionary genetics of the HLA region. They commence with an analysis of the evidence for the existence of recombinational hot spots at various sites within the HLA region on chromosome 6. They then discuss the worldwide allele and haplotype frequency distributions whose analysis has implicated the action of natural selection. Further evidence for natural selection comes from the very different levels of polymorphism within different genes in the HLA-D region. These differences, they argue, are unlikely to be the result of different mutation rates.

The association of certain HLA specificities with diseases has been well documented. The identification of susceptibility genes for such diseases has, however, been very difficult, largely because such genes must have low penetrance. The authors survey some of the statistical techniques that have been used in attempting to deal with this problem.

The Bodmers conclude their paper with an analysis of the possible evolutionary history of the HLA system, suggesting that all the genes in the Class I and Class II region have a common origin by duplication perhaps 700 million years ago. The possible role of gene conversion and related phenomena in producing new variants is also discussed. New population genetic models are required to study the patterns of variation, produced by gene conversion and duplication, that emerge from the molecular data.

CHAPTER EIGHT

On the Theory of Evolution
under Genetic and Cultural Transmission
with Application to the Lactose
Absorption Problem

Marcus W. Feldman and
Luigi L. Cavalli-Sforza

Introduction

It is perhaps surprising that the evolutionary theory of discrete valued phenotypic variation which does not obey simple Mendelian rules of inheritance has received less quantitative study than has continuous or quantitative variation. The reason probably is the wide acceptance of a genetic basis for continuous variation in terms of many loci, each of small effect, whose aggregate produces a continuous distribution of phenotypes (Fisher 1918). This hypothesis may lead to the distribution most amenable to dynamic analysis, which is the Gaussian. For discrete valued or for qualitative traits, the most common assumption is that the variants are themselves different genotypes. For medical applications where the trait dichotomy is diseased versus healthy, and familial analysis suggests something more complicated than simple Mendelian inheritance, a continuous "liability to disease" trait has been postulated, with a threshhold above which disease occurs (Falconer 1965).

Our work (Feldman and Cavalli-Sforza 1976 et seq.) has focused on the role of phenotype-to-phenotype transmission as a force in the evolution of discrete valued traits. In these studies genetic variation influenced the phenotypic transmission from parent to child (*vertical* transmission) or among others in the population (i.e., *oblique* or *horizontal* transmission). This genetic variation could be manifest at the parental level ("teaching") or at the level of reception by offspring (which in the cultural context would be called learning). These forms of phenotypic transmission are described in detail in Cavalli-Sforza and Feldman (1981).

Throughout most of our previous studies (Feldman and Cavalli-Sforza 1976, 1984, 1986; Cavalli-Sforza and Feldman 1983) in which evolution is studied at both genotypic and phenotypic levels the (Darwinian) fitness differences have been primarily between phenotypes, with the genotypic effects on the transmission efficiency. Although there are no *direct* fitness effects of the genotype, the resulting genotypic selection is "induced" by that on the phenotypes. That this induced selection is not the same as direct Darwinian selection is exemplified by the fact that heterozygote superiority in the transmission of an advantageous phenotype is not sufficient for genetic polymorphism (Feldman and Cavalli-Sforza 1976). It is also the case that nongenetic transmission of altruism interferes with the evolutionary rules for the initial increase of the latter due to Hamilton (1964) (see Feldman, Cavalli-Sforza, and Peck 1985).

The type of cultural transmission may also have an important effect on the evolution of a given set of pheno-genotypes. For example, in a simple haploid genetic model with selection on a phenotype whose transmission was determined by the genotype at one locus, Feldman and Cavalli-Sforza (1986) found that oblique transmission could not produce genetic polymorphism, while horizontal transmission could.

The dynamics of our models are expressed in terms of the frequencies of the pheno-genotypes as they change over time. Since our introduction of these models, other workers, notably Boyd and Richerson (1985) and Aoki (1986, 1987), have made simplifying assumptions that reduce the dimensionality of the problem. Generally, these involve the omission of genotype-dependent vertical transmission with the consequence that the gene and phenotype frequencies fully describe the dynamics. In fact, this results in the absence of gene-culture disequilibrium (Feldman and Cavalli-Sforza 1984).

In this paper we investigate a class of models in which the selection on a dichotomous phenotype depends on the genotype, and where both vertical and oblique transmission rates also vary with the genotype of the transmitter or the receiver. This might be regarded as another dimension of phenotypic variation, where the Darwinian fitness of a trait depends on its carrier genotype. One of the most interesting evolutionary examples in which this may have occurred is the phenotype of milk use or nonuse and the genetic variant that controls the ability of adult humans to absorb lactose. It is reasonable to suppose that both Darwinian selection based on nutritional properties of milk and the cultural transmission of milk use have been influenced by the lactose-absorbing genotype. Before investigating this as a diploid genetic model, we discuss analogous haploid genetic models.

Haploid Asexual Model: Construction

There are two alleles, A and a, and two phenotypes labeled 1 and 2. The frequencies of the four pheno-genotypes A_1, a_1, A_2, a_2 are u_A, u_a, n_A, n_a. We assume that phenotype 1 is acquired by copying or learning from a parent who is of this phenotype. The rate of learning is less than 100% and depends on the genotype. Thus A acquires phenotype 1 from parents at rate δ_1 and a at rate δ_2. In each case the remaining offspring are of phenotype 2.

After this vertical transmission, the pheno-genotype frequencies are u_A^v, u_a^v, n_A^v, n_a^v, namely,

$$u_A^v = \delta_1 u_A \tag{8.1a}$$

$$u_a^v = \delta_2 u_a \tag{8.1b}$$

$$n_A^v = n_A + (1 - \delta_1)u_A \tag{8.1c}$$

$$n_a^v = n_a + (1 - \delta_2)u_a. \tag{8.1d}$$

Following vertical transmission, oblique transmission occurs by random contact between offspring of phenotype 2 and members of the parental generation with phenotype 1. The rates of transmission are f_1 and f_2 for A and a, respectively. After oblique transmission, the new frequencies are $u_A^0, u_a^0, n_A^0, n_a^0$ with

$$u_A^0 = u_A^v + f_1 n_A^v(u_A + u_a) \tag{8.2a}$$

$$u_a^0 = u_a^v + f_2 n_a^v(u_A + u_a) \tag{8.2b}$$

$$n_A^0 = n_A^v[1 - f_1(u_A + u_a)] \tag{8.2c}$$

$$n_a^0 = n_A^v[1 - f_2(u_A + u_a)]. \tag{8.2d}$$

Throughout the discussion, phenotype 1 is viewed as the one which involves the positive acquisition of something while phenotype 2 is the "clean-slate" type. We have in the past referred to 1 as "skilled" and 2 as "naive," and we have assigned fitnesses to the two phenotypes independently of the genotype. Here our goal is a diploid lactose intolerance model, and for comparison, the haploid model is formulated in a similar way with genotypic effects on fitness as well as transmission. The fitnesses of A_1, a_1, A_2, a_2 are, respectively, $1 + s_1, 1 - s_2, 1, 1$. In other words, the acquisition of trait 1 is disadvantageous to genotype a when $s_2 > 0$. We shall see later that in the analogous diploid model $s_2 > 0$ reflects any disadvantage to lactose nonabsorbers when they drink milk. After selection, we have

$$\bar{w}u'_A = (1 + s_1)\{\delta_1 u_A + f_1[n_A + (1 - \delta_1)u_A](u_A + u_a)\} \tag{8.3a}$$

$$\bar{w}u'_a = (1 - s_2)\{\delta_2 u_a + f_2[n_a + (1 - \delta_2)u_a](u_A + u_a)\} \tag{8.3b}$$

$$\bar{w}n'_A = [n_A + (1 - \delta_1)u_A][1 - f_1(u_A + u_a)] \tag{8.3c}$$

$$\bar{w}n'_a = [n_a + (1 - \delta_2)u_a][1 - f_2(u_A + u_a)] \tag{8.3d}$$

$$\bar{w} = 1 + s_1\{\delta_1 u_A + f_1[n_A + (1 - \delta_1)u_A](u_A + u_a)\}$$
$$- s_2\{\delta_2 u_a + f_2[n_a + (1 - \delta_2)u_a](u_A + u_a)\}. \tag{8.3e}$$

Haploid Asexual Model: Analysis

Throughout what follows we shall assume $\delta_1 < 1$, $\delta_2 < 1$, $f_1 < 1$, $f_2 < 1$. These assumptions preclude fixation on phenotype 1. We may use a geometric analogy familiar to students of two-locus population genetics and regard the frequency vector as varying within a tetrahedron as drawn in Figure 8.1.

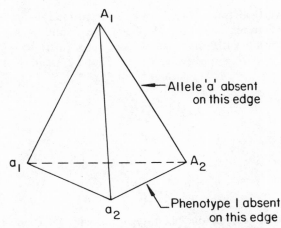

Figure 8.1. Tetrahedron representing the frequencies of the four phenogenotypes, A_1, A_2, a_1, a_2. On the edge where allele a is absent, $u_a = n_a = 0$. On the edge where phenotype 1 is absent, $u_A = u_a = 0$.

Two gene fixation edges and the edge where phenotype 1 is absent are shown in Figure 8.1. (For u_A to reach 1 we must have either $\delta_1 = 1$ or $f_1 = 1$). We consider first the stability properties of these three boundaries of the frequency simplex. On the gene fixation edge where $u_a = n_a = 0$, there may be an equilibrium between u_A and n_A. In fact, if

$$(1 + s_1)(f_1 + \delta_1) > 1, \tag{8.4}$$

there is a single valid equilibrium point given by the admissible root of the quadratic equation,

$$s_1 f_1 \delta_1 u_A^2 - u_A[s_1(f_1 + \delta_1) + (1 + s_1)f_1\delta_1]$$
$$+ (1 + s_1)(f_1 + \delta_1) - 1 = 0 \quad (8.5)$$

(see Feldman and Cavalli-Sforza 1986). Under (8.4) this root is locally stable in the $u_A - n_A$ edge. When (8.4) is reversed, there is no boundary equilibrium with A fixed.

In the same way, if

$$(1 - s_2)(f_2 + \delta_2) > 1, \quad (8.6)$$

there is a locally stable phenotypic polymorphism on the $u_a - n_a$ edge, given by a quadratic analogous to (8.5).

Finally, every point on the phenotype-fixation edge $u_A = u_a = 0$ is an equilibrium. This edge is a neutral equilibrium curve insofar as evolution of A and a is concerned in the absence of phenotype 1.

Initial Increase of the Rare Phenotype

First we investigate the initial increase properties of a *mutant phenotype 1* introduced near fixation of the population on phenotype 2, where the frequencies of A and a prior to the arrival of phenotype 1 were $\hat{n}_A$ and $\hat{n}_a$. When u_A and u_a are small enough so that their quadratic and higher powers may be ignored, we have the linear dynamics

$$u_A' = (1 + s_1)[\delta_1 u_A + f_1 \hat{n}_A(u_A + u_a)] \quad (8.7a)$$

$$u_a' = (1 - s_2)[\delta_2 u_a + f_2 \hat{n}_a(u_A + u_a)]. \quad (8.7b)$$

It is the eigenvalues of the matrix on the right side of (8.7) that determine the increase of phenotype 1. These are given by the roots of the characteristic quadratic,

$$g(\lambda) = \lambda^2 - \lambda[(1 + s_1)(\delta_1 + f_1\hat{n}_A) + (1 - s_2)(\delta_2 + f_2\hat{n}_a)]$$
$$+ (1 + s_1)(1 - s_2)(\delta_1\delta_2 + \delta_1 f_2 \hat{n}_a + \delta_2 f_1 \hat{n}_A). \quad (8.8)$$

Note that if $f_1 = f_2 = 0$, the increase of u_A is independent of that of u_a.

The initial increase of u_A and u_a is governed by $g(1)$ and $g'(1)$. If $g(1) < 0$, then they increase; if $g(1) > 0$ and $g'(1) > 0$, then phenotype 1 cannot initially increase at a geometric rate. Exploration of these conditions is facilitated by the expression,

$$g(1) = [(1 + s_1)\delta_1 - 1][(1 - s_2)(\delta_2 + f_2) - 1]\hat{n}_a$$
$$+ [(1 - s_2)\delta_2 - 1][(1 + s_1)(\delta_1 + f_1) - 1]\hat{n}_A. \quad (8.9)$$

The stability analysis can then be decomposed into four cases:

Case 1. $(1 - s_2)(f_2 + \delta_2) < 1, (1 + s_1)(f_1 + \delta_1) < 1$

No gene fixation equilibria exist, $g(1) > 0$ and $g'(1) > 0$. Hence both roots of (8.8) are less than unity and the $n_A - n_a$ edge is most likely to be stable.

Case 2. $(1 - s_2)(f_2 + \delta_2) > 1, (1 + s_1)(f_1 + \delta_1) > 1$

In this case it is obvious that if $(1 + s_1)\delta_1 < 1$, $g(1) < 0$ (since $\delta_2 < 1$) and the edge is unstable. If $(1 + s_1)\delta_1 > 0$ and $g(1) > 0$, then we examine $g'(1)$. Since $g(1) > 0$ we have from (8.9)

$$[(1 - s_2)(\delta_2 + f_2) - 1]\hat{n}_a$$
$$> \frac{[1 - (1 - s_2)\delta_2][(1 + s_1)(f_1 + \delta_1) - 1]}{(1 + s_1)\delta_1 - 1}\hat{n}_A. \quad (8.10)$$

Rewrite $g'(1)$ as

$$g'(1) = 2 - (1 + s_1)(\delta_1 + f_1\hat{n}_A) - (1 - s_2)(\delta_2 + f_2\hat{n}_a)$$
$$= \hat{n}_A[1 - (1 + s_1)(\delta_1 + f_1)] + \hat{n}_a[1 - (1 + s_1)\delta_1]$$
$$+ \hat{n}_A[1 - (1 - s_2)\delta_2] + \hat{n}_a[1 - (1 - s_2)(\delta_2 + f_2)]$$
$$< \hat{n}_A[1 - (1 - s_2)\delta_2] - \hat{n}_A[1 - (1 - s_2)\delta_2]\left\{\frac{(1 + s_1)(f_1 + \delta_1) - 1}{(1 + s_1)\delta_1 - 1}\right\}$$
$$< 0,$$

since $f_1 > 0$.

Since $g'(1) < 0$, both eigenvalues are greater than unity and the $n_A - n_a$ edge is unstable.

Case 3. $(1 + s_1)(f_1 + \delta_1) < 1, (1 - s_2)(f_2 + \delta_2) > 1$

The edge equilibrium $u_A - n_A$ does not exist. We use the same kind of argument as in case 2:

$$g'(1) = \hat{n}_A[1 - (1 + s_1)(f_1 + \delta_1)] + \hat{n}_a[1 - (1 + s_1)\delta_1]$$
$$+ \hat{n}_A[1 - (1 - s_2)\delta_2] + \hat{n}_a[1 - (1 - s_2)(f_2 + \delta_2)].$$

But if $g(1) > 0$, then

$$\hat{n}_A[1 - (1 - s_2)\delta_2] > \frac{[1 - (1 + s_1)\delta_1][(1 - s_2)(f_2 + \delta_2) - 1]\hat{n}_a}{[1 - (1 + s_1)(f_1 + \delta_1)]}$$
$$> [(1 - s_2)(f_2 + \delta_2) - 1]\hat{n}_a,$$

so that $g'(1) > 0$. The only possible instability condition then is $g(1) < 0$. Clearly, from (8.9), if $\hat{n}_a$ is close to 1 (and $\hat{n}_A$ is small) this condition $g(1) < 0$ may hold; and by the same token, if $\hat{n}_A$ is close to 1, $g(1) > 0$. Thus, if $\hat{n}_a$

is large enough, the user phenotype may increase, whereas if $\hat{n}_a$ is too small, it will be eliminated.

Case 4. $(1 + s_1)(f_1 + \delta_1) > 1, (1 - s_2)(f_2 + \delta_2) < 1$

If $(1 + s_1)\delta_1 > 1$, then from (8.9) it is clear that $g(1) < 0$ and for all $\hat{n}_A$ the user phenotype increases. If $(1 + s_1)\delta_1 < 1$ and $g(1) > 0$, then $g'(1) > 0$. Again, the only way that the edge could be unstable is if $g(1) < 0$. Thus, if $(1 + s_1)\delta_1 < 1$ and $\hat{n}_a$ is small, the $n_A - n_a$ edge is unstable, whereas if $\hat{n}_a$ is large, it will be stable.

To summarize, suppose that phenotypic polymorphism in the A-fixation edge is stable. Then phenotype 1 increases when rare if $\hat{n}_A$ is large enough. Corresponding results apply if the a-fixation edge is stable. Otherwise, the initial increase properties for u_A on the A-fixation edge and u_a on the a-fixation edge extend to the whole of the $n_A - n_a$ fixation edge.

Initial Increase of Rare Alleles

First consider the gene fixation edge where A is absent from the population. If $(1 - s_2)(f_2 + \delta_2) > 1$, then the equilibrium that is the unique valid root of the quadratic

$$s_2 f_2 \delta_2 u_a^2 + u_a[(1 - s_2)f_2\delta_2 - s_2(f_2 + \delta_2)]$$
$$+ 1 - (1 - s_2)(f_2 + \delta_2) = 0 \quad (8.11)$$

is stable in this edge. We seek conditions for the rare allele A to increase.

The local stability matrix in the neighborhood of the equilibrium $\hat{u}_a$ that solves (8.11) produces a characteristic polynomial

$$h_a(\lambda) = \lambda^2 - \frac{\lambda}{\hat{\bar{w}}_a}\{(1 + s_1)\delta_1 + 1 + f_1\hat{u}_a[(1 + s_1)(1 - \delta_1) - 1]\}$$
$$+ \frac{\delta_1(1 + s_1)(1 - f_1\hat{u}_a)}{\hat{\bar{w}}_a^2}, \quad (8.12)$$

where

$$\hat{\bar{w}}_a = 1 - s_2\hat{u}_a[\delta_2 + f_2 - f_2\delta_2\hat{u}_a]$$
$$= (1 - s_2)[f_2 + \delta_2 - f_2\delta_2\hat{u}_a]. \quad (8.13)$$

Two cases emerge. First, if $h_a(1) < 0$, then clearly the leading root of (8.12) is greater than unity and the equilibrium is unstable. Second, suppose $h_a(1) > 0$. Rewrite $h_a(1)$ as

$$h_a(1) = (\hat{\bar{w}}_a - 1)[\hat{\bar{w}}_a - \delta_1(1 + s_1) + f_1\hat{u}_a\delta_1(1 + s_1)(\hat{\bar{w}}_a - 1)]$$
$$- s_1\hat{\bar{w}}_a f_1\hat{u}_a. \quad (8.14)$$

But from (8.13)

$$\hat{\bar{w}}_a - 1 = -s_2\hat{u}_a\hat{\bar{w}}_a/(1 - s_2), \tag{8.15}$$

which upon substitution into (8.14) produces

$$h_a(1) = (\hat{\bar{w}}_a - 1)\left[\hat{\bar{w}}_a - \delta_1(1 + s_1) + f_1\hat{u}_a\delta_1(1 + s_1) + \frac{s_1(1 - s_2)}{s_2}f_1\right] > 0.$$

From (8.15) $\hat{\bar{w}}_a < 1$ when $s_2 > 0$. Hence we may assume

$$\hat{\bar{w}}_a - \delta_1(1 + s_1) + f_1\hat{u}_a\delta_1(1 + s_1) + \frac{s_1(1 - s_2)}{s_2}f_1 < 0. \tag{8.16}$$

Now consider $h'_a(1)$. If $h'_a(1) < 0$, then since the local stability matrix is strictly positive, both roots of (8.12) must be larger than unity. Now,

$$\begin{aligned}
h'_a(1) &= 2\hat{\bar{w}}_a - (1 + s_1)\delta_1 - 1 - f_1\hat{u}_a(1 + s_1)(1 - \delta_1) + f_1\hat{u}_a \\
&= [\hat{\bar{w}}_a - (1 + s_1)\delta_1 + f_1\hat{u}_a\delta_1(1 + s_1)] \\
&\quad + [\bar{w}_a - 1 - s_1f_1\hat{u}_a]. \tag{8.17}
\end{aligned}$$

By (8.16) the first term on the right of (8.17) is negative and by (8.15), so is the second. Hence $h'_a(1) < 0$ and the equilibrium is unstable. Thus when $s_1, s_2 > 0$, A always increases when rare.

We now turn to fixation of the A-allele where the unique root of (8.5) is stable in this edge if $(1 + s_1)(f_1 + \delta_1) > 1$. For stability to invasion by a, the characteristic polynomial of the local stability matrix is

$$h_A(\lambda) = \lambda^2 - \frac{\lambda}{\hat{\bar{w}}_A}[(1 - s_2)\delta_2 + f_2(1 - s_2)(1 - \delta_2)\hat{u}_A + (1 - f_2\hat{u}_A)]$$

$$+ \frac{(1 + s_2)\delta_2(1 - f_2\hat{u}_A)}{\hat{\bar{w}}_A^2}, \tag{8.18}$$

where

$$\begin{aligned}
\hat{w}_A &= 1 + s_1\hat{u}_A(\delta_1 + f_1 - \delta_1f_1\hat{u}_A) \\
&= (1 + s_1)(\delta_1 + f_1 - \delta_1f_1\hat{u}_A). \tag{8.19}
\end{aligned}$$

It is obvious that $\hat{\bar{w}}_A > 1$. Write

$$\hat{\bar{w}}_Ah'_A(1) = 2\hat{\bar{w}}_A - \{(1 - s_2)\delta_2(1 - f_2\hat{u}_A) + (1 - s_2f_2\hat{u}_A)\} > 0,$$

since $0 < s_2, \delta_2, f_2, \hat{u}_A < 1$. Thus the only criterion for stability is the sign of $h_A(1)$; if $h_A(1) > 0$, then the equilibrium is stable, if $h_A(1) < 0$, it is unstable.

Now, from (8.18) we have

$$\hat{u}_A = \frac{1 + s_1}{s_1}\frac{(\hat{\bar{w}}_A - 1)}{\hat{\bar{w}}_A}. \tag{8.20}$$

From (8.19) and with the use of (8.20) we obtain

$$h_A(1) = [\hat{\bar{w}}_A - \delta_2(1 - s_2)](\hat{\bar{w}}_A - 1) - f_2\hat{u}_A\hat{\bar{w}}_A[(1 - s_2)(1 - \delta_2) - 1]$$
$$- f_2\hat{u}_A\delta_2(1 - s_2)$$
$$= [\hat{\bar{w}}_A - \delta_2(1 - s_2) + f_2\hat{u}_A\delta_2(1 - s_2)](\bar{w}_A - 1) + s_2 f_2\hat{u}_A\hat{\bar{w}}_A$$
$$= \left[\bar{w}_A - \delta_2(1 - s_2) + f_2\hat{u}_A\delta_2(1 - s_2) + \frac{s_2 f_2(1 + s_1)}{s_1}\right](\bar{w}_A - 1).$$

$$(8.21)$$

But from (8.18), $\hat{\bar{w}}_A > 1$, and since $\delta_2 < 1$ both factors on the right side of (8.21) are positive. Thus, when $s_2 > 0$, $s_1 > 0$, we have shown that the equilibrium on the a-fixation edge, if it exists, is unstable while that on the A-fixation edge, when it exists, is locally stable. The latter does not, however, imply that there should be global convergence to the A-fixation edge. We have already seen that in some circumstances part of the $n_A - n_a$ edge may be stable (see cases 3 and 4 of the previous section).

The above results were obtained under the assumption $s_2 > 0$. We now turn to the case $s_2 < 0$. Write $-s_2 = \tilde{s}_2$. Then, under the conditions $(1 + s_1)(f_1 + \delta_1) > 1$, $(1 + \tilde{s}_2)(f_2 + \delta_2) > 1$, the isolated equilibria on the gene fixation edges will be stable in their respective boundaries, and the $n_A - n_a$ edge will be unstable to the introduction of users. If $(1 + s_1)(f_1 + \delta_1) < 1$, $(1 + \tilde{s}_2)(f_2 + \delta_2) < 1$, then, as before, the $n_A - n_a$ edge is locally stable. The cases $(1 + s_1)(f_1 + \delta_1) < 1$, $(1 + \tilde{s}_2)(f_2 + \delta_2) > 1$ and $(1 + s_1)(f_1 + \delta_1) > 1$, $(1 + \tilde{s}_2)(f_2 + \delta_2) < 1$ produce conclusions similar to those of cases 3 and 4 above in that part of the $n_A - n_a$ edge is stable and the other part unstable.

When $\tilde{s}_2 > 0$ the stability configuration of the equilibria on the gene fixation edges involves a more complicated analysis than before. The reason is that on both edges $\bar{w} > 1$, that is, from (8.15) $\hat{\bar{w}}_a > 1$ and from (8.19) $\hat{\bar{w}}_A > 1$. Consider the a-fixation edge, for example, and suppose that $h_a(1) > 0$. Then from (8.15)

$$\hat{\bar{w}}_a - \delta_1(1 + s_1) + f_1\delta_1\hat{u}_a(1 + s_1) > s_1 f_1 \frac{(1 + \tilde{s}_2)}{\tilde{s}_2}.$$

Then, returning to (8.16),

$$h'_a(1) > s_1 f_1 \left[\frac{(1 + \tilde{s}_2)}{\tilde{s}_2} - \tilde{u}_a\right] + \hat{\bar{w}}_a - 1 > 0.$$

Thus the criterion for stability of the equilibrium is the sign of $h_a(1)$. If $h_a(1) < 0$, then it is unstable; if $h_a(1) > 0$, it is stable. The same argument applies on the A-fixation edge. The conclusion is that if $(1 + s_1)(f_1 + \delta_1) > 1$ and $(1 + s_2)(f_2 + \delta_2) > 1$ (so that both gene fixation equilibria exist and are stable in their edges) and $h_a(1) < 0$ and $h_A(1) < 0$, then both gene fixation edges are unstable and so is the phenotype fixation boundary $n_A - n_a$. The inequalities $h_a(1) < 0$ and $h_A(1) < 0$ are, respectively,

$$(1 + \tilde{s}_2)[\delta_2 + f_2 - f_2\delta_2\hat{u}_a] < \delta_1(1 + s_1)$$
$$- f_1\delta_1(1 + s_1)\hat{u}_a + \frac{s_1(1 + \tilde{s}_2)}{\tilde{s}_2} f_1 \quad (8.22)$$

and

$$(1 + s_1)[\delta_1 + f_1 - f_1\delta_1\hat{u}_A] < \delta_2(1 + \tilde{s}_2)$$
$$- f_2\delta_2(1 + \tilde{s}_2)\hat{u}_A + \frac{\tilde{s}_2(1 + s_1)}{s_1} f_2. \quad (8.23)$$

It should be noted that if $s_1 = \tilde{s}_2$, these inequalities are identical to those of Feldman and Cavalli-Sforza (1986) where the fitness depended only on the phenotype. There it was shown that (8.22) held if and only if $\hat{\bar{w}}_a < \hat{\bar{w}}_A$ and (8.23) if and only if $\hat{\bar{w}}_A < \hat{\bar{w}}_a$. In other words, only one of the gene fixation edges was stable under selection on the *phenotype* only. The question remains, if $s_1 \neq \tilde{s}_2$, can both gene fixation edges be unstable?

Rewriting (8.21), we have

$$M\hat{u}_a < \frac{1}{\tilde{s}_2} \{(f_1s_1 - f_2\tilde{s}_2)(1 + \tilde{s}_2) + \tilde{s}_2[\delta_1(1 + s_1) - \delta_2(1 + \tilde{s}_2)]\} \quad (8.22a)$$

and from (8.23)

$$M\hat{u}_a > \frac{1}{s_1} \{(f_1s_1 - f_2\tilde{s}_2)(1 + s_1) + s_1[\delta_1(1 + s_1) - \delta_2(1 + \tilde{s}_2)]\}, \quad (8.23a)$$

where $M = f_1\delta_1(1 + s_1) - f_2\delta_2(1 + \tilde{s}_2)$.

In considering whether both (8.22) and (8.23) can hold, we should also recall that when $f_1 = f_2 = 0$, $\hat{\bar{w}}_a = \delta_2(1 + \tilde{s}_2) > 1$ and $\hat{\bar{w}}_A = \delta_1(1 + s_1) > 1$. Hence from (8.14) and (8.21) it is *impossible* for both boundaries to be stable when $f_1 = f_2 = 0$. An example where both gene fixation edges are unstable in the case and one of the f's is nonzero is the following. Take $\delta_1 = 0.9$, $s_1 = 0.5$, $f_1 = 0$; $\delta_2 = 0.6$, $\tilde{s}_2 = 1$, $f_2 = 0.1$. Then $(1 + s_1)\delta_1 = 1.35 > (1 + s_2)\delta_2 = 1.2 > 1$, so that if f_2 had been zero, there would have been fixation on the A gene. At equilibrium on the A-gene fixation edge,

$$\hat{u}_A = \frac{(1 + \delta_1)s_1 - 1}{s_1\delta_1} = \frac{0.35}{0.45} = 0.777.$$ From (8.22a) and (8.23a), $M = -0.12$

and the right side of (8.23a) is -0.15. Since $(0.777)(-0.12) > -0.15$, (8.23a) holds. Now, $\hat{u}_a$ solves the quadratic

$$0.06u_a^2 + u_a(0.82) - 0.4 = 0,$$

that is, $\hat{u}_a = 0.4715$. The right side of (8.22a) is -0.05. Thus (8.22a) is true if $(0.4715)(-0.12) < -0.05$, which is clearly true. It is worth pointing out that if f_2 is increased to 0.2, then (8.23a) remains true, but (8.22a) fails and the a-fixation edge is stable. Clearly, the general analysis of the two inequalities (8.22), (8.23) promises to be immensely complicated. On the basis of the above numerical work, we might conjecture that the range of parameters that allow both genetic and phenotypic polymorphism will be rather small.

In order to obtain a more precise idea of the likelihood of producing simultaneous genetic and phenotypic polymorphism, a numerical investigation was made. We fixed $s_1 = 1$ and chose $\tilde{s}_2$ uniformly (at random) from $[0, 2]$. The parameters $\delta_1, \delta_2, f_1, f_2$ were all chosen at random uniformly on $[0, 1]$. The parameters chosen in this way were then examined to see whether they satisfied $(1 + s_1)(f_1 + \delta_1) > 1$, $(1 + \tilde{s}_2)(f_2 + \delta_2) > 1$. One thousand parameter sets that statisfied both inequalities were chosen. These allow the existence of both A- and a-gene fixation equilibria. From these one thousand parameter sets we determined how many also satisfied *both* (8.22a) and (8.23a), in which case interior polymorphism exists and would be expected to be stable. We found twenty-two cases out of the one thousand that allowed such polymorphism.

Haploid Sexual Model

Again suppose there are two phenotypes numbered 1 and 2 and two alleles A and a. The fitness values for the four pheno-genotypes and their frequencies are

	A_1	a_1	A_2	a_2
Frequencies:	u_A	u_a	n_A	n_a
Fitnesses:	$1 + s_1$	$1 - s_2$	1	1

We usually assume $s_2 > 0$, but as we shall discuss later there are valid reasons for investigating the case $s_2 < 0$ with $s_1 > |s_2| > 0$. Four transmission parameters describe the model with uniparental transmission. The abbreviations TP for transmitting parent and OP for other parent are used:

		Probability for offspring	
		A_1	a_1
	A_1	δ_1	δ_1^*
TP	a_1	δ_2^*	δ_2
	A_2	0	0
	a_2	0	0

The order of events in the life cycle is as in the asexual case, with mating inserted between selection and vertical transmission, that is, selection, mating, vertical cultural transmission to offspring, oblique transmission, and we take the census before selection. The mating table is presented as Table 8.1 and from it we derive equations (8.24), the frequencies of the phenotypes after vertical transmission. The oblique contact parameters are f_1 for n_A and f_2 for n_a individuals. The complete recursion is (8.25).

Table 8.1
Mating and segregation of sexual haploids.

TP OP	Mating*	u_A	u_a	n_A	n_a
$u_A \times u_A$	$u_A^2(1+s_1)^2$	δ_1	0	$1-\delta_1$	0
$u_A \times u_a$	$u_A u_a(1+s_1)(1-s_2)$	$\delta_1/2$	$\delta_1^*/2$	$\dfrac{1-\delta_1}{2}$	$\dfrac{1-\delta_1^*}{2}$
$u_A \times n_A$	$u_A n_A(1+s_1)$	δ_1	0	$1-\delta_1$	0
$u_A \times n_a$	$u_A n_a(1+s_1)$	$\delta_1/2$	$\delta_1^*/2$	$\dfrac{1-\delta_1}{2}$	$\dfrac{1-\delta_1^*}{2}$
$u_a \times u_A$	$u_A u_a(1+s_1)(1-s_2)$	$\delta_2^*/2$	$\delta_2/2$	$(1-\delta_2^*)/2$	$(1-\delta_2)/2$
$u_a \times u_a$	$u_a^2(1-s_2)^2$	0	δ_2	0	$1-\delta_2$
$u_a \times n_A$	$u_a n_A(1-s_2)$	$\delta_2^*/2$	$\delta_2/2$	$(1-\delta_1^*)/2$	$(1-\delta_2)/2$
$u_a \times n_a$	$u_a n_a(1-s_2)$	0	δ_2	0	$1-\delta_2$
$n_A \times u_A$	$u_A n_A(1+s_1)$	0	0	1	0
$n_A \times u_a$	$u_a n_A(1-s_2)$	0	0	1/2	1/2
$n_A \times n_A$	n_A^2	0	0	1	0
$n_A \times n_a$	$n_A n_a$	0	0	1/2	1/2
$n_a \times u_A$	$u_A n_a(1+s_1)$	0	0	1/2	1/2
$n_a \times u_a$	$u_a n_a(1-s_2)$	0	0	0	1
$n_a \times n_A$	$n_A n_a$	0	0	1/2	1/2
$n_a \times n_a$	n_a^2	0	0	0	1

* TP is transmitting parent, OP is the other parent.

After vertical transmission,

$$Tu_A^v = \delta_1 u_A(1 + s_1)\left[u_A(1 + s_1) + \frac{u_a(1 - s_2)}{2} + n_A + \frac{n_a}{2}\right]$$

$$+ \frac{\delta_2^*}{2} u_a(1 - s_2)[u_A(1 + s_1) + n_A] \qquad (8.24a)$$

$$Tu_a^v = \frac{\delta_1^*}{2} u_A(1 + s_1)[u_a(1 - s_2) + n_a]$$

$$+ \delta_2 u_a(1 - s_2)\left[u_a(1 - s_2) + n_a + \frac{u_A(1 + s_1)}{2} + \frac{n_A}{2}\right] \qquad (8.24b)$$

$$Tn_A^v = [(1 - \delta_1)u_A(1 + s_1) + n_A]\left[u_A(1 + s_1) + \frac{u_a(1 - s_2)}{2} + n_A + \frac{n_a}{2}\right]$$

$$+ \left[\frac{n_a}{2} + \frac{1 - \delta_2^*}{2} u_a(1 - s_2)\right][u_A(1 + s_1) + n_A] \qquad (8.24c)$$

$$Tn_a^v = \left[\frac{1 - \delta_1^*}{2} u_A(1 + s_1) + \frac{n_A}{2}\right][u_a(1 - s_2) + n_a] + [n_a + (1 - \delta_2)$$

$$\times u_a(1 - s_2)]\left[u_a(1 - s_2) + n_a + \frac{u_A(1 + s_1)}{2} + \frac{n_A}{2}\right], \qquad (8.24d)$$

where

$$T = [u_A(1 + s_1) + u_a(1 - s_2) + n_A + n_a]^2 = T_1^2. \qquad (8.24e)$$

After selection,

$$u_A' = u_A^v + f_1 n_A^v \left(\frac{u_A(1 + s_1) + u_a(1 - s_2)}{T_1}\right) \qquad (8.25a)$$

$$u_a' = u_A^v + f_2 n_A^v \left(\frac{u_A(1 + s_1) + u_a(1 - s_2)}{T_1}\right) \qquad (8.25b)$$

$$n_A' = n_A^v \left(1 - f_1 \frac{u_A(1 + s_1) + u_a(1 - s_2)}{T_1}\right) \qquad (8.25c)$$

$$n_a' = n_A^v \left(1 - f_2 \frac{u_A(1 + s_1) + u_a(1 - s_2)}{T_1}\right). \qquad (8.25d)$$

When A is fixed, there is a unique valid equilibrium of the quadratic

$$s_1^2 u_A^2 + u_A\{2s_1 - (1 + s_1)[\delta_1 s_1 + f_1 s_1 - f_1 \delta_1(1 + s_1)]\}$$
$$+ 1 - (\delta_1 + f_1)(1 + s_1) = 0, \qquad (8.26)$$

which is stable in the A-fixation boundary provided

$$(1 + s_1)(f_1 + \delta_1) > 1. \tag{8.27}$$

Similarly, there is a unique valid root of the quadratic

$$s_2^2 u_a^2 + u_a\{-2s_2 + (1 - s_2)[\delta_2 s_2 + f_2 s_2 + f_2 \delta_2(1 - s_2)]\}$$
$$+ 1 - (\delta_2 + f_2)(1 - s_2), \tag{8.28}$$

which is stable in the a-fixation edge if

$$(1 - s_2)(f_2 + \delta_2) > 1. \tag{8.29}$$

Initial Increase of Rare Phenotype

As in the haploid asexual case, there can be no equilibrium on the edge where all individuals are of phenotype 1. It is conceivable that the other three edges have equilibria. First consider the $n_A - n_a$ edge, which, as in the haploid asexual case, is neutral. Near the point on this edge specified by $\hat{n}_A$ and $\hat{n}_a = 1 - \hat{n}_A$ the characteristic polynomial of the local stability matrix is

$$G(\lambda) = \lambda^2 - \lambda\left\{(1 + s_1)\left[\delta_1 \frac{(1 + \hat{n}_A)}{2} + f_1 \hat{n}_A\right] + (1 - s_2)\right.$$

$$\times \left[\delta_2 \frac{(1 + \hat{n}_a)}{2} + f_2 \hat{n}_a\right]\right\} + (1 + s_1)(1 - s_2)\left\{\left[\frac{\delta_1}{2} + \hat{n}_A\left(\frac{\delta_1}{2} + f_1\right)\right]\right.$$

$$\times \left[\frac{\delta_2}{2} + \hat{n}_a\left(\frac{\delta_2}{2} + f_2\right)\right] - \hat{n}_A \hat{n}_a\left(\frac{\delta_2^*}{2} + f_1\right)\left(\frac{\delta_1^*}{2} + f_2\right)\right\}. \tag{8.30}$$

Again if $G(1) < 0$, then the phenotype 1 will invade. If $G(1) > 0$ and $G'(1) > 0$, then it cannot invade. Reorganizing (8.30) we have

$$G(1) = \hat{n}_A\left[(1 - s_2)\frac{\delta_2}{2} - 1\right][(1 + s_1)(\delta_1 + f_1) - 1]$$

$$+ \hat{n}_a\left[(1 + s_1)\frac{\delta_1}{2} - 1\right][(1 - s_2)(\delta_2 + f_2) - 1]$$

$$+ \hat{n}_A \hat{n}_a(1 + s_1)(1 - s_2)$$

$$\times \left[\frac{\delta_1 \delta_2}{4} - \frac{\delta_1^* \delta_2^*}{4} + \frac{f_1}{2}(\delta_2 - \delta_1^*) + \frac{f_2}{2}(\delta_1 - \delta_2^*)\right]. \tag{8.31}$$

Obviously, if $\delta_1^* = \delta_2$ and $\delta_1 = \delta_2^*$, the form of (8.31) is extremely similar to (8.9) in the haploid case. But this is exactly the case where the transmission probability depends on the offspring's genotype—not on that of the parent (i.e., on the learner's genotype and not the teacher's). It is interesting that the alternative case in which the transmitting parent's genotype determines the rate of transmission to the offspring does not simplify so elegantly. In this case, $\delta_1 = \delta_1^*$, $\delta_2 = \delta_2^*$, and the last term of (8.31) becomes $\hat{n}_A \hat{n}_a (1 + s_1)(1 - s_2) \dfrac{(f_1 - f_2)}{2} (\delta_2 - \delta_1)$, which does not vanish unless $f_1 = f_2$ or $\delta_2 = \delta_1$. The latter is, of course, uninteresting.

As in the haploid asexual case, if $(1 + s_1)(\delta_1 + f_1) > 1$ and $(1 - s_2)(\delta_2 + f_2) < 1$, we expect that part of the $n_A - n_a$ edge close to $\hat{n}_a = 1$ is stable and that part close to $\hat{n}_A = 1$ is unstable. The sexual case does provide some interesting contrasts with the asexual model. For example, it is possible for the $n_A - n_a$ edge to be unstable in the asexual case and stable in the sexual case even when $\delta_1^* = \delta_2^* = 0$. Set $\hat{n}_A = \hat{n}_a = 0.5$; $s_1 = 0.1$, $\delta_1 = 0.5$, $f_1 = 0.5$; $s_2 = 0.1$, $\delta_2 = 0.6$, $f_2 = 0.6$. Then $(1 + s_1)(f_1 + \delta_1) = 1.1$ and $(1 - s_2)(f_2 + \delta_2) = 1.08$. Both gene fixation edges are stable in their boundaries and, from (8.9), the $n_A - n_a$ edge is unstable. In fact, the analysis of the haploid asexual model I indicates that the equilibrium on the a-fixation edge is locally unstable, and that on the A-fixation edge is locally stable. Now, with these numbers substituted into (8.31), the value of $G(1)$ is positive and from (8.30), $G'(1) > 0$. For this set of transmission and selection parameters from initial conditions near $\hat{n}_A = 1$ or $\hat{n}_a = 1$, phenotype 1 will invade but there will be at least one interval near the middle of the edge from which invasion will not occur. Numerical iteration confirms these predictions. If $\delta_1^* = \delta_2^* = 0$, however, and $(1 + s_1)(\delta_1 + f_1) < 1$ and $(1 - s_2)(\delta_2 + f_2) < 1$, then neither asexual nor sexual models allow the $n_A - n_a$ edge to be invaded by phenotype 1.

When δ_1^*, $\delta_2^* \neq 0$, it is clearly possible that the same $s_1, \delta_1, f_1, s_2, \delta_2, f_2$ values that ensure the stability of the whole $n_A - n_a$ edge in the asexual case do not do so in the sexual case. That is, some values of $\hat{n}_A$ and $\hat{n}_a$ will be invasible when δ_1^*, δ_2^* are large enough.

Initial Increase of Rare Alleles

Here we are interested in the fate of A introduced near the equilibrium specified by (8.26) and a introduced near that specified by (8.28). The analysis of the local stability of these gene fixation equilibrium with $f_1 \neq 0$ and $f_2 \neq 0$ is algebraically prohibitive, so our conclusions concern the situation $f_1 = f_2 = 0$. Suppose that $\hat{u}_A$ is the edge equilibrium where A is fixed, that is, $\hat{u}_A$ is the admissible root of (8.26). Then $\hat{u}_A$ can be shown to be stable or unstable to invasion by a according to whether

$$\hat{u}_A \left[s_1(1 + s_1 \hat{u}_A) - \frac{\delta_2 s_1 (1 - s_2)}{2} + \frac{s_2 \delta_1^* (1 + s_1)}{2} \right] \qquad (8.32)$$

is positive or negative, respectively.

The first point to note is that when $s_2 > 0$, $1 > \delta_2(1 - s_2)/2$ so that (8.32) is positive and the A-fixation equilibrium is stable. Or course, when $s_2 > 0$ there is no a-fixation equilibrium. The case $s_2 = -\tilde{s}_2 < 0$ is more interesting. Here A-fixation is unstable if

$$s_1(1 + s_1\hat{u}_A) - \frac{\delta_2 s_1(1 + \tilde{s}_2)}{2} - \frac{\tilde{s}_2 \delta_1^*(1 + s_1)}{2} < 0, \qquad (8.33)$$

and a-fixation is unstable if

$$\tilde{s}_2(1 + \tilde{s}_2\hat{u}_a) - \frac{\delta_1 \tilde{s}_2(1 + s_1)}{2} - \frac{s_1 \delta_2^*(1 + \tilde{s}_2)}{2} < 0, \qquad (8.34)$$

where, for existence of $\hat{u}_A$ and $\hat{u}_a$, we require $(1 + s_1)\delta_1 > 1$ and $(1 + \tilde{s}_2)\delta_2 > 1$.

The inequalities (8.33) and (8.34) should be compared to (8.22a) and (8.23a), the corresponding inequalities in the asexual haploid case. Recall that with $f_1 = f_2 = 0$, (8.22a) and (8.23a) could not both be valid.

Rewritting (8.33) and (8.34), for simultaneous instability of both gene fixation edge equilibria, it is necessary that

$$\frac{s_1 \delta_2(1 + \tilde{s}_2)}{2} + \frac{\tilde{s}_2 \delta_1^*(1 + s_1)}{2} > s_1 \qquad (8.35)$$

and

$$\frac{\tilde{s}_2 \delta_1(1 + s_1)}{2} + \frac{s_1 \delta_2^*(1 + \tilde{s}_2)}{2} > \tilde{s}_2. \qquad (8.36)$$

When (8.35) and (8.36) hold, then for (8.33) and (8.34) to hold it is sufficient that

$$\hat{u}_A < \frac{1}{s_1^2}\left[\frac{s_1 \delta_2(1 + \tilde{s}_2)}{2} + \frac{\tilde{s}_2 \delta_1^*(1 + s_1)}{2} - s_1\right] \qquad (8.33a)$$

and

$$\hat{u}_a < \frac{1}{\tilde{s}_2^2}\left[\frac{\tilde{s}_2 \delta_1(1 + s_1)}{2} + \frac{s_1 \delta_2^*(1 + \tilde{s}_2)}{2} - \tilde{s}_2\right], \qquad (8.34a)$$

respectively. Substitution of (8.33a) and (8.34a) into (8.26) and (8.28), respectively (with $f_1 = f_2 = 0$), produces the conditions

$$\frac{s_1 \delta_2(1 + \tilde{s}_2)}{2} + \frac{\tilde{s}_2 \delta_1^*(1 + s_1)}{2} > s_1 \delta_1(1 + s_1) \qquad (8.37)$$

and

$$\frac{\tilde{s}_2 \delta_1(1 + s_1)}{2} + \frac{s_1 \delta_2^*(1 + \tilde{s}_2)}{2} > \tilde{s}_2 \delta_2(1 + \tilde{s}_2) \qquad (8.38)$$

for instability of $\hat{u}_A$ and $\hat{u}_a$, respectively. Several sets of parameters that satisfy the conditions for instability of $\hat{u}_A$ and $\hat{u}_a$ have been found. It is of interest to note that in the case $\delta_1^* = \delta_2^* = 0$, which might be the most directly comparable to the haploid case, if $\delta_2(1 + \tilde{s}_2) > 2$ and $\delta_1(1 + s_1) > 2$, then polymorphism can be maintained. It is also worth noting that if $s_1 = \tilde{s}_2$, so that selection is on the phenotype only (with no genetic effects), the conditions (8.37) and (8.38) become

$$\delta_1^*/2 > \delta_1 - \delta_2/2, \qquad \delta_2^*/2 > \delta_2 - \delta_1/2. \qquad (8.39)$$

In order to obtain some indication of the likelihood of polymorphism here, we set $f_1 = f_2 = 0$, $s_1 = 1$ and chose $\tilde{s}_2$ at random uniformly on $[0, 2]$. $\delta_1, \delta_2, \delta_1^*$ and δ_2^* were chosen at random uniformly on $[0, 1]$. One hundred thousand such sets of parameters were chosen and eleven sets of parameters that satisfied (8.37) and (8.38) with $(1 + s_1)\delta_1 > 1$, $(1 + \tilde{s}_2)\delta_2 > 1$ were revealed from the 100,000. We conclude that even in the sexual haploid model, the chance of genetic polymorphism is very low. Nevertheless, it is possible without oblique transmission, contrary to the asexual case.

A Diploid Model for the Lactose Absorption Polymorphism

Consider a single locus with alleles A and a and genotypes AA, Aa, aa. The situation of two phenotypes, that is, six pheno-genotypes, was studied by Feldman and Cavalli-Sforza (1976) in the framework of genetic variability in rates of transmission of the phenotype. The rule of transmission is defined by the parameters $\beta_{ghi,lm}$ where g, h, and i refer to the mother's, father's, and child's genotypes, and l and m are mother's and father's phenotypes. Here we use the numerical convention

genotype	AA	Aa	aa
number	1	2	3

and l, m are 1 or 2. The parameter $\beta_{ghi,lm}$ is the probability that a child of genotype i is of phenotype 1 if its mother and father have genotypes g and h and phenotypes l and m, respectively.

Our intention here is to construct a model in which the phenotypic dichotomy is between milk use and nonuse by adults, and where the gene influences the ability to hydrolyze the milk sugar, lactose, into its constituent monosaccharides glucose and galactose. In most mammals the production of lactase I, which mediates this hydrolysis, declines at an early age. Thus after weaning, most humans, for example, are unable to catabolize large amounts of ingested lactose. Such people are called lactose malabsorbers, and suffer from a number of familiar symptoms such as flatulence, intestinal cramps, and diarrhea. An interesting exception to this common state of affairs occurs in human populations that have high frequencies of lactose absorbers. The latter individuals are able as adults to metabolize sugar completely without exhibiting the intolerance symptoms. (It should be noted that intolerance to lactose may depend on other factors so that the correlation between lactose absorption and intolerance is not complete.)

Simoons (1978) comprehensively reviewed the data on population frequencies of absorption and malabsorption around the world. He concluded that unmixed ethnic groups can be classified either as absorbers (with 0–30% malabsorbers) or as malabsorbers (60–100% malabsorbers). Durham (1988) considered a subsample of sixty populations from Simoons' compilation and found that among hunters and gatherers the frequency of lactose absorbers is low (average 12.6%), as it is among recently dairying agriculturalists (11.9%) and nondairying pastoralists (15.5%). Among the dairying populations of Northern Europe, the frequency averages 91.5%.

This correlation between lactose absorption and the history of dairying was first observed by Simoons (1969) and McCracken (1971). They proposed a *coevolutionary* hypothesis for the evolution of the biochemical trait of lactose absorption. Lactose absorption segregates in families from Mexican, Nigerian, Jewish, Pima Indian, Asian, and Northern European populations as an autosomal dominant gene (Johnson et al. 1977). Both Simoons and McCracken suggest that the cultural trait, dairying, provided a positive selective pressure for lactose absorbing genotypes by enabling lactose absorbers to add fresh milk to their food supply. In times of food shortage, lactose absorbers would have an additional source of nutrition that would not be available to malabsorbers, and would have increased fitness as a result. This is called the "culture historical" hypothesis for the association between dairying and lactose absorption.

In the situation where absorbers benefit by milk use, malabsorbers may also increase their food supply by processing milk in such a way that lactose is broken down. This would entail a general fitness advantage to "users" over "nonusers."

Aoki (1986) developed a stochastic model for the joint evolution of lactose tolerance and milk use using two-dimensional diffusion approximations. Using an estimate of about 6,000 years ago for the advent of dairying, he concluded that with effective population sizes of five hundred or larger, selection coefficients larger than 5% for lactose absorbers using milk must be invoked to explain present frequencies of lactose absorption in dairying populations. In a second study Aoki (1987) has extended Fisher's (1937) theory for the wave of advance for an advantageous gene to a pair of interdependent waves, one for the frequency of the lactose absorption allele and the other for the frequency of milk use. He was not satisfied that reasonable parameters produced speeds of the traveling waves compatible with archeological estimates of the rate of spread of farming by Ammerman and Cavalli-Sforza (1984). One reason for this, as noted by Aoki, may be an increasing selective advantage to absorbers at northern latitudes.

In the treatment we now present, genotype-related differences in vertical transmission of the trait of milk use are included; this was omitted from Aoki's models. In addition, genetic differences in the rate of oblique transmission are considered. The model is in fact a diploid extension of the haploid sexual model with uniparental transmission that we have discussed above. The two phenotypes 1 and 2 are now called "user" and "nonuser" of milk (or dairying). There are three genotypes, AA, Aa, aa, with A dominant to a such that AA and Aa are absorbers, with the same vertical transmission rates. We include AA and Aa in the symbol $\bar{A}$ and aa is written $\bar{a}$. Transmission is uniparental (for example, from just the mother); an offspring's probability of becoming a user depends only on whether the mother is a milk user and if she is an absorber. The other parent plays no role in phenotypic transmission. The transmission rates are summarized in the table below.

Transmitting parent	Relative transmission rates for offspring			
	$u_{\bar{A}}$	$u_{\bar{a}}$	$n_{\bar{A}}$	$n_{\bar{a}}$
$u_{\bar{A}}$	δ_1	δ_1^*	$1 - \delta_1$	$1 - \delta_1^*$
$u_{\bar{a}}$	δ_2^*	δ_2	$1 - \delta_2^*$	$1 - \delta_2$
$n_{\bar{A}}$	0	0	1	1
$n_{\bar{a}}$	0	0	1	1

Superimposed on the entries in the table must be the relevant probabilities for genetic transmission.

Now, let u_1, u_2, u_3 be the frequencies of the user phenotype of genotypes AA, Aa, and aa, respectively, with n_1, n_2, n_3 the corresponding frequencies of nonusers. Selection occurs prior to mating, and the fitness differentials also reflect the genetic dominance. Thus users of genotype AA and Aa each have fitness $1 + s_1$ relative to $1 - s_2$ for users of genotype aa and 1 for all nonusers. After mating, vertical transmission from the transmitting parent occurs, followed by oblique transmission. As before, frequencies are censused after oblique transmission and before selection. The frequencies after vertical transmission are written with a tilde, while primes indicate the completed recursion for variables after oblique transmission. Then we have

$$T^2\tilde{u}_1 = \delta_1[u_1(1 + s_1) + u_2(1 + s_1)/2]\{u_1(1 + s_1) + n_1 \\ + [u_2(1 + s_1) + n_2]/2\}. \tag{8.40a}$$

$$T^2\tilde{u}_2 = \delta_1\{[u_1(1 + s_1) + u_2(1 + s_1)][u_2(1 + s_1) + n_2]/2 \\ + u_1(1 + s_1)[u_3(1 - s_2) + n_3] + u_2(1 + s_1)[u_1(1 + s_1) + n_1]/2 \\ + u_2(1 + s_1)[u_3(1 - s_2) + n_3]/2\} \\ + \delta_2^*\left\{u_3(1 - s_2)\left[u_1(1 + s_1) + n_1 + \frac{u_2(1 + s_1)}{2} + \frac{n_2}{2}\right]\right\}. \tag{8.40b}$$

$$T^2\tilde{u}_3 = \delta_1^*\{u_2(1 + s_1)[u_2(1 + s_1) + n_2)]/4 \\ + u_2(1 + s_1)[u_3(1 - s_2) + n_3]/2\} \\ + \delta_2\{u_3(1 - s_2)[u_2(1 + s_1) + n_2]/2 \\ + u_3(1 - s_2)[u_3(1 - s_2) + n_3]\}. \tag{8.40c}$$

$$T^2\tilde{n}_1 = (1 - \delta_1)[u_1(1 + s_1) + u_2(1 + s_1)/2]\{u_1(1 + s_1) + n_1 \\ + [u_2(1 + s_1) + n_2]/2\} + (n_1 + n_2/2)\{u_1(1 + s_1) + n_1 \\ + [u_2(1 + s_1) + n_2]/2\}. \tag{8.40d}$$

$$T^2\tilde{n}_2 = (1 - \delta_1)\{[u_1(1 + s_1) + u_2(1 + s_1)][u_2(1 + s_1) + n_2]/2 \\ + u_1(1 + s_1)[u_3(1 - s_2) + n_3] \\ + u_2(1 + s_1)[u_1(1 + s_1) + n_1]/2 \\ + u_2(1 + s_1)[u_3(1 - s_2) + n_3]/2\} \\ + (1 - \delta_2^*)\left\{u_3(1 - s_2)\left[u_1(1 + s_1) + n_1 + \frac{u_2(1 + s_1)}{2} + \frac{n_2}{2}\right]\right\} \\ + \left(n_1 + \frac{n_2}{2}\right)\left[u_3(1 - s_2) + n_3 + \frac{u_2(1 + s_1)}{2} + \frac{n_2}{2}\right] \\ + \left(n_3 + \frac{n_2}{2}\right)\left[u_1(1 + s_1) + n_1 + \frac{u_2(1 + s_1)}{2} + \frac{n_2}{2}\right]. \tag{8.40e}$$

$$T^2\tilde{n}_3 = (1 - \delta_1^*)\{u_2(1 + s_1)[u_2(1 + s_1) + n_2]/4 \\ + u_2(1 + s_1)[u_3(1 - s_2) + n_3]/2\} \\ + (1 - \delta_2)\{u_3(1 - s_2)[u_2(1 + s_1) + n_2]/2 \\ + u_3(1 - s_2)[u_3(1 - s_2) + n_3]\} \\ + \left(n_3 + \frac{n_2}{2}\right)\left[u_3(1 - s_2) + n_3 + \frac{u_2(1 + s_2)}{2} + \frac{n_2}{2}\right], \tag{8.40f}$$

where

$$T = (u_1 + u_2)(1 + s_1) + u_3(1 - s_2) + n_1 + n_2 + n_3. \qquad (8.40\text{g})$$

After vertical transmission, offspring contact adults of the previous generation and may then be "converted" from nonuser to user status at rates f_1, f_2, f_3 for genotypes AA, Aa, and aa. The frequencies are then

$$u_1' = \tilde{u}_1 + f_1 \tilde{n}_1 [(u_1 + u_2)(1 + s_1) + u_3(1 - s_2)]/T \qquad (8.41\text{a})$$

$$u_2' = \tilde{u}_2 + f_2 \tilde{n}_2 [(u_1 + u_2)(1 + s_1) + u_3(1 - s_2)]/T \qquad (8.41\text{b})$$

$$u_3' = \tilde{u}_3 + f_3 \tilde{n}_3 [(u_1 + u_2)(1 + s_1) + u_3(1 - s_2)]/T \qquad (8.41\text{c})$$

$$n_1' = \tilde{n}_1 \{ 1 - f_1 [(u_1 + u_2)(1 + s_1) + u_3(1 - s_2)]/T \} \qquad (8.41\text{d})$$

$$n_2' = \tilde{n}_2 \{ 1 - f_2 [(u_1 + u_2)(1 + s_1) + u_3(1 - s_2)]/T \} \qquad (8.41\text{e})$$

$$n_3' = \tilde{n}_3 \{ 1 - f_3 [(u_1 + u_2)(1 + s_1) + u_3(1 - s_2)]/T \}. \qquad (8.41\text{f})$$

Initial Increase of Milk Use

As in the haploid models with $\delta_1, \delta_2 < 1$, the state of fixation on the user phenotype cannot be stable. We first seek conditions that allow the users to invade the population when they arise at very low frequency near fixation of the nonusers. Suppose the nonusers are at frequencies $\hat{n}_1, \hat{n}_2, \hat{n}_3$, prior to the introduction of users AA, Aa, aa at frequencies $\varepsilon_1, \varepsilon_2, \varepsilon_3$. The local linear analysis in the neighborhood of $\hat{n}_1, \hat{n}_2, \hat{n}_3$ produces the matrix transformation $\boldsymbol{\varepsilon}' = \mathbf{A}\boldsymbol{\varepsilon}$ where $\boldsymbol{\varepsilon} = (\varepsilon_1, \varepsilon_2, \varepsilon_3)^T$ is the vector of small frequencies of users of genotype AA, Aa, aa, respectively, $\hat{p}_A = \hat{n}_1 + \hat{n}_2/2 = 1 - \hat{p}_a$ and

$$\mathbf{A} = \begin{bmatrix} (\delta_1 \hat{p}_A + f_1 \hat{n}_1)(1 + s_1) & \left(\dfrac{\delta_1 \hat{p}_A}{2} + f_1 \hat{n}_1 \right)(1 + s_1) & f_1 \hat{n}_1(1 - s_2) \\[2ex] (\delta_1 \hat{p}_a + f_2 \hat{n}_2)(1 + s_1) & \left(\dfrac{\delta_1}{2} + f_2 \hat{n}_2 \right)(1 + s_1) & (\delta_2^* \hat{p}_A + f_2 \hat{n}_2)(1 - s_2) \\[2ex] f_3 \hat{n}_3(1 + s_1) & \left(\dfrac{\delta_1^* \hat{p}_a}{2} + f_3 \hat{n}_3 \right)(1 + s_1) & (\delta_2 \hat{p}_a + f_3 \hat{n}_3)(1 - s_2). \end{bmatrix}$$

For comparison with the analyses of the haploid models, we first set $f_i = 0$ ($i = 1, 2, 3$). The matrix $\mathbf{A}$ remains three-dimensional even under this simplification. The characteristic polynomial reduces to $Q_D(\lambda)$ with

$$Q_D(\lambda) = -\lambda^3 + C_2 \lambda^2 - C_1 \lambda + C_0, \qquad (8.42)$$

where

$$C_2 = (1 + s_1)\delta_1\hat{p}_A + \frac{(1 + s_1)\delta_1}{2} + (1 - s_2)\delta_2\hat{p}_a,$$

$$C_1 = (1 + s_1)(1 - s_2)\delta_1\delta_2\hat{p}_a(1/2 + \hat{p}_A) + (1 + s_1)^2\delta_1^2\hat{p}_A^2/2$$
$$- \delta_1^*\delta_2^*(1 + s_1)(1 - s_2)\hat{p}_A\hat{p}_a/2,$$

$$C_0 = (1 + s_1)^2(1 - s_2)\delta_1\hat{p}_A^2\hat{p}_a(\delta_1\delta_2 - \delta_1^*\delta_2^*)/2.$$

As in the haploid models, we may examine the stability at the ends of the $(\hat{n}_1, \hat{n}_2, \hat{n}_3)$ edge, namely, where $\hat{n}_1 + \frac{\hat{n}_2}{2} = \hat{p}_A = 1$ and $\hat{n}_3 + \frac{\hat{n}_2}{2} = \hat{p}_a = 1$.

When $\hat{p}_A = 1$, the eigenvalues of $\mathbf{A}$ are 0, $\frac{(1 + s_1)\delta_1}{2}$, $(1 + s_1)\delta_1$. When $\hat{p}_a = 1$, the eigenvalues are 0, $\frac{(1 + s_1)\delta_1}{2}$, $(1 - s_2)\delta_2$. If $\frac{(1 + s_1)\delta_1}{2} > 1$, then for values of $\hat{p}_A$ near zero and one, the user phenotype will increase when rare. The same is true if $(1 + s_1)\delta_1 > 1$ and $(1 - s_2)\delta_2 > 1$; the latter of course requires $s_2 = -\tilde{s}_2 < 0$. In the haploid sexual model, we considered the example with $\delta_1^* = \delta_2^* = 0$, $(1 + s_1)(f_1 + \delta_1) = 1.1$, $(1 - s_2)(f_2 + \delta_2) = 1.08$. Let us use the same example; take $\delta_1^* = \delta_2^* = 0$, $f_1 = f_2 = f_3 = 0$, $(1 + s_1)\delta_1 = 1.1$, $(1 + \tilde{s}_2)\delta_2 = 1.08$. Then at $\hat{p}_A = 1$ and at $\hat{p}_a = 1$, users will increase when rare. But when $\hat{p}_A = 0.5$, all three roots of $Q_D(\lambda)$ are positive and less than unity, so that users will *not* increase when rare. Once again the success of the phenotypic innovation depends in an interesting way on the initial gene frequency.

Initial Increase of Rare Alleles

We now turn to the boundaries where the alleles are fixed. On the A-fixation boundary there is a single possible equilibrium given by the valid root of the quadratic equation (8.26). This is stable in the boundary if $(1 + s_1)(f_1 + \delta_1) > 1$. On the a-fixation boundary, a corresponding equilibrium exists and is stable in the boundary if $(1 + \tilde{s}_2)(f_3 + \delta_2) > 1$. For comparison with the analysis in the haploid sexual model set, $f_1 = f_2 = f_3 = 0$. Then the equilibrium on the A-fixation boundary is $\hat{u}_1 = [\delta_1(1 + s_1) - 1]/s_1$. In view of the dominance of A with respect to a we do not expect a geometric increase of a near $\hat{u}_1$. In fact, the characteristic quadratic for the local stability of $\hat{u}_1$ to invasion by a has unity as its largest root.

On the other gene fixation boundary near where a is fixed at the equilibrium $\hat{u}_3 = \delta_2(1 + \tilde{s}_2)$, the local linear analysis for the initial increase of

A produces a condition identical to (8.38). If (8.38) holds, then A increases in this diploid model as well. Clearly, if we had not assumed genetic dominance, the equivalence of AA and Aa for transmission and selection, a condition analogous to (8.37) would have sufficed for the increase of a when introduced near fixation of A at $\hat{u}_1$.

Numerical Analysis of the Diploid Model

It is obvious that the dynamics of the full recursion system (8.41) will be complicated. The range of complications was explored numerically in the following way. A rather high value of s_1 was chosen, $s_1 = 1.0$, and values of s_2 were chosen from $s_2 = 0.0$ to $s_2 = -2.0$ with steps of -0.2. With $s_2 > 0$, fixation of A is always stable, so the case $s_2 \leq 0$ was regarded as more interesting. Three initial vectors of $u_1, u_2, u_3, n_1, n_2, n_3$ were chosen as follows:

A: $u_1 = 0.008$, $u_2 = 0.007$, $u_3 = 0.005$, $n_1 = 0.95$, $n_2 = 0.02$, $n_3 = 0.01$

B: $u_1 = 0.007$, $u_2 = 0.015$, $u_3 = 0.008$, $n_1 = 0.24$, $n_2 = 0.50$, $n_3 = 0.23$

C: $u_1 = 0.008$, $u_2 = 0.005$, $u_3 = 0.007$, $n_1 = 0.01$, $n_2 = 0.02$, $n_3 = 0.95$.

Each starting condition is close to the edge of fixation in nonusers, so that the dynamics are those that follow the introduction of a low frequency of users with a high (A), intermediate (B), and low (C) frequency of the tolerance gene. For each value of s_2 and these initial vectors, 100 sets of transmission parameters $\delta_1, \delta_1^*, \delta_2, \delta_2^*, f_1, f_2, f_3$ were chosen at random, uniformly on $(0, 1)$, and (8.41) iterated in double precision either to equilibrium or for 1,000 generations, whichever came first. The following is a list of the outcomes of this survey, in terms of the sorts of equilibria reached, or what happened after 1,000 generations.

Outcome I: Complete genetic and phenotypic polymorphism, the same from A, B, C

Outcome II: Fixation on the tolerance gene A from A, B, C

Outcome IIa: After 1,000 generations, close to fixation in allele A from A, B, C

Outcome III: Fixation on allele a from A, B, C

Outcome IIIa: After 1,000 generations, close to fixation in allele a from A, B, C

Outcome IV: Fixation on A or a depending on the starting condition

Outcome IVa: Fixation on A from some starting condition and, from another starting condition, close to fixation on a after 1,000 generations
Outcome IVb: Fixation on a from some starting condition and, from another starting condition, close to fixation on A after 1,000 generations
Outcome V: Fixation on the nonuser phenotype from A, B, C
Outcome VI: Complete polymorphism from some starting condition, fixation on nonusers from others
Outcome VII: Fixation on A from some starting point condition, fixation on nonusers from others
Outcome VIIIa: Fixation on A, complete polymorphism from different starting conditions
Outcome VIIIb: Fixation on a, complete polymorphism from different starting conditions
Outcome IX: Two different stable polymorphisms

In Table 8.2 the results of the hundred tests for each s_2 value are presented.

The main features of Table 8.2 are the shift away from polymorphism and fixation on A to fixation on a as the fitness of aa increases. More surprising, however, are the entries in the right-most third of the table where the dependence on initial conditions is revealed. The possibility of two stable pheno-genotypic polymorphisms and the coexistence and stability of fixations and polymorphisms suggest that if there is interaction between genes and culture, the success of a new cultural variant may depend on the genetic makeup of the population in which it arises. And by

Table 8.2
Numerical survey of the diploid model.

s_2	I	II	IIa	III	IIIa	IV	IVa	IVb	V	VI	VII	VIIIa	VIIIb	IX
							Outcome							
0	32	21	11	0	0	0	0	0	3	15	18	0	0	0
−0.2	35	25	18	0	0	0	0	0	4	5	8	5	0	0
−0.4	36	19	15	1	0	4	0	0	4	9	4	7	0	1
−0.6	40	18	8	5	0	13	0	0	2	4	5	5	0	0
−0.8	36	7	7	14	0	23	0	0	3	2	1	7	0	0
−1.0	21	8	3	26	5	22	2	1	3	3	2	3	1	0
−1.2	21	3	5	40	1	21	0	0	2	2	2	1	2	0
−1.4	8	3	4	57	2	16	1	1	3	2	1	1	1	0
−1.6	13	3	0	52	0	23	1	1	0	4	1	0	2	0
−1.8	7	2	0	64	1	25	0	1	0	0	0	0	0	0
−2.0	7	0	2	68	1	17	1	2	1	1	0	0	0	0

NOTE: The outcomes are as listed in the text.

the same token, the success of a new mutant allele may depend on the frequency distribution of the cultural traits when it arises.

Numerical Results for Lactose Absorption

The results of the previous section involve a selection coefficient $s_1 = 1$, which is unreasonable for the lactose absorption trait. We have therefore examined three values of s_1 in the more realistic range of the advantage for milk-using lactose absorbers, $s_1 = 0.05$, $s_1 = 0.1$, $s_1 = 0.2$. In addition, the fitness of milk-using malabsorbers was constrained so that s_2 was -0.1, 0.0 or $+0.1$. The transmission parameters are just guesses, but we suspect that δ_1 should be the highest of the vertical transmission parameters. We chose $\delta_1 = 0.9$, $\delta_1^* = 0.6$. The smaller value for δ_1^* reflects the idea that a malabsorbing offspring is less likely to become a user. The rates of transmission from malabsorbing parents were varied; we chose three sets, $\delta_2 = \delta_2^* = 0.5$; $\delta_2 = 0.5$, $\delta_2^* = 0.8$; $\delta_2 = \delta_2^* = 0.8$. These are lower than the transmission rates for an absorbing mother with an absorbing child. Oblique transmission rates were set at $f_1 = f_2 = 0.8$, $f_3 = 0.2$, reflecting the greater likelihood of acceptance of milk use by absorbers on contact with users. The initial conditions were A, B, and C of the previous section. The frequencies of the lactose absorbing allele A and the milk-use phenotype were taken at 300 generations and 1,000 generations, if equilibrium had not been achieved earlier. In almost every case there was little difference between the frequencies at 300 and at 1,000 generations. The estimated time since the advent of dairying is 300 generations. Note that for all starting conditions milk use is initially very low in frequency. Table 8.3 presents the results.

The first fact to emerge from Table 8.3 is that a 5% selective advantage to lactose-absorbing milk users is usually not sufficient to ensure the increase of this allele, nor indeed will it ensure success of milk use. Under the most favorable transmission regime, δ_C, the frequency of users increases initially but by 50 generations it has started to decrease to the 35–36% reported in the table. The decrease continues to 1,000 generations. With increasing s_1 it becomes easier to realize high frequencies of allele A, although with $s_1 = 0.1$ the frequency of allele A does not reach a value characteristic of Northern European populations when its initial frequency is low (i.e., from starting condition C) until more than 600 generations. With an advantage of 20% to absorbers who use milk from an initial frequency of about $2\frac{1}{2}\%$, high enough frequencies of A and milk use are possible after 300 generations.

There is an important effect of the initial frequency of the absorbing allele. Thus, if milk use arose in a population where allele A was at roughly

Table 8.3
Dynamics of lactose absorption milk use[a]

| Parameters[b] | | | A | | B | | C | |
s_1	s_2		g	u	g	u	g	u
0.05	−0.1	δ_A	0.014	0	0.014	0	0.013	0
		δ_B	0.013	0	0.013	0	0.012	0
		δ_C	0.006	0.359	0.004	0.355	0.002	0.353
0.05	0.0	δ_A	0.017	0	0.017	0	0.015	0
		δ_B	0.016	0	0.016	0	0.014	0
		δ_C	0.424	0.812	0.035	0.206	0.009	0.064
0.05	0.1	δ_A	0.824	0.962	0.020	0	0.017	0
		δ_B	0.830	0.964	0.020	0	0.016	0
		δ_C	0.836	0.966	0.007	0	0.007	0
0.1	−0.1	δ_A	0.013	0	0.013	0	0.013	0
		δ_B	0.027	0.004	0.013	0	0.013	0
		δ_C	0.228	0.673	0.149	0.584	0.127	0.556
0.1	0.0	δ_A	0.805	0.958	0.017	0	0.014	0
		δ_B	0.811	0.961	0.016	0	0.014	0
		δ_C	0.813	0.963	0.805	0.961	0.412	0.805
0.1	0.1	δ_A	0.888	0.972	0.019	0	0.017	0
		δ_B	0.888	0.972	0.888	0.972	0.019	0
		δ_C	0.889	0.973	0.888	0.973	0.007	0
0.2	−0.1	δ_A	0.869	0.970	0.865	0.969	0.014	0
		δ_B	0.869	0.971	0.866	0.970	0.012	0
		δ_C	0.867	0.971	0.865	0.971	0.861	0.970
0.2	0.0	δ_A	0.911	0.975	0.910	0.974	0.014	0
		δ_B	0.911	0.975	0.910	0.975	0.014	0
		δ_C	0.911	0.975	0.910	0.975	0.907	0.975
0.2	0.1	δ_A	0.933	0.976	0.932	0.976	0.016	0
		δ_B	0.933	0.977	0.932	0.976	0.016	0
		δ_C	0.934	0.977	0.932	0.977	0.016	0

[a] Frequencies are reported after 300 generations. The starting conditions A, B, C are from the previous section. $g = u_1 + n_1 + (u_2 + n_2)/2$ is the gene frequency of A, the allele for the dominant genetic trait of absorption, and $u = u_1 + u_2 + u_3$ is the frequency of milk users.
[b] $\delta_1 = 0.9$, $\delta_1^* = 0.6$; $f_1 = f_2 = 0.8$; $f_3 = 0.2$. The parameter choices are δ_A: $\delta_2 = \delta_2^* = 0.5$; δ_B: $\delta_2 = 0.5$, $\delta_2^* = 0.8$; δ_C: $\delta_2 = \delta_2^* = 0.8$.

50% (starting condition B), then a 10% advantage to absorbers is sufficient to bring the frequency of A to more than 80% and that of milk use to over 90% in 300 generations.

The role of s_2 is of interest because there is some ambiguity over whether it should be positive or negative. It is conceivable that intolerant milk users gain some nutritional advantage either by processing milk or through other dietary concomitants of dairying; this would correspond to $s_2 < 0$. It might be expected that $s_2 > 0$ should entail a higher frequency of the absorbing allele. This is not always true from starting condition C, where the frequency of A is higher with $s_2 = 0$ than with $s_2 = -0.1$ or $s_2 = 0.1$. One explanation for this is that the higher frequency of users with $s_2 = 0$ creates a stronger frequency dependent selection on A. It certainly suggests some manifestation of gene-culture disequilibrium.

Discussion

Throughout our discussion, both vertical and oblique transmission are essentially phenotype-to-phenotype. Genetics has a dual role in our schema. First, it affects this phenotype-to-phenotype transmission. Second, and this constitutes a departure from most of our previous analyses, the selection on the phenotypes is influenced by the genotype in the sense that the pheno-genotype of the transmitting parent influences the vertical transmission rate. Of course this could be regarded as another level of phenotypic variation except that we are interested in the evolution of both the gene and the phenotype. It is natural, therefore, to maintain the distinction between genetic and phenotypic variation.

A key feature of our treatment is the emphasis on vertical transmission. For simplicity we have modeled this as uniparental. In a recent study of the AKA Pygmies, vertical transmission of foraging techniques was shown to be by far the most important mechanism of information transfer (Hewlett and Cavalli-Sforza 1986). The inclusion of vertical transmission has the effect of increasing the dimensionality of the joint dynamics of the gene and phenotype. Both the analytical results and the numerical findings of Table 8.3 demonstrate the sensitivity of the dynamics to the rate of vertical transmission. It is not surprising that the weaker the selection on the phenotype, the greater is the sensitivity to the transmission rule. Our emphasis on the role of vertical transmission should be recognized as distinct from Aoki's (1986, 1987) models in which, without vertical phenotypic transmission, the gene and user frequencies suffice to describe the evolution.

Even so there is a convergence of results between the two treatments in the following sense. For the increase of lactose absorption from a few percent to frequencies characteristic of northern European populations in a reasonable time, both treatments need to invoke selection coefficients higher than a few percent.

The analysis of each model and the numerical treatments illustrate the importance of what we call in population genetics "history"—that is, the initial conditions play an important role in the evolutionary outcome. The fate of a phenotypic innovation clearly depends on the allele frequency among the resident population. In addition, multiple polymorphic equilibria and simultaneous stability of polymorphism and gene or phenotype fixation may occur. These are possible because the realized selection on the genotypes is frequency-dependent and of a rather complex form.

It is worth comparing the above dynamics with those of a simple dominant gene with a fitness of $1 + s_1$ for the dominant phenotype and $1 - s_2$ for the homozygous recessive. Of course there is no qualitative dependence in the initial conditions, and there is rapid progress toward high gene frequencies of the dominant type except when $-s_2 > s_1$ (which is equivalent to the first case $s_1 = 0.05$, $s_2 = -0.1$ of Table 8.3). There is no acceleration of the selection process under the cultural conditions envisaged here, but the latter complicate the selection process to the extent that the outcome may differ from that expected under purely genetic transmission and, in some cases, be reversed.

Acknowledgment

The authors thank Dr. K. Aoki for valuable criticism of an early draft.

References

Ammerman, A. J., and Cavalli-Sforza, L. L. 1984. *The Neolithic Transition and the Genetics of Populations in Europe.* Princeton University Press, Princeton, N.J.

Aoki, K. 1986. A stochastic model of gene-culture coevolution suggested by the "culture historical hypothesis" for the evolution of adult lactose absorption in humans. *Proc. Natl. Acad. Sci. USA.* 83: 2929–2933.

Aoki, K. 1987. Gene-culture waves of advance. J. Math. Biol. 25: 453–464

Boyd, R., and Richerson, P. J. 1985. *Culture and the Evolution Process.* University of Chicago Press, Chicago.

Cavalli-Sforza, L. L., and Feldman, M. W. 1981. *Cultural Transmission and Evolution.* Princeton University Press, Princeton, N.J.

Cavalli-Sforza, L. L., and Feldman, M. W. 1983. Cultural versus genetic adaptation. *Proc. Natl. Acad. Sci. USA.* 79: 1331–1335.

Durham, W. H. 1988. *Coevolution*: Genes, Culture and Human Diversity. Stanford University Press, Stanford, Calif. To appear.

Falconer, D. S. 1965. The inheritance of liability to certain diseases, estimated from the incidence among relatives. *Ann. Hum. Genet.* 29: 51–76.

Feldman, M. W., and Cavalli-Sforza, L. L. 1976. Cultural and biological evolutionary processes; selection for a trait under complex transmission. *Theor. Pop. Biol.* 9: 238–259.

Feldman, M. W., and Cavalli-Sforza, L. L. 1984. Cultural and biological evolutionary processes: Gene-culture disequilibrium. *Proc. Natl. Acad. Sci. USA.* 81: 1604–1607.

Feldman, M. W., and Cavalli-Sforza, L. L. 1986. Towards a theory for the evolution of learning. In *Evolutionary Processes and Theory*, pp. 725–741. Ed. S. Karlin and E. Nevo. Academic Press, New York.

Feldman, M. W.; Cavalli-Sforza, L. L.; and Peck, J. R. 1985. Gene-culture coevolution: Models for the evolution of altruism with cultural transmission. *Proc. Natl. Acad. Sci. USA.* 82: 5814–5818.

Fisher, R. A. 1918. The correlation between relatives on the supposition of Mendelian inheritance. *Trans. Roy. Soc. Edinburgh* 52: 399–433.

Fisher, R. A. 1937. Wave of advance of advantageous genes. *Ann. Eugenics* 7: 355–369

Hamilton, W. D. 1964. The genetic evolution of social behavior, I. *J. Theor. Biol.* 7: 1–52.

Hewlett, B. S., and Cavalli-Sforza, L. L. 1986. Cultural transmission among AKA Pygmies. *Am. Anthropologist* 88: 922–934.

Johnson, J. D., et al. 1977. Lactose malabsorption among the Pima Indians of Arizona. *Gastroenterology* 73: 1299–1304.

McCracken, R. D. 1971. Lactose deficiency: An example of dietary evolution. *Current Anthropology* 12: 479–517.

Simoons, F. J. 1969. Primary adult lactose intolerance and the milking habit: A problem in biological and cultural interrelations. *Am. J. Digestive Diseases* 14: 819–836.

Simoons, F. J. 1978. The geographic hypotheses and lactose malabsorption. *Am. J. Digestive Diseases* 23: 963–980.

CHAPTER NINE

Two-Locus Models of Kin Selection among Haplodiploids: Effects of Recombination and Epistasis on Relatedness

Marcy K. Uyenoyama

Introduction

Hamilton's (1964a,b) theory of kin selection proposes that natural selection favors genes that promote altruism among genetically related individuals, provided that relatedness between donor and recipient exceeds the ratio of the cost to the benefit associated with the behavior. Studies of dynamic recursion systems that explicitly model the processes of genetic transmission, zygote formation, and interactions among relatives have restricted attention for the most part to single loci that modify the rate of performance of altruism (see reviews in Charlesworth 1980; Uyenoyama and Feldman 1980; Michod 1982). The multilocus models of kin selection that have been analyzed to date fall into two major categories distinguished by both technical approach and intent. Yokoyama and Felsenstein (1978) cast the propensity for altruistic behavior as a quantitative character. They demonstrated that the condition for the increase of the mean of this character *within* the span of a generation corresponds to Hamilton's one-locus criterion with the phenotypic correlation between interactants assuming the role of relatedness. Their criterion for increase in the mean *between* generations is obtained by modifying the phenotypic correlation by the heritability. This study showed that the appealing heuristic that summarizes one-locus kin selection theory can be generalized in a simple and elegant way to accommodate a large number of independent (uncorrelated) genetic loci. Engels (1983) regarded the cost-benefit ratio itself as an evolving quantitative character. The second major category includes two-locus models that explicitly describe the processes of transmission and selection (Aoki and Moody 1981; Feldman and Eshel 1982; Mueller and

Feldman 1985). Models of this kind have addressed the development of genetic associations in response to coevolutionary interactions between the first locus, which controls the propensity to perform altruism, and the second locus, which modifies brood sex ratio (Aoki and Moody 1981) or parental interference (Feldman and Eshel 1982). Mueller and Feldman (1985) studied the modification of the propensity for altruistic behavior by both loci.

My approach to multilocus kin selection theory departs from previous work, which has sought to demarcate the purview of Hamilton's rule. I define relatedness such that the rule holds, and explore the implications of this definition. By construction, this study does not constitute a test of Hamilton's theory; rather, it permits the investigation of the meaning of relatedness in multilocus systems and, in particular, the effects of recombination and epistasis on the adaptive value of altruism.

I consider evolutionary changes at two sex-linked loci influencing the performance of altruism among brothers and among sisters, and report conditions governing the introduction of alleles at one or two loci in populations at monomorphic or polymorphic equilibria. First, I partition the variation in altruistic propensity in populations in linkage disequilibrium into components that are due to the additive effect of alleles, dominance, and epistasis. Because linkage disequilibrium implies correlation among the causes of variation, the order in which the components of variation are removed affects their relative magnitudes (see Kempthorne 1955). Even in populations in linkage equilibrium, a complete partitioning requires eight orthogonal scales, the definition of which (apart from orthogonality) is arbitrary from a mathematical point of view (Cockerham 1954; see Ewens 1979, Chap. 2). However, the choice of scales in any proposed multilocus generalization of the cost-benefit criterion is not arbitrary: the central question is whether any scale has the desired properties. A two-locus extension of the covariance method (Li 1967, 1976, Chap. 21; Price 1970, 1972) suggests that the appropriate scale attributes the variation in altruistic propensity primarily to the invading alleles.

Second, I study a two-locus expression for relatedness in populations in linkage disequilibrium. One-locus relatedness measures incorporate genotypic variances and covariances between donors and recipients of altruism (Orlove and Wood 1978; Michod and Hamilton 1980; Seger 1981; Uyenoyama and Feldman 1981). In multilocus systems, linkage disequilibrium appears in the covariances among relatives (Cockerham 1956; Gallais 1974; Weir and Cockerham 1977). Mueller and Feldman (1985) produced a partitioning that assigns additive effects to chromosomes, but found that Hamilton's rule holds only in the absence of epistasis or recombination. By construction, the allelic (rather than chromosomal) partitioning described here and its associated measure of relatedness preserve

Hamilton's rule without restriction on the rate of recombination, epistasis, or the intensity of selection.

Third, in order to place one- and two-locus models within a common conceptual framework, I establish a correspondence between the local stability conditions and a local analogue of Hamilton's rule. Near equilibrium states characterized by zero additive genetic variance in altruistic propensity, relatedness between sisters assumes its one-locus value of 3/4; the dependence of the local stability conditions on epistasis and the recombination rate derives from their influence on the average effect of substitution alone. Near equilibrium states characterized by positive additive genetic variance, relatedness between sisters departs from its one-locus value under all intensities of selection. In their study of parental manipulation, Feldman and Eshel (1982) observed that, in the absence of recombination, the conditions for initial increase near the two kinds of equilibrium states differ. In the present study, elements of the full two-locus structure influence both the adaptiveness of altruistic behavior and the average effect of substitution of new alleles near equilibrium states characterized by positive additive genetic variance for altruistic propensity.

Hamilton's Rule

THE COVARIANCE METHOD

Characters that are positively correlated with fitness increase under natural selection (Li 1967, 1976; Price 1970, 1972). This approach has provided an appealing heuristic framework for the study of the evolution of altruistic behavior (Seger 1981; Uyenoyama, Feldman, and Mueller 1981; Crow and Aoki 1982; Uyenoyama 1984; Wade 1985). Define fitness for a given genotype as the proportional change in genotypic frequency from birth to reproductive age. Under the assumption of additivity of effects associated with performance and receipt of altruism, fitness can be represented as a variable of the form

$$F - E(F) = [1 + \beta B - \gamma C] - [1 + \beta E(B) - \gamma E(C)]$$
$$= \beta[B - E(B)] - \gamma[C - E(C)], \tag{9.1}$$

where B represents the level of altruism received by the individual in question; C altruism expressed by the individual itself; β and γ benefit and cost parameters which convert the character values into units of fitness; and E the expectation operator. For genotype i, define fitness as the excess (see Fisher 1941), given by

$$F_i - E(F) = T(u_i' - u_i^*)/u_i^*, \tag{9.2}$$

where u_i^* represents the frequency of the ith genotype at birth, u_i' the frequency after selection, and T the normalizer that ensures that the u_i' sum to unity. For example, under control of altruistic propensity by two sex-linked biallelic loci, nine genotypes occur among females and four (AB, Ab, aB, and ab) among males. Unlike more conventional measures of fitness, (2) attributes offspring to the fitness of their own genotypes, rather than to their parents' genotypes (see Denniston 1978).

To the ith genotype associate the character value Q_i. The covariance between fitness and the scale Q is

$$\text{Cov}(FQ) = \sum u_i^* Q_i T(u_i' - u_i^*)/u_i^* \qquad (9.3a)$$

$$= T[\text{E}'(Q) - \text{E}^*(Q)], \qquad (9.3b)$$

where $\text{E}'(Q)$ denotes the average value of Q after selection and $\text{E}^*(Q)$ the average value at birth (Robertson 1966; Falconer 1966; Li 1967, 1976; Price 1970, 1972; Crow and Nagylaki 1976). Equation (9.3) indicates that the mean value of a character that is positively correlated with fitness increases within generations. The fundamental theorem of natural selection examines the change in mean fitness, with Q defined as fitness itself (see Fisher 1958, Chap. 2). In the presence of dominance and epistasis, the change in mean fitness resulting from processes inherent in genetic transmission (including recombination) depends on the effects of those processes on associations within and between loci (Kimura 1958, 1965; Nagylaki 1976; Crow and Nagylaki 1976).

With respect to kin selection, the genotypic value of an individual reflects its propensity for altruistic behavior relative to the population mean. For example, the ith genotype, which performs altruism at the level g_i, has genotypic value $[g_i - \text{E}^*(g)]$, in which $\text{E}^*(g)$ represents the mean within the population at birth. Using (9.1), in which the variable B corresponds to the genotypic values of associates (siblings, in the cases to be considered) and C the individual's own genotypic value, (9.3a) can be expressed as

$$\text{Cov}[(\beta D_G - \gamma R_G)R_Q] = \beta \, \text{Cov}(D_G R_Q) - \gamma \, \text{Cov}(R_G R_Q), \qquad (9.4)$$

where D_G denotes the genotypic value of donors of altruism, R_G the individual's (recipient's) own genotypic value, and R_Q the individual's own value on the scale Q. If Q can be defined to be uncorrelated with all other components of variance within individuals, then $\text{Cov}(R_G R_Q)$ reduces to $\text{Var}(Q)$, the variance of Q. Combining (9.3) and (9.4) produces

$$\text{Cov}(FQ) = T[\text{E}'(Q) - \text{E}^*(Q)] = \text{Var}(Q)[\beta b_{D \to R} - \gamma], \qquad (9.5)$$

where the relatedness between donor and recipient with respect to the scale Q is given by

$$b_{D \to R} = \text{Cov}(D_G R_Q)/\text{Var}(Q) \qquad (9.6)$$

(compare Yokoyama and Felsenstein 1978; Orlove and Wood 1978; Michod and Hamilton 1980). Equation (9.5) expresses Hamilton's rule: the character Q is positively associated with fitness if the cost incurred by altruistic actions is less than the benefit derived by recipients, discounted by the relatedness between donor and recipient.

Because (9.3) and (9.5) describe changes within a generation, they hold independently of the definition of the scale Q and the genetic basis of the character under consideration (Price 1970, 1972). The extension of this description over a full generation cycle requires investigation of the relationship between $E^*(Q)$ and $E(Q)$, the mean value among parents in the preceding generation. Define the scale Q such that

$$E^*(Q) = E(Q) \qquad (9.7)$$

holds without restriction on the genetic structure of the population. Provided such a scale exists, (9.3), (9.5), and (9.7) indicate that Hamilton's rule determines the change in mean value of Q over a full generation.

ONE-LOCUS KIN SELECTION

Under determination of altruistic propensity by a single multiallelic locus, the appropriate scale is simply the additive genotypic value (Uyenoyama, Feldman, and Mueller 1981). To genotype A_iA_j, associate additive genotypic value ($\alpha_i + \alpha_j$), where the average effects (α_i) are obtained by minimization of the dominance variance (see Fisher 1941). For sex-linked loci, assign males of genotype A_i additive genotypic value α_i. The average additive genotypic value among offspring at birth is a function of the α_i and the gene frequencies alone. At an autosomal locus controlling altruism among siblings irrespective of sex, the gene frequencies among parents and their offspring are identical, satisfying (9.7). Under determination by a sex-linked locus, the average additive genotypic value at birth is

$$E^*(Q) = \sum \alpha_i(2f_i^* + m_i^*), \qquad (9.8)$$

where f_i^* denotes the frequency of A_i among females at birth and the m_i^* the frequency among males at birth (see Uyenoyama, Feldman, and Mueller 1981). In the absence of selection, male and female gene frequencies initiated at different values oscillate toward equality; however, the invariance of the weighted average of gene frequencies in (9.8) (see, for example, Cavalli-Sforza and Bodmer 1971, Chap. 2), ensures that (9.7) holds.

Under random mating, the dominance component of the donor is uncorrelated with the additive component of the recipient, and (6) reduces to the linear regression coefficient of the additive genotypic value of the recipient on that of the donor (Uyenoyama and Feldman 1981; Uyenoyama, Feldman, and Mueller 1981). However, under inbreeding (see Uyenoyama 1984) those two quantities are in general correlated between relatives even though by definition the additive and dominance components are uncorrelated within individuals (Kempthorne 1955; Harris 1964).

TWO-LOCUS KIN SELECTION

In their study of altruistic propensity determined by two autosomal loci, Mueller and Feldman (1985) assigned additive values to chromosomes (rather than alleles), and showed that Hamilton's rule fails except in the absence of recombination, epistasis, or genotypic linkage disequilibrium. This result derives from the violation of (9.7), which requires

$$E^*(Q) - E(Q) = \sum \alpha_i(x_i^* - x_i) = 0, \tag{9.9}$$

where x_i^* and x_i denote the frequencies of the ith chromosome (AB, Ab, aB, or ab) at birth and among adults of the preceding generation, respectively. Recombination in the presence of genotypic linkage disequilibrium among parents causes the chromosomal frequencies among offspring and their parents to differ.

Returning to the definition of the scale Q in terms of effects of alleles rather than chromosomes (see Fisher 1958; Kimura 1958) avoids violation of (9.7). (Both the genic and chromosomal partitionings studied by Kimura [1965], which he showed to be equivalent, assign effects to alleles.) For example, assign to diploid individuals additive genotypic values of the form $(\alpha_i + \alpha_j + \beta_i + \beta_j)$, where the first two terms represent the effects of the alleles at the A locus and the remaining terms the effects of the alleles at the B locus; to hemizygous individuals, assign additive genotypic values of the form $(\alpha_i + \beta_j)$. Under sex-linkage, (9.7) requires that

$$\sum \alpha_i(2f_{Ai}^* + m_{Ai}^*) + \sum \beta_i(2f_{Bi}^* + m_{Bi}^*) = \sum \alpha_i(2f_{Ai} + m_{Ai}) + \sum \beta_i(2f_{Bi} + m_{Bi}), \tag{9.10}$$

where the subscripts index alleles at locus A and locus B. An analogous equation holds for autosomal loci. Because (9.10) involves only gene frequencies, it holds without restriction on recombination, epistasis, or linkage disequilibrium. Alternatively, in cases in which the evolutionary changes of interest are occurring at only one of the loci, it is convenient to restrict the scale Q to the additive genotypic value associated with the particular locus in question. For example, the change in the average additive genotypic value at the B locus can be studied by assigning females additive genotypic values of the form $(\beta_i + \beta_j)$ and males β_j. This approach will be used in subsequent sections to interpret the local stability conditions governing the initial increase of an allele at a locus linked to a distinct locus held in polymorphic equilibrium.

The Partitioning of Variance and the Definition of Relatedness under Linkage Disequilibrium

Kempthorne (1955) has summarized the conceptual difficulties that confront attempts to partition the total phenotypic variance into components attributed to correlated factors: for example, the partitioning itself is not unique and the magnitude of the components of variance attributed to the factors depends on the order in which they are isolated. Even if the effects associated with the additive, dominance, and epistatic components are defined with respect to populations in linkage equilibrium, the expressions for the covariances among relatives in the presence of linkage disequilibrium are extraordinarily complex (Gallais 1974; Weir and Cockerham 1977). Equation (9.10), together with (9.3) and (9.5), indicates that Hamilton's rule holds for two loci (and, in fact, for any number of loci), provided that the appropriate additive genotypic scale can always be defined. This section describes the partitioning and its associated covariances that permit the extension of Hamilton's rule to multiple epistatic loci under linkage disequilibrium.

CHROMOSOMAL AND GENOTYPIC FREQUENCIES

Associate indices 1 through 4 to the four chromosomal types (AB, Ab, aB, or ab) generated by two biallelic loci. Among males censused after selection, these chromosomes occur in frequencies m_i, for i ranging from 1 through 4 ($\sum m_i = 1$). In the model involving brother-to-brother altruism, the altruistic propensity of genotype i is denoted g_i ($0 \le g_i \le 1$). The genotype formed by chromosomes i and j occurs among adult females with frequency f_{ij} ($\sum f_{ij} = 1$). Under sister-to-sister altruism, individuals having this genotype perform altruism at the rate h_{ij} ($0 \le h_{ij} \le 1$; $h_{14} = h_{23}$). Among adult females, the frequency of chromosome i is

$$f_i = f_{ij} + \sum_{j \ne i} f_{ij}/2. \tag{9.11}$$

In the presence of recombination, the frequency of chromosome i among chromosomes transmitted by females is

$$f_{iT} = f_i + (-1)^{\delta(i)}r(f_{14} - f_{23})/2, \tag{9.12}$$

where $\delta(i)$ equals unity for i equal to 1 and 4, and zero for i equal to 2 and 3. Under genotypic linkage disequilibrium, the frequencies of the two doubly heterozygous arrangements (f_{14} and f_{23}) are unequal. If for all i

$$m_i = f_{iT} \tag{9.13}$$

(the frequency distributions of chromosomes transmitted by males and by females are equal), then among offspring censused at birth (before selection)

$$m_i^* = m_i$$

$$f_{ii}^* = m_i^2 \qquad\qquad\qquad (9.14)$$

$$f_{ij}^* = 2m_i m_j \qquad \text{for} \quad i \neq j.$$

DEFINITION OF COMPONENTS

Under the derivation described in the preceding section, Hamilton's rule requires a scale Q that has two properties: its mean should depend on only allele frequencies and it should be uncorrelated with the remaining components of variance. For each model considered, this additive scale is obtained from the parameters representing altruistic propensity and the genotypic frequencies censused among offspring before the performance of altruism. I adopt the familiar approach of least-squares minimization under the constraint that the average additive effect within each locus equals zero (see Kimura 1958, 1965; Ewens 1979, Section 2.9). A separate minimization isolates the component that is due to dominance interactions from the variance remaining after the removal of the additive component. Finally, the residual variance, attributed to epistasis, can be further decomposed into additive × additive, additive × dominance, and dominance × dominance components (see Cockerham 1954). For example, under altruism among brothers, the additive effects of the alleles are isolated first from the deviations of the altruistic propensities from the mean $[g_i - E^*(g)]$, in which the mean is computed using the genotypic frequencies among altruists at birth (m_i^*).

In subsequent sections, I describe the conditions for the initial increase of alleles introduced in low frequencies at one or both of the loci determining altruistic propensity. A one-gene introduction corresponds to the appearance of an allele (b) in a population at an equilibrium characterized by a stable polymorphism at locus A and monomorphism at locus B. Two-gene introductions occur near monomorphic states for both loci. Under one-gene introductions, the additive effects of the alleles at the locus at which variation is introduced are obtained before the additive effects of the other locus. This procedure attributes the greatest proportion of the variation to the introduced allele, and ensures that the additive components associated with the two loci are uncorrelated with each other, as well as with the components that are due to dominance and epistasis. Under two-gene introductions, the additive effects associated with alleles at the two loci are isolated simultaneously.

Additive, Dominance, and Epistatic Effects in One-Gene Introductions

Under altruism among sisters, two classes of one-locus polymorphic equilibrium states can exist (Uyenoyama and Feldman 1981): the class characterized by zero additive genetic variance was described as "viability-analogous" and the class characterized by positive additive genetic variance as "structural" by Uyenoyama, Feldman, and Mueller (1981). No one-locus polymorphisms exist under altruism among brothers. Consideration of the conditions for initial increase of a new allele (b) is restricted to cases in which the original polymorphism is locally stable to perturbations in frequencies of alleles at the first locus (A). First, the additive effects due to variation at the B locus alone are isolated. Under (9.14), with m_2 and m_4 small and m_1 and m_3 large, the average effect of substitution at the B locus is

$$(\beta_1 - \beta_2) = (m_2 M_{Ab} + m_4 M_{ab})/(m_2 + m_4), \qquad (9.15)$$

where M_{Ab} and M_{ab} represent the marginal deviations in altruistic propensity caused by the new allele (b) in association with the resident alleles (A and a):

$$M_{Ab} = \mathrm{E}(h) - (p_A h_{12} + q_A h_{23}) \qquad (9.16a)$$

$$M_{ab} = \mathrm{E}(h) - (p_A h_{14} + q_A h_{34}), \qquad (9.16b)$$

with p_A the equilibrium frequency of the A allele, $q_A = 1 - p_A$, and $\mathrm{E}(h)$ the average altruistic propensity at the equilibrium. Partitioning the residual variance into components due to additive effects of alleles at the A locus and nonadditive effects at both loci produces

$$(\alpha_1 - \alpha_2) = p_A(h_{11} - h_{13}) - q_A(h_{33} - h_{13}). \qquad (9.17)$$

Expression (9.17) is identical to the additive effect of substitution obtained in the absence of variation at the B locus; it is zero at polymorphisms characterized by zero additive genetic variance and nonzero at polymorphisms characterized by positive additive genetic variance. Appendix 9.1 describes the derivation of the expressions representing dominance and epistasis.

Additive, Dominance, and Epistatic Effects in Two-Gene Introductions

Under brother altruism, the additive effects of the introduced alleles at the A locus (9.18a) and at the B locus (9.18b) are

$$(\alpha_1 - \alpha_2) = (g_1 - g_3) - em_2^* \qquad (9.18a)$$

$$(\beta_1 - \beta_2) = (g_1 - g_2) - em_3^*, \qquad (9.18b)$$

with

$$e = (g_1 - g_2 - g_3 + g_4)m_4^*/[m_3^*m_4^* + m_2^*(m_3^* + m_4^*)]. \qquad (9.18c)$$

Near the resident monomorphism, zero linkage disequilibrium corresponds to the absence of chromosomes carrying both rare alleles ($m_4^* = 0$); the additive effects of substitution reduce in such populations to frequency-independent differences between the altruistic propensities of the resident AB and the rare genotypes carrying exactly one new allele. If the loci are correlated, then interactions among loci contribute to the additive effects through the factor involving the g_i in (9.18c) (compare Fisher's [1918] expression for epistasis). Because males are hemizygous, dominance does not enter into consideration, leaving the residual variance to additive × additive epistasis (see Appendix 9.1).

In the model of sister altruism, the average effects of substitution are obtained by minimizing the nonadditive variance in the same fashion. Near the AB corner of the tetrahedron representing the possible chromosomal frequencies (see Bodmer and Felsenstein 1967), h_{11}, h_{12}, h_{13}, and h_{14} assume the roles of g_1, g_2, g_3, and g_4, respectively. Under (9.14), the expressions for the additive effects are identical to those shown in (9.18), with the g_i replaced by the marginal averages for the chromosomal types (see Kimura 1965; Ewens 1979, Sections 2.9 and 7.4). As in the preceding case, epistasis in the presence of linkage disequilibrium influences the magnitudes of the additive effects attributed to the alleles.

Partitioning of the dominance effects removes no further variance, leaving all of the residual variance to epistasis. Under (9.14), the epistatic parameters associated with genotypes AB/AB, AB/Ab, AB/aB, and AB/ab are identical to those given for genotypes AB, Ab, aB, and ab in the case of brother altruism under the substitutions already noted. This epistatic variance can be entirely attributed to additive × additive epistasis.

Relatedness in One- and Two-Gene Introductions

Relatedness [see (9.6)] depends on the covariance between siblings, and hence the genotypic distribution among parents. As a result of kin selection, the parental genotypic frequencies may exhibit associations between alleles at one or both loci (inbreeding or linkage disequilibria), which can generate correlations between the additive and nonadditive components between siblings (Kempthorne 1955; Harris 1964).

One-Gene Introductions. Under (9.14), relatedness between sisters ($b_{s \to s}$) differs from its one-locus value of 3/4 by

$$b_{s \to s} - 3/4 = (\alpha_1 - \alpha_2)(\beta_1 - \beta_2)D^*/[4 \text{ Var}(A)], \qquad (9.19a)$$

in which

$$D^* = 2(p_A m_4 - q_A m_2) - (1 - 2r)(f_{14} - f_{23}). \qquad (9.19b)$$

In (9.19a), $\text{Var}(A)$ represents the additive genetic variance among sisters at locus B. In general, $\text{Var}(A)$ equals

$$2p_B^* q_B^* (\beta_1 - \beta_2)^2, \qquad (9.20a)$$

with p_B^* the frequency of the B allele among females at birth, and $q_B^* = 1 - p_B^*$. Near the edge representing polymorphism at the A locus and monomorphism at the B locus, (9.20a) reduces under (9.14) to

$$2(m_2 + m_4)(\beta_1 - \beta_2)^2. \qquad (9.20b)$$

Two-Gene Introductions. Under altruism among brothers, the measure of relatedness between brothers ($b_{B \to B}$) near the AB corner can depart from its one-locus value of $1/2$:

$$b_{B \to B} - 1/2$$
$$= r f_{14}[(\alpha_1 - \alpha_2)(g_1 - g_2) + (\beta_1 - \beta_2)(g_1 - g_3)]/[4\,\text{Var}(A)], \quad (9.21)$$

in which $\text{Var}(A)$ represents the additive genetic variance among brothers. In general, this variance is given by

$$p_A^* q_A^* (\alpha_1 - \alpha_2)^2 + p_B^* q_B^* (\beta_1 - \beta_2)^2 + 2(\alpha_1 - \alpha_2)(\beta_1 - \beta_2)D^*, \quad (9.22a)$$

in which p_A^* and p_B^* represent the frequencies of the A and B alleles among males at birth; $q_A^* = 1 - p_A^*$; $q_B^* = 1 - p_B^*$; and D^* is the gametic linkage disequilibrium ($m_1^* m_4^* - m_2^* m_3^*$). Near the AB corner, $\text{Var}(A)$ reduces to

$$(\alpha_1 - \alpha_2)^2 (m_3^* + m_4^*)$$
$$+ (\beta_1 - \beta_2)^2 (m_2^* + m_4^*) + (\alpha_1 - \alpha_2)(\beta_1 - \beta_2)(1 - r)f_{14}. \quad (9.22b)$$

Expressions (9.21) and (9.22) do not require (9.14).

The deviation between sister-to-sister relatedness and its one-locus value of $3/4$ is identical to the right side of (9.21) under (9.14), with h_{11}, h_{12}, and h_{13} replacing g_1, g_2, and g_3, and with $\text{Var}(A)$ representing the additive genetic variance among females at birth. For general frequency distributions, this variance equals

$$2p_A^* q_A^* (\alpha_1 - \alpha_2)^2 + 2p_B^* q_B^* (\beta_1 - \beta_2)^2$$
$$+ 2(\alpha_1 - \alpha_2)(\beta_1 - \beta_2)(4D^* - \Delta^*), \quad (9.23)$$

in which p_A^*, q_A^*, p_B^*, and q_B^* represent frequencies among females at birth; D^* the gametic linkage disequilibrium ($f_1^* f_4^* - f_2^* f_3^*$); and Δ^* the genotypic linkage disequilibrium $(f^*_{14} - f^*_{23})$. Near the AB corner, (9.23) reduces to twice (9.22b) under (9.14).

Summary of Effects of Disequilibria and Nonadditive Components on Relatedness. Expression (9.19) indicates that associations between additive components at different loci can cause relatedness between sisters to depart from the one-locus value of 3/4. Near polymorphisms characterized by zero additive genetic variance at the A locus (viability-analogous), the effect of substitution at the A locus is zero (see [9.17]), implying that $b_{s \to s}$ equals 3/4 irrespective of the level of epistasis, dominance, or genotypic associations in the parental or offspring generations. Near all other equilibria under consideration, relatedness may fall above or below the corresponding one-locus values depending on the nature of the associations between loci.

As recombination diminishes, two-locus systems approach one-locus systems in structure. Under tight linkage (r near 0), relatedness converges to its one-locus value in all two-gene introductions [see (9.21)]. However, in one-gene introductions [see (9.19)], $b_{s \to s}$ may depart from 3/4 even if linkage is complete; this feature derives from the isolation of the components that are due to variation at the B locus first (chromosomes Ab and ab), followed by the other components.

The Covariance Criterion and the Conditions for Local Stability

Equations (9.3) and (9.5) can be expressed in matrix form by defining a column vector $\mathbf{X}$ with elements representing the frequencies of all genotypes, censused at adulthood and listed in the following order: $f_{1j}, j = 1$ through 4; $f_{2j}, j = 1$ through 4; etc.; $m_j, j = 1$ through 4. The vector is constrained to lie in the valid region, which is characterized by nonnegative elements which sum to unity within sexes. Let $\mathbf{M}$ represent the matrix that describes the transformation of $\mathbf{X}$ over a full generation, with T the normalizing scalar that ensures that $\mathbf{X}'$, the vector in the next generation, lies in the valid region. The change in $\mathbf{X}$ over one generation is given by

$$T \Delta \mathbf{X} = (\mathbf{M} - T\mathbf{I})\mathbf{X}, \tag{9.24}$$

in which $\mathbf{I}$ denotes the identity matrix. The matrix $(\mathbf{M} - T\mathbf{I})$ contains elements similar to (9.1), in which B represents the average altruistic propensity among the siblings of a given individual and C the individual's own propensity for altruism. Let $\mathbf{A}$ represent a row vector with elements corresponding to the additive genotypic values of the elements of $\mathbf{X}$. In one-gene introductions, the additive genotypic value associated with the nearly monomorphic locus alone contributes to $\mathbf{A}$; in two-gene introductions, $\mathbf{A}$ incorporates the full two-locus additive genotypic value. The analogue of (9.3) and (9.5) is

$$\mathbf{A}T\,\Delta\mathbf{X} = \mathbf{A}(\mathbf{M} - T\mathbf{I})\mathbf{X} = \beta\,\mathrm{Cov}(D_G R_A) - \gamma\,\mathrm{Var}(A). \qquad (9.25)$$

Consider $\mathbf{M_S}$, the reduced version of $\mathbf{M}$ corresponding to the linearized transformation that holds approximately near one- and two-gene introductions. Genotypes occurring in frequencies of the second order in perturbations from the equilibrium point need not be considered in the local analysis. The constraint that the genotypic frequencies sum to unity within sexes requires that one additional variable in each sex be excluded from the analysis. The choice of these variables is arbitrary; for convenience, I exclude one of the common genotypes in each sex. A sufficient condition for local stability (all eigenvalues of $\mathbf{M_S}$ less than unity in absolute value) is that the determinants of all principal minors of the matrix $(T\mathbf{I} - \mathbf{M_S})$, where T is the mean fitness at the equilibrium, be positive (see Gantmacher 1959, p. 71). These criteria apply to all nonnegative $\mathbf{M_S}$, including reducible matrices.

Consider the case in which $\mathbf{M_S}$ is an $n \times n$ irreducible matrix. (Appendix 9.2 includes the argument for nonnegative reducible matrices.) Let $\mathbf{V}(\lambda)$ denote the eigenvector corresponding to the maximal eigenvalue (λ), which is positive and unique (Perron-Frobenius theorem). The $n - 1$ elements of $\mathbf{V}$ in terms of the last element, V_n, are obtained by solving

$$(\lambda T\mathbf{I} - \mathbf{M_S})^{(n-1)}\mathbf{V}^{(n-1)}(\lambda) = -V_n\mathbf{L}^{(n-1)}, \qquad (9.26)$$

in which the first factor denotes the matrix $(\lambda T\mathbf{I} - \mathbf{M_S})$ with its nth row and column deleted, the second the eigenvector with its nth element deleted, and $\mathbf{L}^{(n-1)}$ the nth column of the matrix $(\lambda T\mathbf{I} - \mathbf{M_S})$ with its nth element deleted. The uniqueness of the maximal eigenvalue guarantees the positivity of the determinant of the matrix $(\lambda T\mathbf{I} - \mathbf{M_S})^{(n-1)}$. All elements of the eigenvector can be expressed in terms of the eigenvalue and the last element (V_n), which can be chosen to be positive. Using these expressions, the characteristic equation, $C(\lambda)$, may be obtained by applying the last row of the matrix $(\lambda T\mathbf{I} - \mathbf{M_S})$ to the eigenvector:

$$(\lambda T\mathbf{I} - \mathbf{M_S})_n\mathbf{V}(\lambda) = V_n C(\lambda)/\mathrm{Det}[(\lambda T\mathbf{I} - \mathbf{M_S})^{(n-1)}], \qquad (9.27)$$

where the denominator of the right side is the determinant of the $(n - 1)$th principal minor.

Define $\mathbf{V}(1)$ as the vector that satisfies (9.26) with unity substituted for the dominant eigenvalue λ. This vector will differ from $\mathbf{V}(\lambda)$, unless the dominant eigenvalue equals unity. Further, if the dominant eigenvalue departs from unity, the expressions for the elements of $V(1)$ depend on the ordering of the rows of M_S. Under all orderings of rows, the positivity of the determinants of all $n - 1$ lower-order principal minors of $(TI - M_S)$ guarantees that the elements of $V(1)$ are nonnegative. The positivity of the determinants of all $n - 1$ lower-order principal minors implies that unity exceeds the dominant eigenvalue of the nonnegative $(n - 1) \times (n - 1)$ matrix $M_S^{(n-1)}$. For all values l greater than the dominant eigenvalue of $M_S^{(n-1)}$, the adjoint matrix $B(l)$ [the ijth element of which is the algebraic complement of the ijth element in the determinant of the matrix $(lTI - M_S)^{(n-1)}$] is nonnegative (see Gantmacher 1959, p. 66). Nonnegativity of the adjoint matrix implies that the elements of $V(1)$ are nonnegative. Because the element of $V(1)$ depends on the ordering of the rows, my examination of relatedness and other components of selection will address qualitative properties that hold for all nonnegative $V(1)$.

The condition that the determinant of $(TI - M_S)$ be positive can be related to the covariance criterion (9.25) by applying $(M_S - TI)$ to $V(1)$. The first $n - 1$ terms of the vector $(M_S - TI)V(1)$ vanish by definition [see (9.26)]. The nth term, from (9.27), is equal to

$$-V_n C(1)/\text{Det}[(TI - M_S)^{(n-1)}]. \qquad (9.28)$$

Multiplication on the left by A produces

$$A(M_S - TI)V(1) = -A_n V_n C(1)/\text{Det}[(TI - M_S)^{(n-1)}] \qquad (9.29a)$$

$$= [\beta \, \text{Cov}(D_G R_A) - \gamma \, \text{Var}(A)]|_{V(1)}, \qquad (9.29b)$$

where A_n is the additive genotypic value of the nth genotype, and (9.29b) is evaluated at $V(1)$.

The stability conditions phrased in terms of the determinants of the principal minors can be restated using (9.29) as

$$\text{Det}[(TI - M_S)^{(n-1)}] > 0 \qquad \text{for} \quad i = 1 \text{ through } n - 1 \qquad (9.30a)$$

$$-A_n[\beta \, \text{Cov}(D_G R_A) - \gamma \, \text{Var}(A)]|_{V(1)} > 0. \qquad (9.30b)$$

Inequality (9.30b) expresses the local analogue of Hamilton's rule: the equilibrium is stable if the invading genotypes are less altruistic (A_n negative) and altruism is favored (the expression in brackets is positive), or if both conditions are reversed. The relatedness measure relevant to (9.30b) is evaluated at $V(1)$, the elements of which are nonnegative if (9.30a) holds. For one- and two-gene introductions, the analogue of (9.14) holds when calculated at $V(1)$ under altruism by sisters; hence, the expressions for the variances and relatedness coefficients given in (9.19) through (9.23) apply, all evaluated at $V(1)$.

Altruism among Brothers

GENERAL INITIAL INCREASE CONDITIONS

A monomorphic population resists invasion by alleles a and b when introduced in low frequencies if all three of the following conditions are met:

$$(g_1 - g_2)(\beta/2 - \gamma) > 0$$

$$(9.31a)$$

$$(g_1 - g_3)(\beta/2 - \gamma) > 0 \qquad (9.31b)$$

$$(1 - r)(g_1 - g_4)(\beta/2 - \gamma) + 2rT$$
$$+ r(1 - r)(g_1 - g_2 - g_3 + g_4)\beta/2 > 0, \qquad (9.31c)$$

in which $T = [1 + g_1(\beta - \gamma)]$ is the mean fitness among residents. In the absence of recombination ($r = 0$), the inequalities in (9.31) reduce to the one-locus conditions, which ensure that each of three rare alleles (haplotypes Ab, aB, and ab) fail to increase when introduced by itself. Mueller and Feldman (1985) obtained similar results in their study of sibling altruism determined by two autosomal loci.

The reducibility of the stability matrix introduces the possibility that both genes may eventually exit the population even when (9.31c) is violated (see Karlin and Feldman 1968). For example, while the simultaneous introduction of the alleles may initially result in their increase, the population may converge to a face (corresponding to the loss of one of the new alleles) and then return to the corner along an edge.

Conditions (9.31a) and (9.31b) correspond to (9.30a), which involves the determinants of the principal minors, and (9.31c) corresponds to (9.30c), the local analogue of Hamilton's rule. One-locus kin selection theory adequately accounts for the first two conditions in (9.31). Under the assumption that both (9.31a) and (9.31b) are satisfied, local expressions for the additive genetic variance and the covariance between brothers are required for the interpretation of the remaining condition in terms of Hamilton's rule. Substitution of the elements of $\mathbf{V}(1)$ for the genotypic frequencies in (9.18) produces the average effects of substitution at the two loci. In (9.30b), the negative of the additive genotypic value of the individual carrying both rare alleles ($-A_n$) is

$$(\alpha_1 - \alpha_2) + (\beta_1 - \beta_2). \qquad (9.32a)$$

From (9.21), the second bracket in (9.30b) evaluated at $\mathbf{V}(1)$ is

$$(\beta/2 - \gamma)/\text{Var}(A)$$
$$+ \beta r f_{14}[(\alpha_1 - \alpha_2)(g_1 - g_2) + (\beta_1 - \beta_2)(g_1 - g_3)]/4. \qquad (9.32b)$$

Expression (9.32b) reflects the failure of the two-locus measure of relatedness between brothers to converge to the one-locus value of $1/2$ under recombination, even in the absence of epistasis. In agreement with the analysis of the preceding section, the product of (9.32a) and (9.32b) is proportional to (9.31c).

While (9.31c) holds for general parameter sets, its relationship to Hamilton's rule emerges most clearly in simple cases. In order to elucidate the effects of epistasis on the additive genotypic values and relatedness, I restrict consideration to the nonepistatic case and the "phenotypic" model studied by Mueller and Feldman (1985).

THE NONEPISTATIC CASE

In the absence of epistasis ($g_1 - g_2 - g_3 + g_4 = 0$), the average effects of substitution at each locus assume values expected on the basis of one-locus theory:

$$(\alpha_1 - \alpha_2) = (g_1 - g_3) \tag{9.33a}$$

$$(\beta_1 - \beta_2) = (g_1 - g_2). \tag{9.33b}$$

The local stability of the equilibrium to the initial increase of the new alleles when introduced separately [see (9.31a) and (9.31b)] requires that (9.33a) and (9.33b) have the same sign. Recombination permits the generation of individuals carrying both rare alleles, which have similar effects on altruistic propensity. As a result of the correlation in effect between loci, relatedness between brothers is greater in the presence of recombination than in its absence [see (9.21)]. While relatedness exceeds $1/2$, the cost-benefit ratio never lies between these two values, which implies that the departure of relatedness from its one-locus value does not qualitatively affect the adaptiveness of altruism. In agreement, the first two conditions in (9.31) imply the third, indicating that the stability of the equilibrium to the separate introduction of each new allele ensures stability to their simultaneous introduction.

EPISTASIS BETWEEN ALLELES HAVING IDENTICAL EFFECTS SEPARATELY

Consider the special case in which $g_2 = g_3$, which implies that the separate introduction of either new allele generates the same effect on altruistic propensity. This case corresponds to the "phenotypic" model of Mueller and Feldman (1985), in which the altruistic propensity of an individual depends on the number of heterozygous loci (compare Wright 1952; Lewontin and Kojima 1960). Under this restriction, the average effects of substitution of the rare alleles when introduced together are equal [see (9.18) with $m_2^* = m_3^*$] and proportional to

$$m_2^*(g_1 - g_2) + m_4^*(g_1 - g_4). \tag{9.34a}$$

Epistasis causes a departure from the one-locus values given in (9.33).

For the purposes of illustration, assume that both factors in (9.31a) are positive (similar results hold for the case in which both are negative). This assumption implies that at each locus considered separately, altruism is favored (the cost-benefit ratio is less than $1/2$) and that the rare allele reduces the propensity for altruism ($g_1 > g_2$). Upon the simultaneous introduction of the rare alleles, the average effect of substitution ($\alpha_1 - \alpha_2$) is proportional to

$$L - \beta r(g_1 - g_2)/2, \tag{9.34b}$$

in which L represents the left side of (9.31c). A sufficient condition for the local stability of the equilibrium is that the rare alleles reduce the propensity for altruism when introduced together [(9.34a) positive]. This condition holds, for example, if the rate shown by the resident exceeds the rates of individuals bearing rare alleles ($g_1 > g_2, g_4$).

If g_4 exceeds g_1 by a sufficient margin, then ($\alpha_1 - \alpha_2$) is negative, indicating that the rare alleles acting in concert can increase the propensity for altruism. Were relatedness between brothers equal to or greater than the one-locus value of $1/2$ (which exceeds the cost-benefit ratio), one would expect that the negativity of ($\alpha_1 - \alpha_2$) would ensure the invasion of the rare alleles. However, because relatedness exceeds $1/2$ only if ($\alpha_1 - \alpha_2$) is positive [see (9.19)], the two-locus analogue of Hamilton's rule (9.30b) suggests that the magnitude of relatedness relative to the cost-benefit ratio also affects the stability conditions. For cost-benefit ratios close to $1/2$ and recombination rates in an intermediate range, relatedness falls below the cost-benefit ratio (altruism is disfavored) and the equilibrium is locally stable. For other parameter values, relatedness lies below $1/2$, but above the cost-benefit ratio (altruism is favored), and the new alleles increase when rare. Mueller and Feldman (1985) obtained similar results, but did not seek to reconcile the results with their two-locus version of the cost-benefit criterion, which holds only in the absence of epistasis. Bodmer and Parsons (1962, Section VIIC) observed concordant behavior under frequency-independent viability selection: alleles that are favorable in combination but not singly increase when rare only if linkage is sufficiently tight.

Epistasis can cause departures from expectations based on one-locus theory through two effects. First, interactions between loci can reverse the expression of the alleles when present in the same genome relative to their effects when introduced separately. For example, the ordering $g_4 > g_1 > g_2 = g_3$ implies that the two alleles have an opposite effect when acting together. Second, epistasis can reverse the adaptive value of altruism through its influence on relatedness. In the presence of recombination, correlations between the additive and nonadditive components between siblings generate departures of the relatedness measures from their one-locus values.

Altruism among Sisters

In this section, I describe the analysis of one-gene introductions near equilibria characterized by zero and positive additive genetic variance. Because the initial increase conditions near corner equilibria (two-gene introductions) are qualitatively similar to those obtained for the model of brother altruism, I dispense with their presentation.

INITIAL INCREASE CONDITIONS NEAR EDGE EQUILIBRIA CHARACTERIZED BY ZERO ADDITIVE GENETIC VARIANCE

The equilibrium state prior to the introduction of the new allele is characterized by

$$p_A/q_A = (h_{33} - h_{13})/(h_{11} - h_{13}), \tag{9.35a}$$

with $E(h)$, the average altruistic propensity, equal to

$$E(h) = (h_{11}h_{33} - h_{13}^2)/(h_{11} + h_{33} - 2h_{13}) \tag{9.35b}$$

(Michod and Abugov 1980; Uyenoyama and Feldman 1981). This equilibrium is valid and locally stable with respect to perturbations within the edge corresponding to polymorphism at the A locus if h_{13} lies outside the range bounded by h_{11} and h_{33}, and if

$$(3\beta/4 - \gamma)(h_{13} - h_{11}) > 0. \tag{9.36}$$

Relatedness between sisters maintains its one-locus value of 3/4 regardless of the effects of alleles at the B locus, the recombination rate, or associations between loci [see (9.19)]. The two-locus analogue of the cost-benefit criterion (9.30b) suggests that the two-locus structure influences the conditions for initial increase only through the average effect of substitution $(-A_n)$. I describe the effect of recombination in the presence of epistasis and associations between loci in two special cases: (1) a family of altruistic propensity matrices in which the new allele at the B locus expresses identical effects when in association with either allele at the A locus; and (2) a perturbation of one of the matrices in the first class.

Equal Effects of Chromosomes Ab and ab. For matrices of the h_{ij} in which M_{Ab} and M_{ab} (9.16) are equal, the average effect of substitution associated with the B locus assumes a value that is independent of the frequencies of genotypes carrying the rare allele. A number of well-known matrices studied in models of two-locus viability selection belong to this family (see Bodmer and Felsenstein 1967, Section 7).

Under additivity between loci, one-locus genotypes *AA*, *Aa*, and *aa* are associated with positive parameters a_i (i = 1, 2, 3), and *BB*, *Bb*, and *bb* with b_i (i = 1, 2, 3), with the altruistic propensity of any two-locus genotype determined as the sum of the corresponding one-locus parameters. This set satisfies $M_{Ab} = M_{ab}$, with the average effect of substitution at the B locus (9.15) reducing to $(b_1 - b_2)$. This stability condition suggested by (9.30b) is

$$(3\beta/4 - \gamma)(b_1 - b_2) > 0. \tag{9.37a}$$

Multiplicative interactions between loci imply that the altruistic propensity of a two-locus genotype is determined by the product of the one-locus parameters. This construction satisfies $M_{Ab} = M_{ab}$, with the average effect of substitution (9.15) reducing to $E(a)(b_1 - b_2)$, where $E(a)$ represents the average parameter associated with the A locus ($p_A a_1 + q_A a_2 = p_A a_2 + q_A a_3$). Condition (9.37a) applies to this case as well.

The analogue of the Lewontin-Kojima model (1960) assumes that among individuals with a given genotype at the one locus, the two homozygotes at the other locus have identical propensities for altruism. This assumption implies $h_{11} = h_{33}, p_A = q_A$, and $h_{12} = h_{34}$, which satisfies $M_{Ab} = M_{ab}$. (The "phenotypic" model studied by Mueller and Feldman [1985] corresponds to a subset of the Lewontin-Kojima model, with the additional restriction that $h_{13} = h_{12} = h_{34}$.) The stability condition suggested by (9.30b) in this case is

$$(3\beta/4 - \gamma)(h_{11} - h_{12} + h_{13} - h_{14}) > 0, \tag{9.37b}$$

in which the second factor is proportional to the average effect of substitution at the B locus (9.15).

The assumption that all individuals heterozygous at the B locus share the same propensity for altruism ($h_{12} = h_{14} = h_{34}$) also satisfies $M_{Ab} = M_{ab}$. In this case, (9.30b) suggests

$$(3\beta/4 - \gamma)(E(h) - h_{12}) > 0, \tag{9.37c}$$

in which $E(h)$ is defined in (9.35b).

Examination of the full set of stability conditions governing one-gene introductions confirms that for each of the special cases described here, the corresponding condition given in (9.37) determines the local stability of the resident population to the initial increase of the b allele. These results resemble the conditions derived by Bodmer and Parsons (1962, Section VIIC) and Bodmer and Felsenstein (1967, Section 7) for models of viability selection controlled by two autosomal loci. The additional factor $(3\beta/4 - \gamma)$ determines whether altruism is favored.

Near Equality among B Locus Heterozygotes. In order to explore more general altruistic propensity matrices, I studied a perturbation of a matrix belonging to the family of special cases studied. Under the assignment

$$h_{12} = E(h) - \varepsilon$$

$$h_{14} = E(h) \quad (9.38)$$

$$h_{34} = E(h) + \varepsilon,$$

the average effect of substitution at the B locus (9.15) reduces to

$$(\beta_1 - \beta_2) = \varepsilon(p_A m_2 - q_A m_4)/(m_2 + m_4). \quad (9.39)$$

In the absence of recombination ($r = 0$), the edge equilibrium resists invasion by chromosomes Ab and ab if both

$$(3\beta/4 - \gamma)M_{Ab} > 0 \quad (9.40a)$$

$$(3\beta/4 - \gamma)M_{ab} > 0. \quad (9.40b)$$

Because M_{Ab} and M_{ab} have opposite signs under (9.38), the inequalities in (9.40) cannot be satisfied simultaneously. This result implies that for tight linkage (r positive but small), the edge equilibrium is unstable.

If ε is assumed to be very small and r is chosen to be large relative to ε, then the sole condition for local stability of the edge equilibrium is

$$\varepsilon(h_{11} - h_{33}) > 0. \quad (9.41)$$

For parameter sets satisfying (9.41), the new allele at the B locus fails to increase under loose linkage (r large) but invades the population under tight linkage (r small). Because relatedness equals 3/4 regardless of the values of the h_{ij}, this result illustrates the effects of recombination in the presence of linkage disequilibrium and epistasis on the average effect of substitution (9.15) alone. Qualitatively similar conclusions were obtained by Bodmer and Felsenstein (1967, see IIc and IIIc in their Table 5), who showed that tight linkage promotes instability under general forms of frequency-independent viability selection controlled by two autosomal loci.

INITIAL INCREASE CONDITIONS NEAR EDGE EQUILIBRIA
CHARACTERIZED BY POSITIVE ADDITIVE GENETIC VARIANCE

Dominance permits the existence of a second one-locus equilibrium class (structural), in which the additive genetic variance is positive. This equilibrium class has no analogue in models of two-locus frequency-independent viability selection in randomly mating populations, in which the equilibrium additive genetic variance is always zero (Ewens 1976; Ewens and Thomson 1977). The frequencies of the A allele at the two equilibrium states in this class correspond to the valid roots of the quadratic given in Equation (11) of Uyenoyama and Feldman (1981), in which h_1, h_2, and h_3 correspond to h_{11}, h_{13}, and h_{33}, respectively. Of these two equilibria, at most one and possibly neither are locally stable. Appendix 9.3 presents the conditions for existence and stability of this equilibrium class. Under

$$h_{11} > h_{13} > h_{33} \tag{9.42a}$$

$$h_{11} + h_{33} > 2h_{13} \tag{9.42b}$$

$$3\beta/4 > \gamma, \tag{9.42c}$$

the A allele increases when rare and when close to fixation; for intermediate frequencies, the population converges to the equilibrium at which the frequency of the A allele lies between 0 and 1/2. Condition (9.42a) implies that the average effect of substitution at the A locus ($\alpha_1 - \alpha_2$) is positive. I obtained conditions for the initial increase of the new allele at the B locus (one-gene introduction) near this point.

The departure of the average effect of substitution at the A locus from zero implies that relatedness between sisters (9.19) differs from 3/4. Consequently, elements of the two-locus structure (recombination, epistasis, disequilibria) affect both the average effect of substitution of the rare allele on the propensity for altruism and the magnitude of relatedness relative to the cost-benefit ratio. I restrict consideration to four special cases: (1) no epistasis, (2) no recombination, (3) near equality of rates of altruism among individuals carrying the rare allele, and (4) weak selection at the B locus.

Absence of Epistasis. Near the edge, associations between loci may arise in the absence of epistasis. Both gametic linkage disequilibrium among offspring ($p_A m_4 - q_A m_2$) and genotypic linkage disequilibrium among parents ($f_{14} - f_{23}$) evaluated at $V(1)$ [see (9.29)] differ from zero unless $b_1 = b_2$ (BB and Bb have identical effects on the propensity for altruism). If b_1 exceeds b_2, then the average effect of substitution (9.15) is positive and the edge equilibrium resists invasion of the b allele for all values of the recombination fraction, even though correlations between loci affect the magnitude of relatedness (9.19) relative to the cost-benefit ratio. Because 3/4 exceeds the cost-benefit ratio (9.42c), one would expect this result on the basis of one-locus theory alone.

Absence of Recombination. The Ab chromosome fails to increase when rare if

$$(h_{33} - h_{13})(h_{11} - h_{12}) + (h_{11} - h_{13})(h_{13} - h_{14}) > 0, \qquad (9.43a)$$

and the ab chromosome fails to increase when rare if

$$(h_{33} - h_{13})(h_{13} - h_{14}) + (h_{11} - h_{13})(h_{33} - h_{34}) > 0. \qquad (9.43b)$$

Under (9.42), these conditions are identical to those in (9.40), which govern the local stablility of the edge equilibrium characterized by zero additive genetic variance, even though that equilibrium class does not exist under (9.42).

The average effect of the AB chromosome on altruistic propensity corresponds to M_{Ab} in (9.16a); M_{ab} in (9.16b) assumes this role for the ab chromosome. Were relatedness equal to 3/4, one would expect on the basis of one-locus theory that (9.42c) implies the local stability of the edge equilibrium whenever M_{Ab} and M_{ab} are positive (invading chromosomes decrease the rate of altruism). This expectation is not always borne out, indicating that epistasis can cause relatedness to fall below the cost-benefit ratio (altruism is disfavored). To illustrate the effect of epistasis on the magnitude of relatedness relative to the cost-benefit ratio, I describe results obtained under the assumptions that altruistic propensity is determined multiplicatively by the two loci and that individuals heterozygous at the B locus share the same propensity for altruism ($h_{12} = h_{14} = h_{34}$).

Under the multiplicative altruistic propensity matrix, both inequalities in (9.43) reduce under (9.42) to

$$(b_1 - b_2)(a_1 a_3 - a_2^2) > 0. \qquad (9.44)$$

The positivity of $(b_1 - b_2)$ is sufficient to ensure that M_{ab} is positive, given (9.42b). In this case, if a_2 falls between the arithmetic and geometric means of a_1 and a_3, then (9.42b) is satisfied and (9.44) violated. Chromosome ab can increase when rare, even though it decreases the rate of altruism and 3/4 exceeds the cost-benefit ratio. This result reflects that relatedness falls below both the cost-benefit ratio and 3/4.

Under the assumption $h_{12} = h_{14} = h_{34}$, both chromosomes Ab and ab increase when introduced separately, provided that

$$h_{12} > E(h), \qquad (9.45)$$

with $E(h)$ defined in (9.35b). As in the preceding case, the signs of M_{Ab} and M_{ab} are not sufficient to determine local stability. For example, if h_{13}

falls between the arithmetic and geometric means of h_{11} and h_{33}, then (9.42) is satisfied and E(h) is negative, which indicates, from (9.45), that the new chromosomes increase when rare regardless of the signs of M_{Ab} and M_{ab}.

Near Equality among Individuals Heterozygous at the B Locus. To demonstrate that the effects of epistasis on the average effect of substitution and the magnitude of relatedness relative to the cost-benefit ratio are not restricted to cases involving absolute linkage, I describe results obtained under the assumption that the altruistic propensity parameters of individuals heterozygous at the B locus (h_{12}, h_{14}, h_{34}) are nearly identical. Assume that

$$h_{12} = E(h) + \varepsilon_1 \tag{9.46a}$$

$$h_{14} = E(h) + \varepsilon_2 \tag{9.46b}$$

$$h_{34} = E(h) + \varepsilon_3, \tag{9.46c}$$

in which E(h) is defined in (9.35b) and the ε_i are small. The positivity of E(h) requires that h_{13} falls below the geometric mean of h_{11} and h_{33}. For small ε_i, M_{Ab} and M_{ab} are positive and nearly equal, indicating that the new allele decreases the propensity for altruism [see (9.15)].

Because 3/4 exceeds the cost-benefit ratio (9.42c), intuition based on one-locus theory suggests that the edge equilibrium should be locally stable. However, analysis of the stability conditions indicates that the equilibrium is unstable for all ε_i small relative to the recombination rate. Under absolute linkage ($r = 0$), the ε_i can be chosen to ensure the local stability of the equilibrium point. These results imply that epistasis can cause relatedness to fall below the cost-benefit ratio even under loose linkage. In contrast, tight linkage promotes instability near edge equilibria characterized by zero additive genetic variance under a similar assignment of altruistic propensity parameters [see (9.38)].

Weak Selection at the B Locus. The initial increase conditions depend on the recombination fraction under general parameter sets; in particular, strong selection is not required. Assume that the new allele only weakly modifies the altruistic propensity of its carriers:

$$h_{12} = h_{11} + \varepsilon_1 \tag{9.47a}$$

$$h_{14} = h_{13} + \varepsilon_2 \tag{9.47b}$$

$$h_{34} = h_{33} + \varepsilon_3, \tag{9.47c}$$

where the ε_i are small relative to all other parameters including the recombination fraction. The average effect of substitution at the B locus

(9.15) is proportional to

$$(\alpha_1 - \alpha_2)(p_A m_4 - q_A m_2), \qquad (9.48)$$

ignoring terms of order ε_i and higher. The expression of the A locus dominates the effects attributed to the B locus through the correlation between loci (gametic linkage disequilibrium). Because this correlation is on the order of the average effect of substitution, weak selection does not imply convergence of relatedness (9.19) to its one-locus value of 3/4. Analysis of the successive principal minors described in (9.30) indicates that the sole condition for local stability is given by (9.30b), the local analogue of Hamilton's rule (see Appendix 9.4).

Because the complexity of this condition obscures the effect of recombination, I restrict the analysis still further. Under

$$\varepsilon_1 = -\varepsilon_3 = \varepsilon \text{ and } \varepsilon_2 = 0, \qquad (9.49a)$$

the equilibrium is locally stable only if ε is positive, regardless of the recombination fraction. For positive ε, the average effect of substitution at the B locus (9.48) is positive (alleles a and b are positively correlated), implying that the new allele reduces altruistic propensity.

Assume that the ε_i are positive and satisfy

$$(\varepsilon_1 - \varepsilon_2)(h_{33} - h_{13}) = (\varepsilon_3 - \varepsilon_2)(h_{11} - h_{13}). \qquad (9.49b)$$

This family of parameter sets includes the case in which the ε_i are proportional to the corresponding h_{ij}. If ε_2 lies above the arithmetic mean of ε_1 and ε_3, the edge equilibrium is unstable for all recombination rates. If ε_2 lies between the arithmetic and geometric means, the equilibrium is locally stable for all recombination rates. The condition for local stability depends upon the recombination fraction as well as the ε_i in the remaining case (ε_2 below the geometric mean). These results indicate that the behavior of the two-locus system departs from its one-locus counterpart even under weak selection.

Discussion

HAMILTON'S RULE EXTENDED TO MULTIPLE LOCI

Yokoyama and Felsenstein (1978) showed that the change in the mean level of altruism within a generation as a result of selection within cooperating assemblages depends on the cost of altruism borne by donors, the benefit to recipients, and a phenotypic correlation that assumes the role of relatedness. In terms of the notation used in preceding sections,

$$T(P' - P^*) = \beta \text{ Cov}(D_P R_P) - \gamma \text{ Var}(P), \qquad (9.50)$$

where P denotes the character of altruism, and P' and $P*$ the mean value of this character after selection and at birth, respectively. [The character value P corresponds to G in (9.5).] The extension of this description over a full generation requires determination of the response of the mean among offspring $(P*' - P*)$ to the selection differential among their parents $(P' - P*)$. In the absence of selection, these two quantities are related by the heritability (h^2), defined as the linear regression coefficient of offspring phenotype on parental phenotype:

$$(P*' - P*) = h^2(P' - P*). \tag{9.51}$$

Yokoyama and Felsenstein obtained their polygenic version of Hamilton's rule by substituting (9.51) into (9.50) to produce

$$T(P*' - P*) = h^2 \, \text{Var}(P)[\beta b_{D \to R} - \gamma]. \tag{9.52}$$

Implicit in the application of (9.51) to populations under selection is the assumption that the regression is linear throughout its range (see Kempthorne 1957, p. 327); in particular, it implies that dominance and epistasis have negligible effects.

The simple derivation of the measure of relatedness makes this approach appealing. However, because this analysis ignores nonadditive components, the study of their effects on relatedness and the adaptive value of altruism requires the analysis of explicit multilocus models.

Equation (9.5) [compare (9.50)] indicates that the change in mean scale (Q) within generations is determined by the cost-benefit criterion, where Q is defined to be uncorrelated with all other components of variance within individuals. Definition of Q in terms of effects of alleles permits extension of (9.5) across a full generation by ensuring that the means among offspring and their parents are identical [compare (9.7) and (9.51)]. Under this scale, the difference between relatedness and the cost-benefit ratio is proportional to a function involving changes in gene frequencies [see (9.25)]. Under one-gene introductions (invasion of an allele at a monomorphic locus linked to a locus held in stable polymorphism), Q represents the sum of the additive effects of the alleles held at the formerly monomorphic locus alone. In this case, (9.25) is proportional to the change in gene frequency of the new allele. Under two-gene introductions (simultaneous invasion of a monomorphic population by alleles at both loci), Q represents the full two-locus additive genotypic value, expressed as the sum of the additive effects of all four alleles. By construction, (9.25) indicates that Hamilton's rule holds for two loci (or any number of loci) under these definitions for the additive scale Q.

Correspondence between the Covariance Approach and Local Stability Analysis

Taylor (1987) studied the relationship between inclusive fitness arguments and the expected change in gene frequency derived from explicit one-locus genetic models. He examined the direction and magnitude of gene frequency change in the neighborhood of the neutral distribution, and showed that explicit genetic models and inclusive fitness arguments (which incorporate neutral values for relatedness coefficients) produce approximately equivalent results under weak selection. Under weak selection, the dominant eigenvalue of the local stability matrix $\mathbf{M_S}$ lies near unity, which implies that the conditions (9.30a) involving the lower order principal minors are satisfied. The vector $\mathbf{V}(1)$ in (9.29) differs only slightly from the eigenvector associated with the dominant eigenvalue, indicating that (9.30b) corresponds to Taylor's result.

Limiting expressions for relatedness near equilibrium states in one-locus kin selection models involving general selection intensities have been obtained by evaluating relatedness with respect to the genotypic distribution corresponding to the dominant eigenvector of the local stability matrix (Uyenoyama and Feldman 1981; Uyenoyama, Feldman, and Mueller 1981). Eshel and Feldman (1984) demonstrated that the elements of the leading eigenvector represent the appropriate weights used in determining the average fitness among carriers of the new allele in general models of two-locus viability selection. The approach described in preceding sections departs from earlier work in that the relatedness measures are evaluated at $\mathbf{V}(1)$, the vector that satisfies the eigenvector equations with unity substituted for the eigenvalue. Explicit expressions for the elements of this vector are readily obtained and do not require knowlege of the eigenvalue. By suggesting local analogues of the average effect of substitution and relatedness, (9.30) unifies the interpretation of conditions for local stability obtained from one- and two-locus models of kin selection among siblings.

Introductions near Equilibrium States Characterized by Zero Additive Genetic Variance in Altruistic Propensity

Two-gene introductions involve monomorphic populations having no genetic variance of any kind. Mueller and Feldman (1985) obtained conditions similar to (9.31) and observed that the departure of the results from expectations based on one-locus theory requires epistasis. Under weak selection, the average effects of substitution at both loci are small and the measures of linkage disequilibrium are of the same order of magnitude, implying that relatedness (9.21) remains near its one-locus value. Strong epistatic selection can influence the magnitude of relatedness relative to the cost-benefit ratio and the average effects of substitution on altruistic propensity.

Near edge equilibria characterized by zero additive genetic variance (one-gene introductions), relatedness between sisters retains its one-locus value of 3/4, irrespective of dominance, epistasis, recombination, or the intensity of selection [see (9.19)]. The local analogue of Hamilton's rule (9.30b) suggests that the average effect of substitution alone should determine local stability for a given cost-benefit ratio. A number of simple cases were described to illustrate the influence of recombination and epistasis on the average effect of substitution. Under altruistic propensity matrices for which the marginal deviations of the rare chromosomes [see (9.16)] are equal, the average effect of substitution and the local stability conditions (9.37) are independent of the recombination fraction. This result also arises under frequency-independent viability selection (Bodmer and Parsons 1962; Bodmer and Felsenstein 1967). Under a form of weak selection (9.38), for which the new allele at the B locus has equal but opposite effects when in combination with the two homozygotes at the A locus, tight linkage promotes instability.

Introductions near Equilibrium States Characterized by Positive Additive Genetic Variance in Altruistic Propensity

Equilibrium states with positive additive genetic variance arise under the frequency-dependence inherent in models of kin selection (Uyenoyama and Feldman 1981); these states have no analogue under frequency-independent selection, under which the additive component of variance is zero at all equilibrium states for arbitrary numbers of loci (Ewens 1976; Ewens and Thomson 1977). Under inbreeding, all polymorphic equilibrium states arising under kin selection maintain positive additive genetic variance (Uyenoyama 1984). Feldman and Eshel (1982) showed that the initial increase conditions obtained in the absence of recombination in their two-locus model of parental manipulation differ between the two equilibrium classes.

The effect of recombination in the presence of epistasis was explored through the analysis of special cases. Under the assignment of altruistic propensities given in (9.46), loose linkage (r large relative to the ε_i) promotes instability even though the new allele reduces the rate of altruism and the cost-benefit ratio falls below 3/4. Weak selection (9.47) does not ensure the convergence of relatedness to its one-locus value. Near equilibria characterized by positive additive genetic variance, recombination affects both the average effect of substitution and relatedness.

Acknowledgment

This study was supported by Public Health Service grants HD-17925 and GM-37841. I thank M. W. Feldman and L. D. Mueller, who collaborated on the early stages of this project.

Appendix 9.1. The Partitioning of Variance

One-Gene Introduction under Sister Altruism. Expression (9.15) is obtained by minimizing the variance remaining after the additive effects of the B locus are subtracted:

$$u_B^*[(h_{BB} - \text{E*}(h)) - 2\beta_1]^2 + w_B^*[(h_{bb} - \text{E*}(h)) - 2\beta_2]^2$$
$$+ v_B^*[(h_{Bb} - \text{E*}(h)) - 2(\beta_1 + \beta_2)]^2, \quad (9A1.1)$$

in which u_B^*, v_B^*, and w_B^* denote the frequencies at birth of genotypes BB, Bb, and bb; h_{BB}, h_{Bb}, and h_{bb} the average altruistic propensities among individuals bearing those genotypes; and $\text{E*}(h)$ the average propensity among offspring at birth.

Extraction of the dominance parameters by minimization of the residual variance as before produces

$$\text{d}(AA) = q^2(h_{11} + h_{33} - 2h_{13})$$
$$\text{d}(Aa) = -p\text{d}(AA)/q$$
$$\text{d}(aa) = p^2\text{d}(AA)/q^2 \qquad (9A1.2)$$
$$\text{d}(Bb) = (\alpha_1 - \alpha_2)(p_A m_4 - q_A m_2)/(m_2 + m_4)$$

for the effects due to dominance expressed by genotypes AA, Aa, aa, and Bb. The dominance effect for BB is zero, and the effect for bb is not required because such individuals occur in negligible frequencies. Epistatic effects for each genotype are assigned as the residual obtained by subtracting the additive and dominance components from the genotypic values $[h_{ij} - \text{E*}(h)]$.

Two-Gene Introduction under Brother Altruism. Extraction of the additive effects entails minimization of

$$m_1^*\{[g_1 - \text{E*}(g)] - (\alpha_1 + \beta_1)\}^2 + m_2^*\{[g_2 - \text{E*}(g)] - (\alpha_1 + \beta_2)\}^2$$
$$+ m_3^*\{[g_3 - \text{E*}(g)] - (\alpha_2 + \beta_1)\}^2 + m_4^*\{[g_4 - \text{E*}(g)] - (\alpha_2 + \beta_2)\}^2. \quad (9A1.3)$$

The expressions in (9.18) are obtained under the assumption

$$p_A^*\alpha_1 + q_A^*\alpha_2 = p_B^*\beta_1 + q_B^*\beta_2 = 0, \qquad (9A1.4)$$

in which

$$p_A^* = m_1^* + m_2^* \tag{9A1.5a}$$

$$p_B^* = m_1^* + m_3^*. \tag{9A1.5b}$$

Subtraction of the additive genotypic values $(\alpha_i + \beta_j)$ from the genotypic values $[g_i - E^*(g)]$ produces the epistatic effects

$$e(Ab) = -em_3^* \tag{9A1.6a}$$

$$e(aB) = -em_2^* \tag{9A1.6b}$$

$$e(ab) = em_2^* m_3^*/m_4^*, \tag{9A1.6c}$$

in which e is given in (9.18c). The epistatic effect associated with the common genotype AB is zero.

Appendix 9.2. Reducible Stability Matrices

Near edge equilibria (one-gene introductions), the full stability matrix resolves into two blocks: the first describes changes within the edge itself and the second departs from the edge. By assumption, the equilibrium resists perturbations within the edge, necessitating consideration of only the latter block. The argument for irreducible matrices can be applied to this block, which is irreducible in the presence of recombination. Because the several common genotypes share the same additive genotypic value with respect to the B locus, they do not contribute to the variance or covariance; consequently, they can be ignored in (9.30). In the absence of recombination, the transformation describes movement along the two faces adjacent to the edge on which the equilibrium lies. The argument again applies to each of the irreducible matrices corresponding to these faces.

Near the corners (two-gene introductions), $\mathbf{M_S}$ includes two blocks corresponding to the one-locus subspaces, in which the dynamics are governed by one-locus theory. If changes in the interior of the tetrahedron (rather than along the edges) tend to dominate the transformation, then all four possible haplotypes will be generated. In this case, the rows and columns of $\mathbf{M_S}$ and the elements of $\mathbf{X}$ can be renamed with the frequency of the genotype bearing two rare alleles corresponding to the nth dimension. An argument identical to that for irreducible matrices applies; the assumption that the eigenvalue corresponding to the nth dimension exceeds those corresponding to the one-locus subspaces guarantees that the determinants of the principal minors [see (9.30a)] are positive.

Appendix 9.3. Conditions for Existence and Internal Stability of Equilibria Having Positive Additive Genetic Variance

Uyenoyama and Feldman (1981) reported that either zero or two structural equilibria occur in one-locus models. A necessary condition for the existence of this class is

$$(\beta 3/4 - \gamma)(h_{11} + h_{33} - 2h_{13}) > 0, \tag{9A3.1}$$

which ensures the instability of the equilibrium characterized by zero additive genetic variance [see (9.35)]. Two valid equilibrium points exist for β lying between $\gamma 4/3$ and one of the roots of the quadratic

$$[(\beta - \gamma)2(h_{33} - h_{13}) + \beta(h_{11} - h_{33})/4]^2$$
$$- 4(h_{11} + h_{33} - 2h_{13})(\beta 3/4 - \gamma)[1 + h_{33}(\beta - \gamma)] = 0. \quad (9A3.2)$$

The root of interest is the one that lies between

$$\gamma 4/3 \text{ and } 2\gamma \text{ if } (h_{11} + h_{33} - 2h_{13}) > 0 \qquad (9A3.3a)$$

$$\gamma \text{ and } \gamma 4/3 \text{ if } (h_{11} + h_{33} - 2h_{13}) < 0. \qquad (9A3.3b)$$

Of these two equilibria, at most one and possibly neither are locally stable. If the viability-analogous equilibrium (9.35) lies between the gene frequencies associated with the equilibria described here, then all three polymorphisms are unstable.

Appendix 9.4. Condition for Local Stability of the Structural Equilibrium under Weak Selection

Under (9.47), the sole condition for local stability is given by (9.30b), which reduces to

$$\beta(\alpha_1 - \alpha_2)[(\beta 3/4 - \gamma) - 2r(1 - r)(\beta - \gamma)]X/4$$
$$- (\beta/4)^2(h_{11} + h_{33} - 2h_{13})Y$$
$$+ 2r(1 - r)(\varepsilon_1 + \varepsilon_3 - 2\varepsilon_2)[(\beta - \gamma)(\alpha_1 - \alpha_2)]^2 > 0, \quad (9A4.1)$$

in which

$$X = (h_{11} - h_{13})(\varepsilon_3 - \varepsilon_2) - (h_{33} - h_{13})(\varepsilon_1 - \varepsilon_2) \qquad (9A4.2a)$$

$$Y = (h_{11} - h_{13})\varepsilon_3 + (h_{33} - h_{13})\varepsilon_2 p_A X. \qquad (9A4.2b)$$

References

Aoki, K., and Moody, M. 1981. One- and two-locus models of the origin of worker behavior in Hymenoptera. *J. Theor. Biol.* 89: 449–474.

Bodmer, W. F., and Felsenstein, J. 1967. Linkage and selection: Theoretical analysis of the deterministic two-locus random mating model. *Genetics* 57: 237–265.

Bodmer, W. F., and Parsons, P. A. 1962. Linkage and recombination in evolution. *Adv. Genetics* 11: 1–100.

Cavalli-Sforza, L. L., and Bodmer, W. F. 1971. *The Genetics of Human Populations.* W. H. Freeman, San Francisco.

Cavalli-Sforza, L. L., and Feldman, M. W. 1978. Darwinian selection and "altruism." *Theor. Pop. Biol.* 14: 268–280.

Charlesworth, B. 1980. Models of kin selection. In *Evolution of Social Behavior: Hypotheses and Empirical Tests*, pp. 11–26. Ed. H. Markl. Verlag Chemie, Weinheim, F.R.G.

Cockerham, C. C. 1954. An extension of the concept of partitioning hereditary variance for the analysis of covariances among relatives when epistasis is present. *Genetics* 39: 859–882.

Cockerham, C. C. 1956. Effects of linkage on the covariances between relatives. *Genetics* 41: 138–141.

Crow, J. F., and Aoki, K. 1982. Group selection for a polygenic behavioral trait: A differential proliferation model. *Proc. Natl. Acad. Sci. USA.* 79: 2628–2631.

Crow, J. F., and Nagylaki, T. 1976. The rate of change of a character correlated with fitness. *Amer. Natur.* 110: 207–213.

Denniston, C. 1978. An incorrect definition of fitness revisited. *Ann. Hum. Genet.* 42: 77–85.

Engels, W. R. 1983. Evolution of altruistic behavior by kin selection: An alternative approach. *Proc. Natl. Acad. Sci. USA.* 80: 515–518.

Eshel, I., and Feldman, M. W. 1984. Initial increase of new mutants and some continuity properties of ESS in two-locus systems. *Amer. Natur.* 124: 631–640.

Ewens, W. J. 1976. Remarks on the evolutionary effect of natural selection. *Genetics* 83: 601–607.

Ewens, W. J. 1979. *Mathematical Population Genetics.* Springer-Verlag, New York.

Ewens, W. J., and Thomson, G. 1977. Properties of equilibria in multi-locus genetic systems. *Genetics* 87: 807–819.

Falconer, D. S. 1966. Genetic consequences of selection pressure. In *Genetic and Environmental Factors in Human Ability*, pp. 219–232. Ed. J. E. Meade and A. S. Parkes. Oliver and Boyd, Edinburgh.

Feldman, M. W., and Cavalli-Sforza, L. L. 1981. Further remarks on Darwinian selection and "altruism." *Theor. Pop. Biol.* 19: 251–260.

Feldman, M. W., and Eshel, I. 1982. On the theory of parent-offspring conflict: A two-locus genetic model. *Amer. Natur.* 119: 285–292.

Fisher, R. A. 1918. The correlation between relatives on the supposition of Mendelian inheritance. *R. Soc. (Edinburgh) Trans.* 52: 399–433.

Fisher, R. A. 1941. Average excess and average effect of a gene substitution. *Ann. Eugen.* 11: 53–63.

Fisher, R. A. 1958. *The Genetical Theory of Natural Selection.* 2nd ed. Dover, New York.

Gallais, A. 1974. Covariances between arbitrary relatives with linkage and epistasis in the case of linkage disequilibrium. *Biometrics* 30: 429–446.

Gantmacher, F. R. 1959. *The Theory of Matrices.* Vol II, Chelsea, New York.

Hamilton, W. D. 1964a. The genetical evolution of social behaviour, I. *J. Theor. Biol.* 7: 1–16.

Hamilton, W. D. 1964b. The genetical evolution of social behaviour, II. *J. Theor. Biol.* 7: 17–52.

Harris, D. L. 1964. Genotypic covariances between inbred relatives. *Genetics* 50: 1319–1348.

Karlin, S., and Feldman, M. W. 1968. Analysis of models with homozygote × heterozygote matings. *Genetics* 59: 105–116.

Kempthorne, O. 1955. The correlations between relatives in random mating populations. *Cold Spring Harbor Symp. Quant. Biol.* 20: 60–75.

Kempthorne, O. 1957. *An Introduction to Genetic Statistics.* John Wiley and Sons, New York.

Kimura, M. 1958. On the change of population fitness by natural selection. *Heredity* 12: 145–167.

Kimura, M. 1965. Attainment of quasi linkage equilibrium when gene frequencies are changing by natural selection. *Genetics* 52: 875–890.

Lande, R., and Arnold, S. J. 1983. The measurement of selection on correlated characters. *Evolution* 37: 1210–1226.

Lewontin, R. C., and Kojima, K. 1960. The evolutionary dynamics of complex polymorphisms. *Evolution* 14: 458–472.

Li, C. C. 1967. Fundamental theorem of natural selection. *Nature* 214: 505–506.

Li, C. C. 1976. *First Course in Population Genetics.* Boxwood Press, Pacific Grove, Calif.

Maynard Smith, J. 1980. Models of the evolution of altruism. *Theor. Pop. Biol.* 18: 151–159.

Michod, R. E. 1982. The theory of kin selection. *Ann. Rev. Ecol. Syst.* 13: 23–55.

Michod, R. E., and Abugov, R. 1980. Adaptive topographies in family-structured models of kin selection. *Science* 210: 667–669.

Michod, R. E., and Hamilton, W. D. 1980. Coefficients of relatedness in sociobiology. *Nature* 288: 694–697.

Mueller, L. D., and Feldman, M. W. 1985. Population genetic theory of kin selection: A two-locus model. *Amer. Natur.* 125: 535–549.

Nagylaki, T. 1976. The evolution of one- and two-locus systems. *Genetics* 83: 583–600.

Orlove, M. J., and Wood, C. J. 1978. Coefficients of relationship and coefficients of relatedness in kin selection: A covariance form for the rho formula. *J. Theor. Biol.* 73: 679–686.

Price, G. R. 1970. Selection and covariance. *Nature* 227: 520–521.

Price G. R. 1972. Fisher's "fundamental theorem" made clear. *Ann. Hum. Genet., Lond.* 36: 126–140.

Robertson, A. 1966. A mathematical model of the culling process in dairy cattle. *Anim. Prod.* 8: 95–108.

Seger, J. 1981. Kinship and covariance. *J. Theor. Biol.* 91: 191–213.

Taylor, P. D. 1988. Inclusive fitness models with two sexes. *Theor. Pop. Biol.* To appear.

Uyenoyama, M. K. 1984. Inbreeding and the evolution of altruism under kin selection: Effects on relatedness and group structure. *Evolution* 38: 778–795.

Uyenoyama, M. K., and Feldman, M. W. 1980. Theories of kin and group selection: A population genetics perspective. *Theor. Pop. Biol.* 17: 380–414.

Uyenoyama, M. K., and Feldman, M. W. 1981. On relatedness and adaptive topography in kin selection. *Theor. Pop. Biol.* 19: 87–123.

Uyenoyama, M. K., and Feldman, M. W. 1982. Population genetic theory of kin selection, II. The multiplicative model. *Amer. Natur.* 120: 614–627.

Uyenoyama, M. K.; Feldman, M. W.; and Mueller, L. D. 1981. Population genetic theory of kin selection, I. Multiple alleles at one locus. *Proc. Natl. Acad. Sci. USA.* 78: 5036–5040.

Wade, M. J. 1985. Soft selection, hard selection, kin selection and group selection. *Amer. Natur.* 125: 61–73.

Weir, B. S., and Cockerham, C. C. 1977. Two-locus theory in quantitative genetics. In *Proceedings of the International Conference on Quantitative Genetics*, pp. 247–269. Ed. E. Pollak, O. Kempthorne, and T. B. Bailey, Jr. The Iowa State University Press, Ames.

Wright, S. 1952. The genepics of quantitative variability. In *Quantitative Inheritance*, pp. 5–41. Ed. E.C.R. Reeve and C. H. Waddington. Her Majesty's Stationery Office, London.

Yokoyama, S., and Felsenstein, J. 1978. A model of kin selection for an altruistic trait considered as a quantitative character. *Proc. Natl. Acad. Sci. USA.* 75: 420–422.

Resource Allocation in Mendelian Populations: Further in ESS Theory

Sabin Lessard

10.1. Introduction

Modern ESS (Evolutionarily Stable Strategy) theory began with Maynard Smith and Price's (1973) paper. But its origins can be traced in early frequency-dependent selection models (dealing, e.g., with mimicry or sexual selection) and in sex ratio evolution principles (see, e.g., Fisher 1930). In this respect, Shaw and Mohler (1953) shed some light onto a basic notion, that of reproductive value in sexual populations, with the formula

$$\frac{m_0}{2m} + \frac{f_0}{2f},\tag{10.1.1}$$

where m_0 and f_0 refer to individual fitnesses through male and female functions (or progenies) with mean values m and f in the population. Given a (convex) fitness set for all possible (m_0, f_0), MacArthur (1965) showed that the "optimal strategy" is to maximize the product $m_0 \times f_0$. Charnov, Maynard Smith, and Bull (1976) refined the approach to study conditions for the evolution of dioecy versus hermaphroditism. Charnov (1979, 1982, 1986) applied the product maximization principle to study various aspects of sex allocation and proposed several extensions for structured populations, haplodiploid models, varying environments, and so forth.

ESS theory owes a great deal of its first developments to Hamilton's (1967) work on "unbeatable" sex ratios in nonrandom mating populations, that is, sex ratios that have "selective advantage(s) over any other(s)." Hamilton's concept of an unbeatable sex ratio suggests a global selective advantage. But in the search of ESS sex ratios in Mendelian populations with random mating or not, initial increase properties of mutant alleles have become the predominant criteria in use (see, e.g., Taylor and Bulmer 1980; Uyenoyama and Bengtsson 1979, 1981; Eshel 1975; Eshel and Feldman 1982a,b). As Uyenoyama and Bengtsson (1982) point out, there is some ambiguity in local ESS approaches because some authors look for ESS's as non- invadable population equilibrium strategies and others as ever-invading mutant population strategies. Moreover, there is the need to distinguish between "strong" ESS's (ESS population equilibria whereby any mutant is selected against when it is rare at a geometric rate) and "weak" ESS's (for which the rate of decrease in the frequency of any mutant is best algebraic). These are some of the issues that will be addressed in this article.

Speith (1974) did simulations in order to test the validity of MacArthur's maximization principle in assessing the optimal sex ratio and predicting evolution toward that optimum. With random mating and Mendelian segregation of two alleles at a single locus modifying male and female fitnesses in a diploid population, it is the product of the male and female mean fitnesses $m \times f$ [and not the mean of the products $m_0 \times f_0$ in the notation of (10.1.1)] that appears to be monotonically nondecreasing after the first few generations. However, one probable exception was observed, but without any explanation. This convinced us of the necessity of running our own simulations in a multiallele framework and under various fitness constraints in order to examine in detail several aspects and/or hypotheses concerning the evolution of the product $m \times f$—for example, rate of increase, conditions favoring a decrease if any, conditions promoting polymorphism versus monomorphism, and so on.

An important class of two-sex, one-locus models in diploid random-mating populations were analyzed in Karlin and Lessard (1984). (See also Chap. 7 in Karlin and Lessard 1986.) Assuming a linear relationship between male and female fitnesses [i.e., $f_0 = a + bm_0$ in (10.1.1) for some constants a and b; the case $a = 1$ and $b = -1$ corresponds to pure sex ratio determination models as introduced in Eshel and Feldman (1982a)], it was shown that the product of the male and female mean fitnesses should increase from one equilibrium to the next one attainable following the introduction of a mutant allele (leading, e.g., in the long run to a one-to-one population sex ratio in pure sex ratio determination models; an "attainable" equilibrium is an equilibrium that may be reached if there is convergence). Global convergence according to this scheme was proved for some classes of two-sex, two-allele models including the linear one. Nevertheless, some reserve was expressed about the validity of such an evolutionary property in general two-sex models, but the question was left open. One of our objectives in this paper is to answer this question by analytical and numerical results.

Maynard Smith (1974) was concerned with a pure game-theoretic approach of ESS theory by comparing fitness returns in pairwise contests to identify noninvadable strategies, that is, strategies that have greater fitness than any other once they are adopted by all members of a population. In the original formulation, a strategy is represented by a frequency vector whose components are probabilities of using specific elementary strategies in a contest with an opponent chosen at random in a population consisting essentially of two types, a common type and a rare type. Dynamical studies in polymorphic (and actually autogenous) populations were initiated by Taylor and Jonker (1978), followed by Zeeman (1980), Hines (1980), Akin (1982), and Thomas (1985b), among others. Dynamical properties of ESS's (or ES [Evolutionarily Stable] states in polymorphic populations) in exact diploid one-locus Mendelian populations were first investigated by Maynard Smith (1981) and Eshel (1982) in the case of two-component strategies with no sex differences. In this context, global convergence to an ES state as new alleles are introduced (i.e., "evolutionary attractiveness") was proved by Lessard (1984). (See Thomas 1985a for a formulation of genetical ESS theory extended to the case of multicomponent strategies with no sex differences determined at a single locus.) Evolutionary Genetic Stability (EGS) of ES states, that is, initial increase property of mutant alleles that "render the population strategy closer to the ES state(s)" was shown to hold in one-sex, two-locus systems determining multicomponent strategies (Eshel and Feldman 1984). This property had been previously introduced for one-locus sex ratio determination models (Eshel and Feldman 1982a,b). General two-sex, two-phenotype frequency-dependent selection models in a multiallele one-locus setting were considered in Lessard (1986). (See also Raper 1983 for models more specifically related to sexual selection.) Optimality properties of ES states over successive equilibria as new alleles are introduced can be proved in this context.

In this paper, we present a general theory of resource allocation in two-sex populations (hermaphrodite as well as dioecious populations). The phenotypes are N-dimensional nonnegative resource allocation vectors that are genetically determined at one or several loci. We assume throughout (see Section 10.6 for a discussion of other assumptions) an infinite diploid population with discrete (nonoverlapping) generations and random mating, or random union of gametes, in a constant environment. Frequency-independent as well as frequency-dependent male and female fitness functions are considered. This approach unifies ESS theory and sex-ratio evolution theory in a multilocus Mendelian framework.

10.2. Basic Elements of Resource Allocation Models in Mendelian Populations

10.2.1. GENOTYPE SET

In a diploid population, the genotype of an individual can be represented by a couple $(\mathbf{i}, \mathbf{j})$, where $\mathbf{i}$ and $\mathbf{j}$ denote the genomes of the male and female parental gametes, respectively. If we focus on genes at L specific loci with possible alleles $A_1^{(l)}, \ldots, A_{n_l}^{(l)}$ at the lth locus ($l = 1, \ldots, L$), a genome can be described as a vector of integers $\mathbf{i} = (i_1, \ldots, i_L)$, where i_l is the index of the allele present at the lth locus ($1 \le i_l \le n_l$ for $l = 1, \ldots, L$).

The *genotype set* G will be defined as the set of all possible genotypes $(\mathbf{i}, \mathbf{j})$. Some of these genotypes may not exist in the population at a specific time, but all of them can come into existence by appropriate mutations of existing alleles. Thus G represents the set of genotypes that can originate from any allelic substitutions within a given genetic framework.

In the context of L autosomal loci and n_l possible alleles at locus l for $l = 1, \ldots, L$, we have

$$G = \{(\mathbf{i}, \mathbf{j}): \mathbf{i} = (i_1, \ldots, i_L), \mathbf{j} = (j_1, \ldots, j_L),$$
$$1 \le i_l \le n_l, 1 \le j_l \le n_l, l = 1, \ldots, L\}. \quad (10.2.1)$$

10.2.2. RESOURCE ALLOCATION IN HERMAPHRODITES

Suppose that the various genotypes of a set G account for individual differences in resources allocated by hermaphrodites to a given number N of characters or functions. These may be usual quantitative genetic characters as physical dimensions or amounts of product, or else behavioral traits subject to quantitative variations related, for example, to duration, intensity, and frequency of expression in the population and/or habitat.

The amount of resources allocated to a character k by an individual of genotype $(\mathbf{i}, \mathbf{j})$ in a constant environment is a nonnegative number that will be denoted by $x_k(\mathbf{i}, \mathbf{j})$. For N characters numerated from 1 to N, the nonnegative vector

$$\mathbf{x}(\mathbf{i}, \mathbf{j}) = [x_1(\mathbf{i}, \mathbf{j}), \ldots, x_N(\mathbf{i}, \mathbf{j})] \quad (10.2.2)$$

represents the corresponding *resource allocation vector*. It will be assumed throughout that $\mathbf{x}(\mathbf{i}, \mathbf{j}) = \mathbf{x}(\mathbf{j}, \mathbf{i})$ for all $\mathbf{i}, \mathbf{j}$.

In hermaphrodites, some characters may affect only the male function, others only the female function, others both. In dioecious populations, we would rather resort to a double notation, for example, $\mathbf{x}(\mathbf{i}, \mathbf{j})$ for a male and $\mathbf{y}(\mathbf{i}, \mathbf{j})$ for a female (see Section 10.2.5). In the case of geographically or temporally varying environments with effects on resource allocation, we would use the notation $\mathbf{x}^{(e)}(\mathbf{i}, \mathbf{j})$ where e specifies the environment. It is also conceivable that resource allocation might depend on the distribution of genotypes in the population.

In any circumstances, an N-dimensional nonnegative resource allocation vector will determine a phenotype with respect to N characters or activities.

10.2.3. Resource Allocation Set

A resource allocation vector $\mathbf{x}$ generally belongs to a set X whose properties reflect biological and ecological constraints. A *resource allocation set*, X, has the property that there exist a "cost'" function g and a constant c, such that

$$X = \{\mathbf{x} \geq \mathbf{0} : g(\mathbf{x}) \leq c\}. \tag{10.2.3}$$

(We use the notation $\mathbf{x} \geq \mathbf{0}$ for $x_k \geq 0$ for $k = 1, \ldots, N$ and $(x_1, \ldots, x_N) = \mathbf{x}$.)

The *cost function*, g, is a nonnegative function defined on $\mathbf{x} \geq \mathbf{0}$ that has all positive partial derivatives, that is,

$$\nabla g(\mathbf{x})$$
$$= \left[\frac{\partial}{\partial x_1} g(\mathbf{x}), \ldots, \frac{\partial}{\partial x_N} g(\mathbf{x}) \right] > \mathbf{0} \qquad \text{(every component} > 0), \tag{10.2.4}$$

and goes to infinity as $\sum_{k=1}^{N} x_k \to +\infty$.

Hence a resource allocation set X is bounded above, and the level surface $g(\mathbf{x}) = c$ that bounds X above exhibits only positive normal vectors given by [10.2.4]. Moreover, since g is increasing in each coordinate, $g(\mathbf{x}) \leq c$ entails $g(\mathbf{y}) \leq c$ whenever $\mathbf{0} \leq \mathbf{y} \leq \mathbf{x}$ (i.e., $0 \leq y_k \leq x_k$ for $k = 1, \ldots, N$), that is, every nonnegative vector whose components are less or equal to the components of any vector in X necessarily belongs to X.

In the case of a linear cost function, that is, a cost function in the form

$$g(\mathbf{x}) = \sum_{k=1}^{N} a_k x_k, \tag{10.2.5}$$

the cost per unit of resources allocated to each character is constant [given by $a_k > 0$ for character k in (10.2.5)] and the costs are all cumulative. In such a case the level surface $g(\mathbf{x}) = c$ that bounds X is a hyperplane [perpendicular to $(a_1, \ldots, a_N) = \nabla g$ in (10.2.5)].

10.2.4. Frequency-Dependent Fitness Set

Consider a hermaphrodite population with a given distribution of phenotypes or resource allocation vectors in a set X. In general, the fitness of an individual through the male and female functions may be affected by the population state. We will concentrate on the case where the dependence on the population state is through the mean resource allocation vector denoted by μ. (The dependence on the population state might be more complex in general, involving the whole distribution of phenotypes.)

Let the male and female fitnesses be $m_\mu(\mathbf{x})$ and $f_\mu(\mathbf{x})$, respectively, for an individual of phenotype $\mathbf{x}$ in X in a population with mean phenotype μ. The quantities $m_\mu(\mathbf{x})$ and $f_\mu(\mathbf{x})$ for all $\mathbf{x}$ in the population refer to relative genetic contributions to the next generation through male and female gametes, respectively. They can incorporate viability differentials, effects of sexual selection, and even female fertilities and male virilities if these selective forces operate independently in males and females.

The fitness pair $[m_\mu(\mathbf{x}), f_\mu(\mathbf{x})]$ measures the effectiveness of a resource allocation $\mathbf{x}$ when the resource allocation in the population is μ on the average. The dependence on μ is generally smooth, or at least continuous.

The fitness functions $m_\mu(\mathbf{x})$ and $f_\mu(\mathbf{x})$ are also assumed to be differentiable, nonnegative and nondecreasing in each coordinate with respect to the resource allocation vector $\mathbf{x}$. The nonnegative gradient vectors of $m_\mu(\mathbf{x})$ and $f_\mu(\mathbf{x})$, which indicate the directions of greatest increase, will be denoted by $\nabla m_\mu(\mathbf{x})$ and $\nabla f_\mu(\mathbf{x})$, respectively. It will be assumed throughout that $\nabla m_\mu(\mathbf{x}) + \nabla f_\mu(\mathbf{x})$ is a positive vector for every $\mathbf{x}$.

In the *linear case*, we have

$$m_\mu(\mathbf{x}) = \sum_{k=1}^{N} x_k (\nabla m_\mu)_k,$$

$$f_\mu(\mathbf{x}) = \sum_{k=1}^{N} x_k (\nabla f_\mu)_k, \tag{10.2.6}$$

where the nonnegative vectors ∇m_μ and ∇f_μ are independent of $\mathbf{x}$ and $\nabla m_\mu + \nabla f_\mu$ is positive. In such a case the mean fitnesses in the population through the male and female functions are $m_\mu(\mu)$ and $f_\mu(\mu)$, respectively.

Charnov (1982) has proposed male and female fitness functions independent of μ but not necessarily linear in $\mathbf{x}$ in the form

$$ax_1^t (t > 0) \text{ and } bx_2, \tag{10.2.7}$$

respectively, where x_1 represents the number of male gametes and x_2 the number of female gametes produced with particular reference to hermaphroditic plants (see Charnov 1982 for more details).

Given a resource allocation set X and a mean resource allocation vector μ in a population, the *fitness set* $S(\mu)$ is defined as

$$S(\mu) = \{[m_\mu(\mathbf{x}), f_\mu(\mathbf{x})] : \mathbf{x} \in X\}. \tag{10.2.8}$$

$S(\boldsymbol{\mu})$ is a frequency-dependent bounded set located in the first quadrant that gives all possible fitness pairs with respect to the population state.

10.2.5. RESOURCE ALLOCATION IN DIOECIOUS POPULATIONS

For dioecious populations, the previous formulation applies mutatis mutandis to resources allocated by the mother to her male and female progeny. However, in this case the recurrence equations for the genotype frequencies will slightly differ from those deduced in the next sections.

In the context of offspring allocation with sex differentiation, the resource allocation vector will be denoted by $\mathbf{x}$ for a male and $\mathbf{y}$ for a female. There will be assumed a functional relationship between $\mathbf{x}$ and $\mathbf{y}$ in the form $\mathbf{y} = h(\mathbf{x})$ where h is one-to-one. The vector $h(\mathbf{x})$ represents the phenotype of a female that has the same genotype as a male of phenotype $\mathbf{x}$. For simplicity, all partial derivatives of h are assumed to be positive.

In the above context, the resource allocation sets X and $Y = h(X)$ for males and females, respectively, will be defined with respect to the cost functions g and $g(h^{-1})$, respectively. The corresponding male and female mean resource allocation vectors in the population will be represented by $\boldsymbol{\mu}$ and v. In such a population, the fitness of a male of phenotype $\mathbf{x}$ will be represented by $m_{\boldsymbol{\mu},v}(\mathbf{x})$ and the fitness of a female of phenotype $\mathbf{y}$ by $f_{\boldsymbol{\mu},v}(\mathbf{y})$.

Therefore, the fitness set takes the form

$$S(\boldsymbol{\mu}, v) = \{[m_{\boldsymbol{\mu},v}(\mathbf{x}), f_{\boldsymbol{\mu},v}(\mathbf{y})] : \mathbf{y} = h(\mathbf{x}), \mathbf{x} \in X\}. \qquad (10.2.9)$$

We are in the framework of the previous formulation with fitness functions $m_{\boldsymbol{\mu},v}$ and $f_{\boldsymbol{\mu},v}(h)$ but with a population state described by a pair $(\boldsymbol{\mu}, v)$ of mean resource allocation vectors. In general, $v \neq h(\boldsymbol{\mu})$ unless h is linear. In the case of frequency-independent fitness functions, (10.2.9) can always be written in the form (10.2.8).

10.3. Multiallele One-Locus Resource Allocation Models with Frequency-Independent Fitness Functions

10.3.1. RECURRENCE EQUATIONS FOR GENERAL TWO-SEX, ONE-LOCUS MODELS

Suppose that resource allocation in a hermaphrodite population is controlled at a single autosomal locus with alleles $A_1, \ldots, A_n$ such that the male and female fitnesses for a genotype $A_i A_j$ of phenotype $\mathbf{x}(i, j)$ are given by positive constants $m_{ij}(= m_{ji})$ and $f_{ij}(= f_{ji})$, respectively, for $i, j = 1, \ldots, n$ independently of the population state. Let p_i and q_i be the frequency of

A_i in the male and female gametes, respectively, that participate to the formation of a given generation ($i = 1, \ldots, n$). Assuming random union of gametes in an infinite population with discrete nonoverlapping generations, we will have at the beginning of the next generation

$$
p_i' = \frac{p_i \sum\limits_{j=1}^{n} m_{ij} q_j + q_i \sum\limits_{j=1}^{n} m_{ij} p_j}{2m}, \qquad i = 1, \ldots, n,
$$

$$
q_i' = \frac{p_i \sum\limits_{j=1}^{n} f_{ij} q_j + q_i \sum\limits_{j=1}^{n} f_{ij} p_j}{2f}, \qquad i = 1, \ldots, n,
$$

(10.3.1)

where the normalizing quantities m and f are the mean male and female fitnesses, namely,

$$
m = \sum_{i,j=1}^{n} m_{ij} p_i q_j,
$$

$$
f = \sum_{i,j=1}^{n} f_{ij} p_i q_j.
$$

(10.3.2)

In matrix notation, we have

$$
\mathbf{p}' = \frac{\mathbf{p} \circ \mathbf{Mq} + \mathbf{q} \circ \mathbf{Mp}}{2\langle \mathbf{p}, \mathbf{Mq} \rangle},
$$

$$
\mathbf{q}' = \frac{\mathbf{p} \circ \mathbf{Fq} + \mathbf{q} \circ \mathbf{Fp}}{2\langle \mathbf{p}, \mathbf{Fq} \rangle},
$$

(10.3.3)

where "$\circ$" denotes the Schur product (i.e., the product component by component) and $\langle , \rangle$ the scalar product (i.e., the sum of the products component by component).

The model (10.3.3) is a classical two-sex, one-locus model with male and female fitnesses given by the positive symmetric matrices $\mathbf{M} = \|m_{ij}\|$ and $\mathbf{F} = \|f_{ij}\|$, respectively. It was originally considered in the context of viability selection in dioecious populations and extended later to include multiplicative fecundities (see, e.g., Karlin 1978, Ewens 1979, and references therein). The case $f_{ij} = 1 - m_{ij}$ for all i, j corresponds to pure sex determination models (see, e.g., Eshel and Feldman 1982a).

If ($\mathbf{p}^*$, $\mathbf{q}^*$) is an equilibrium pair of frequency vectors for the model (10.3.3), then its stability with respect to small perturbations on the frequencies of alleles $A_1, \ldots, A_n$ is determined by the gradient matrix, denoted hereafter by $\mathbf{A}$, in the linear approximation of (10.3.3) near the equilibrium. *Stability*, actually local stability with a geometric rate of return to the equilibrium, occurs if and only if the spectral radius of $\mathbf{A}$, $\rho(\mathbf{A})$, is less than 1, that is the modulus of the largest eigenvalue of $\mathbf{A}$ in modulus is less than 1.

10.3.2. INITIAL INCREASE CONDITIONS FOR A MUTANT ALLELE

Let us introduce a new allele A_{n+1} in small frequency into a population at a stable equilibrium $(\mathbf{p}^*, \mathbf{q}^*)$ of (10.3.3). Then the first-order approximation for the transformation of allelic frequencies over two successive generations takes the form

$$
\begin{bmatrix} \mathbf{p}' - \mathbf{p}^* \\ \mathbf{q}' - \mathbf{q}^* \\ p'_{n+1} \\ q'_{n+1} \end{bmatrix} \cong \begin{bmatrix} \mathbf{A}_{2n \times 2n} & \mathbf{B}_{2n \times 2} \\ \mathbf{0}_{2 \times 2n} & \mathbf{C}_{2 \times 2} \end{bmatrix} \begin{bmatrix} \mathbf{p} - \mathbf{p}^* \\ \mathbf{q} - \mathbf{q}^* \\ p_{n+1} \\ q_{n+1} \end{bmatrix},
\tag{10.3.4}
$$

where $\mathbf{0}$ is a matrix with all zero entries and

$$
\mathbf{C} = \begin{bmatrix} \dfrac{m_{n+1}(\mathbf{q}^*)}{2m^*} & \dfrac{m_{n+1}(\mathbf{p}^*)}{2m^*} \\[2ex] \dfrac{f_{n+1}(\mathbf{q}^*)}{2f^*} & \dfrac{f_{n+1}(\mathbf{p}^*)}{2f^*} \end{bmatrix}
\tag{10.3.5}
$$

using the notation m^* and f^* for the male and female mean fitnesses at equilibrium, that is,

$$
m^* = \sum_{i,j=1}^{n} m_{ij} p_i^* q_j^*,
$$
$$
f^* = \sum_{i,j=1}^{n} f_{ij} p_i^* q_j^*,
\tag{10.3.6}
$$

while $m_{n+1}(\mathbf{p}^*)$ and $f_{n+1}(\mathbf{p}^*)$ represent the male and female marginal fitnesses of allele A_{n+1} carried by female gametes at equilibrium, namely,

$$
m_{n+1}(\mathbf{p}^*) = \sum_{j=1}^{n} m_{n+1,j} p_j^*,
$$
$$
f_{n+1}(\mathbf{p}^*) = \sum_{j=1}^{n} f_{n+1,j} p_j^*,
\tag{10.3.7}
$$

and similarly $m_{n+1}(\mathbf{q}^*)$ and $f_{n+1}(\mathbf{q}^*)$ for allele A_{n+1} carried by male gametes.

With the assumption that $\rho(\mathbf{A}) < 1$, the stability property of the equilibrium $(\mathbf{p}^*, \mathbf{q}^*)$ will be preserved following the introduction of A_{n+1} if and only if $\rho(\mathbf{C}) < 1$.

Since $\mathbf{C}$ is a positive matrix, there exist positive vectors $\mathbf{v}$ and $\mathbf{w}$, satisfying

$$\mathbf{u}^T\mathbf{v} = 1 \quad \text{and} \quad \mathbf{v}^T\mathbf{w} = 1 \quad (\mathbf{u} \text{ for the unit vector, } T \text{ for transpose}), \quad (10.3.8)$$

such that

$$\mathbf{Cv} = \rho(\mathbf{C})\mathbf{v} \quad \text{and} \quad \mathbf{w}^T\mathbf{C} = \rho(\mathbf{C})\mathbf{w}^T. \quad (10.3.9)$$

(See, e.g., Gantmacher 1959 for a review of matrix theory.) In particular, we have

$$\rho(\mathbf{C}) = \mathbf{u}^T\mathbf{Cv}$$

$$= v_1\left\{\frac{m_{n+1}(\mathbf{q}^*)}{2m^*} + \frac{f_{n+1}(\mathbf{q}^*)}{2f^*}\right\} + v_2\left\{\frac{m_{n+1}(\mathbf{p}^*)}{2m^*} + \frac{f_{n+1}(\mathbf{p}^*)}{2f^*}\right\}, \quad (10.3.10)$$

where v_1 and v_2 are the components of $\mathbf{v}$ with $v_1, v_2 > 0$ and $v_1 + v_2 = 1$. Denoting the matrix in (10.3.4) by $\mathbf{\Lambda}$, successive iterations give

$$\mathbf{\Lambda}^k = \begin{bmatrix} \mathbf{A}^k & \sum_{r=0}^{k-1} \mathbf{A}^{k-1-r}\mathbf{BC}^r \\ \mathbf{0} & \mathbf{C}^k \end{bmatrix}. \quad (10.3.11)$$

The ergodic theorem for the iterates of a positive matrix ensures

$$\frac{\mathbf{C}^k}{\rho(\mathbf{C})^k} \longrightarrow \mathbf{vw}^T \quad \text{as} \quad k \to \infty \quad (10.3.12)$$

for $\mathbf{v}$ and $\mathbf{w}$ of (10.3.9). Therefore, in the long run near the equilibrium $(\mathbf{p}^*, \mathbf{q}^*)$, the vector of the new components (p_{n+1}, q_{n+1}) in (10.3.4) is in the direction of the leading right eigenvector (v_1, v_2) of $\mathbf{C}$. Moreover,

$$p'_{n+1} + q'_{n+1} \cong p_{n+1}\left\{\frac{m_{n+1}(\mathbf{q}^*)}{2m^*} + \frac{f_{n+1}(\mathbf{q}^*)}{2f^*}\right\}$$

$$+ q_{n+1}\left\{\frac{m_{n+1}(\mathbf{p}^*)}{2m^*} + \frac{f_{n+1}(\mathbf{p}^*)}{2f^*}\right\}$$

$$\cong \rho(\mathbf{C})[p_{n+1} + q_{n+1}] \quad (10.3.13)$$

owing to (10.3.10). Since p_{n+1} and q_{n+1} represent the frequency of A_{n+1} in male and female gametes, respectively, the quantity $(p_{n+1} + q_{n+1})/2$ is the frequency of A_{n+1} at conception. The increase or decrease of this quantity when small but in the long run near an equilibrium (referred to as an *initial increase or decrease* throughout the paper) determines *invasion or extinction*, respectively, of allele A_{n+1} following its introduction. Invasion occurs (at a geometric rate) if $\rho(\mathbf{C}) > 1$ and extinction if $\rho(\mathbf{C}) < 1$.

The interpretation of $\rho(\mathbf{C})$ according to (10.3.10) is that of the mean reproductive value of allele A_{n+1} using the stable distribution of male and female gametes carrying A_{n+1} when rare and the Shaw-Mohler formula for the marginal reproductive value of male and female gametes. Equivalently,

$$\rho(\mathbf{C}) = \frac{m^*_{n+1}}{2m^*} + \frac{f^*_{n+1}}{2f^*}, \tag{10.3.14}$$

where m^*_{n+1} and f^*_{n+1} represent the overall marginal male and female fitnesses of A_{n+1} at equilibrium, that is,

$$\begin{aligned} m^*_{n+1} &= v_1 m_{n+1}(\mathbf{q}^*) + v_2 m_{n+1}(\mathbf{p}^*), \\ f^*_{n+1} &= v_1 f_{n+1}(\mathbf{q}^*) + v_2 f_{n+1}(\mathbf{p}^*), \end{aligned} \tag{10.3.15}$$

with $v_1, v_2 > 0$ satisfying (10.3.9) and $v_1 + v_2 = 1$, compared to the mean male and female fitnesses at equilibrium, m^* and f^*.

For comparisons, the reproductive value of an allele A_i represented at an equilibrium $(\mathbf{p}^*, \mathbf{q}^*)$ of (10.3.3) is

$$\frac{m^*_i}{2m^*} + \frac{f^*_i}{2f^*} = 1, \tag{10.3.16}$$

where

$$\begin{aligned} m^*_i &= \left(\frac{p^*_i}{p^*_i + q^*_i} \right) m_i(\mathbf{q}^*) + \left(\frac{q^*_i}{p^*_i + q^*_i} \right) m_i(\mathbf{p}^*), \\ f^*_i &= \left(\frac{p^*_i}{p^*_i + q^*_i} \right) f_i(\mathbf{q}^*) + \left(\frac{q^*_i}{p^*_i + q^*_i} \right) f_i(\mathbf{p}^*), \end{aligned} \tag{10.3.17}$$

with

$$\begin{aligned} m_i(\mathbf{p}^*) &= \sum_{j=1}^{n} m_{ij} p^*_j, \\ f_i(\mathbf{p}^*) &= \sum_{j=1}^{n} f_{ij} p^*_j, \end{aligned} \tag{10.3.18}$$

and similarly for $m_i(\mathbf{q}^*)$ and $f_i(\mathbf{q}^*)$.

Degenerate Case

If $\rho(\mathbf{C}) = 1$, then the linear approximation (10.3.4) fails to determine the initial fate of allele A_{n+1} following its introduction, and a quadratic analysis is required. This may be the case, for instance, if A_{n+1} is a recessive allele. This also has implications on the stability of nearby equilibria created by bifurcation as the value of $\rho(\mathbf{C})$ crosses 1.

When $\rho(C) = 1$ and $\rho(A) < 1$, we have

$$\Lambda^k \longrightarrow \begin{bmatrix} 0 & (I - A)^{-1}BC^* \\ 0 & C^* \end{bmatrix} = \Lambda^\infty \qquad \text{as} \quad k \to \infty, \qquad (10.3.19)$$

where $C^* = vw^T$. In fact, $\Lambda^\infty = \zeta\omega^T$, where

$$\zeta = \begin{bmatrix} (I - A)^{-1}Bv \\ v \end{bmatrix} \quad \text{and} \quad \omega = \begin{bmatrix} 0 \\ w \end{bmatrix} \qquad (10.3.20)$$

(**I** for the identity matrix and **0** for the zero vector), where ζ and ω are the right and left eigenvectors of Λ corresponding to the leading eigenvalue 1 and satisfying $\zeta^T\omega = 1$.

In the long run, the iterates of (10.3.4) will be oriented in the direction of the leading right eigenvector ζ and the coefficient with respect to ζ will obey the recurrence relationship

$$(w_1 p_{n+1} + w_2 q_{n+1})' \cong (w_1 p_{n+1} + w_2 q_{n+1})$$
$$+ \frac{(w_1 p_{n+1} + w_2 q_{n+1})^2}{2}$$
$$\times [w_1 Q_1(\zeta) + w_2 Q_2(\zeta)], \qquad (10.3.21)$$

where w_1 and w_2 are the components of the leading left eigenvector **w** of C, satisfying (10.3.8) with $w_1, w_2 > 0$, while Q_1 and Q_2 are the quadratic expressions in the Taylor expansions of p'_{n+1} and q'_{n+1} near equilibrium. Explicitly,

$$Q_1(\zeta) = \sum_{i,j=1}^{2n+2} \frac{\partial^2 p'_{n+1}}{\partial z_i \partial z_j} \zeta_i \zeta_j,$$
$$Q_2(\zeta) = \sum_{i,j=1}^{2n+2} \frac{\partial^2 q'_{n+1}}{\partial z_i \partial z_j} \zeta_i \zeta_j, \qquad (10.3.22)$$

where

$$z = (p_1 - p_1^*, \ldots, p_n - p_n^*, q_1 - q_1^*, \ldots, q_n - q_n^*, p_{n+1}, q_{n+1}) \qquad (10.3.23)$$

and the second derivatives of p'_{n+1} and q'_{n+1} are evaluated at $z = 0$.

In this case, there will be convergence to or divergence from the equilibrium $z = 0$ at an algebraic rate of degree 1 following the introduction of A_{n+1} accordingly as

$$Q(\zeta) = w_1 Q_1(\zeta) + w_2 Q_2(\zeta) < 0 \quad (> 0). \qquad (10.3.24)$$

(Compare with Lessard and Karlin 1982).

When a recessive allele A_2 is introduced into a population fixed for an allele A_1, then $m_{12} = m_{11}$, $f_{12} = f_{11}$, and the condition (10.3.24) takes the form

$$\frac{m_{22}}{2m_{11}} + \frac{f_{22}}{2f_{11}} < 1 \quad (>1). \tag{10.3.25}$$

General Case with Multiple Mutant Alleles

Even in the case $\rho(\mathbf{A}) \geq 1$ in (10.3.4), the linear approximation (10.3.13) for the frequency of a new allele A_{n+1} holds as long as the population is near the original equilibrium $(\mathbf{p}^*, \mathbf{q}^*)$. In all cases, initial increase of A_{n+1} at a geometric rate occurs if $\rho(\mathbf{C}) > 1$. In particular, this condition is also valid if A_{n+1} is introduced along with other alleles.

Let us resume.

RESULT 10.1. [See (10.3.15) and (10.3.22) for notation] *Let* $(\mathbf{p}^*, \mathbf{q}^*)$ *be a stable equilibrium of* (10.3.3). *A mutant allele* A_{n+1} *invades at a geometric rate if and only if*

$$\frac{m_{n+1}^*}{2m^*} + \frac{f_{n+1}^*}{2f^*} > 1. \tag{10.3.26}$$

It invades at an algebraic rate of degree 1 if and only if there is equality in (10.3.26) *and*

$$w_1 Q_1(\zeta) + w_2 Q_2(\zeta) > 0. \tag{10.3.27}$$

The condition (10.3.26) *is actually valid at any equilibrium* $(\mathbf{p}^*, \mathbf{q}^*)$.

10.3.3. PRODUCT CHARACTERIZATION FOR INVASION AS A MUTANT AND NONINVADABILITY AS AN EQUILIBRIUM

Assume that all possible (m_{ij}, f_{ij}) in (10.3.1) and (10.3.7) belong to a compact fitness set S as defined in (10.2.8) but independent of the population state. Let $E(S)$ be the set of all convex combinations of points in S, that is,

$$E(S) = \left\{ \sum_{i=1}^r \alpha_i(m_i, f_i) : (m_i, f_i) \in S, \alpha_i \geq 0, \sum_{i=1}^r \alpha_i = 1, r \geq 1 \right\}. \tag{10.3.28}$$

Consider the point $(\hat{m}, \hat{f})$ in $E(S)$ that maximizes the product $m \times f$, that is,

$$\max_{(m, f) \in E(S)} m \times f = \hat{m} \times \hat{f}. \tag{10.3.29}$$

We claim that every (m, f) in $E(S)$ belongs to the half-plane

$$\hat{f}(m - \hat{m}) + \hat{m}(f - \hat{f}) \leq 0, \tag{10.3.30}$$

that is,

$$\frac{m}{2\hat{m}} + \frac{f}{2\hat{f}} \leq 1. \qquad (10.3.31)$$

For every point on the line $t(m, f) + (1 - t)(\hat{m}, \hat{f})$ for $0 \leq t \leq 1$ is in the convex set $E(S)$ if (m, f) is, and the product of the components along this line has $\hat{f}(m - \hat{m}) + \hat{m}(f - \hat{f})$ as a derivative with respect to t at $t = 0$. But such a derivative is ≤ 0 because of (10.3.29). This completes the proof of our claim.

The converse is also true. If (10.3.31) holds for some $(\hat{m}, \hat{f})$ in $E(S)$, then the geometric-arithmetic mean inequality ensures

$$\left(\frac{m \times f}{\hat{m} \times \hat{f}}\right)^{1/2} \leq \frac{m}{2\hat{m}} + \frac{f}{2\hat{f}} \leq 1 \qquad (10.3.32)$$

for every (m, f) in $E(S)$, which entails (10.3.29). Note that the property (10.3.31) has only to hold for all (m, f) in S, since then it necessarily holds for all (m, f) in the convex hull $E(S)$ by linearity.

Moreover, we have equalities in (10.3.32) if and only if $m/\hat{m} = f/\hat{f}$ and $m \times f = \hat{m} \times \hat{f}$. This is possible only in the case $m = \hat{m}$ and $f = \hat{f}$. In particular, this shows that $(\hat{m}, \hat{f})$ is unique.

Another consequence of (10.3.32) is that

$$\frac{\hat{m}}{2m} + \frac{\hat{f}}{2f} \geq \left(\frac{\hat{m} \times \hat{f}}{m \times f}\right)^{1/2} \geq 1 \qquad (10.3.33)$$

for every (m, f) in $E(S)$ with equality if and only if $(m, f) = (\hat{m}, \hat{f})$.

On the other hand, the property

$$\frac{\hat{m}}{2[\hat{m} + t(m - \hat{m})]} + \frac{\hat{f}}{2[\hat{f} + t(f - \hat{f})]} \geq 1, \qquad (10.3.34)$$

that is,

$$\frac{m - \hat{m}}{2[\hat{m} + t(m - \hat{m})]} + \frac{f - \hat{f}}{2[\hat{f} + t(f - \hat{f})]} \leq 0, \qquad (10.3.35)$$

for every (m, f) in $E(S)$ and $0 < t \leq 1$ entails (10.3.30) as $t \to 0$.

In conclusion, we get a twofold characterization for the maximum point of the product $m \times f$ in $E(S)$.

RESULT 10.2. $(\hat{m}, \hat{f})$ in $E(S)$ maximizes $m \times f$ for all (m, f) in $E(S)$ if and only if

$$\frac{m}{2\hat{m}} + \frac{f}{2\hat{f}} \leq 1 \qquad (10.3.36)$$

for all (m, f) in S. Moreover $(\hat{m}, \hat{f})$ is unique. A condition equivalent to (10.3.36) is

$$\frac{\hat{m}}{2m} + \frac{\hat{f}}{2f} \geq 1 \qquad (10.3.37)$$

for every (m, f) in $E(S)$ with equality if and only if $(m, f) = (\hat{m}, \hat{f})$.

There are three possibilities for the position of $(\hat{m}, \hat{f})$ with respect to S, $E(S)$, and the line

$$L = \left\{ (m, f) : \frac{m}{2\hat{m}} + \frac{f}{2\hat{f}} = 1 \right\}. \qquad (10.3.38)$$

These are

(1) $(\hat{m}, \hat{f})$ is in S and the unique point in $L \cap E(S)$.
(2) $(\hat{m}, \hat{f})$ is in S and belongs to a nondegenerate segment $L \cap E(S)$.
(3) $(\hat{m}, \hat{f})$ is not in S and belongs to a nondegenerate segment $L \cap E(S)$.

(See Figure 10.1).

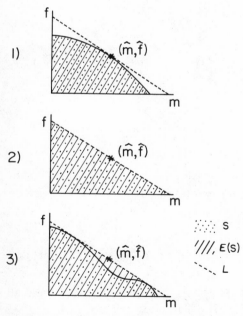

Figure 10.1. Three examples for the location of $(\hat{m}, \hat{f})$ that maximizes $m \times f$ on $E(S)$ with respect to the fitness set S, its convex hull $E(S)$, and the line L of (10.3.38). (See Charnov 1982 for biological interpretations.)

We are now prepared to deduce optimality properties concerning invasion of a mutant allele and noninvadability of an equilibrium. By appeal to Results 10.1 and 10.2, an equilibrium $(\mathbf{p}^*, \mathbf{q}^*)$ of (10.3.3) with fitness set S and mean male and female fitnesses $m^* = \hat{m}$ and $f^* = \hat{f}$, respectively, cannot be invaded geometrically fast by a new allele. There exist no other equilibria having this property since a mutant allele with marginal fitness pair $(m^*_{n+1}, f^*_{n+1}) = (\hat{m}, \hat{f})$ invades geometrically fast any other equilibrium. Besides, the condition

$$\sum_{i,j=1}^{n} \left(\frac{m_{ij}}{2\hat{m}} + \frac{f_{ij}}{2\hat{f}} \right) p_i^* q_j^* = \frac{m^*}{2\hat{m}} + \frac{f^*}{2\hat{f}} = 1 \qquad (10.3.39)$$

compels equality in (10.3.36) for every (m_{ij}, f_{ij}) wherever $p_i^* q_j^* > 0$. Let us summarize.

RESULT 10.3. *The only equilibria* $(\mathbf{p}^*, \mathbf{q}^*)$ *of* (10.3.3) *with fitness set S that cannot be invaded geometrically fast by a new allele have* $(m^*, f^*) = (\hat{m}, \hat{f})$. *Moreover, every* (m_{ij}, f_{ij}) *represented at such an equilibrium belongs to the line L, that is,*

$$\frac{m_{ij}}{2\hat{m}} + \frac{f_{ij}}{2\hat{f}} = 1. \qquad (10.3.40)$$

If $(m^*, f^*) \neq (\hat{m}, \hat{f})$, *then a mutant allele A_{n+1} that has* $(m^*_{n+1}, f^*_{n+1}) = (\hat{m}, \hat{f})$ *invades geometrically fast the equilibrium* $(\mathbf{p}^*, \mathbf{q}^*)$.

10.3.4. DOES THE PRODUCT $m \times f$ INCREASE OVER SUCCESSIVE EQUILIBRIA?

The linear model (10.3.40) was studied in detail in Karlin and Lessard (1983, 1984, 1986) in a context of sex-ratio determination models as introduced in Eshel and Feldman (1982a).

If the fitness set S is restricted to L, the equations (10.3.3) have

$$\frac{\mathbf{M}}{2\hat{m}} + \frac{\mathbf{F}}{2\hat{f}} = \mathbf{U} \qquad \text{(U for the unit matrix).} \qquad (10.3.41)$$

For technical reasons [see (10.3.45) below], it is assumed that $\mathbf{M}$ and all principal submatrices of $\mathbf{M}$ are nonsingular, and similarly for $\mathbf{F}$.

Phenotypic equilibria $(\mathbf{p}^*, \mathbf{q}^*)$ *satisfying* $(m^*, f^*) = (\hat{m}, \hat{f})$, *that is,* $\langle \mathbf{p}^*, \mathbf{M}\mathbf{q}^* \rangle = \hat{m}$ *and* $\langle \mathbf{p}^*, \mathbf{F}\mathbf{q}^* \rangle = \hat{f}$, *are characterized by the equation*

$$\rho[\text{diag}(\mathbf{M}\mathbf{q}^*) + \text{diag}(\mathbf{q}^*)\mathbf{M}] = 2\hat{m}, \qquad (10.3.42)$$

with corresponding right eigenvector $\mathbf{p}^*$, or equivalently,

$$\rho[\text{diag}(\mathbf{F}\mathbf{p}^*) + \text{diag}(\mathbf{p}^*)\mathbf{F}] = 2\hat{f}, \tag{10.3.43}$$

with corresponding right eigenvector $\mathbf{q}^*$ [$\text{diag}(\mathbf{x})$ for diagonal matrix with $\mathbf{x}$ on the main diagonal], owing to the Perron-Frobenius theory for the spectral radius of nonnegative matrices. [See (10.3.48) below for the existence of such equilibria.]

Apart from phenotypic equilibria, there are *genotypic equilibria* $(\tilde{\mathbf{p}}, \tilde{\mathbf{p}})$ with $(\tilde{m}, \tilde{f}) \neq \hat{m}, \hat{f})$ also called *symmetric* (Eshel and Feldman 1982a), satisfying

$$\tilde{\mathbf{p}} = \frac{\tilde{\mathbf{p}} \circ \mathbf{M}\tilde{\mathbf{p}}}{\langle \tilde{\mathbf{p}}, \mathbf{M}\tilde{\mathbf{p}} \rangle}. \tag{10.3.44}$$

To avoid degeneracies in linear approximations for stability analyses, it is assumed that the eigenvalues of the gradient matrix at every genotypic equilibrium are all $\neq 1$.

If a *polymorphic* (i.e., with all alleles represented) genotypic equilibrium exists, it is unique under the assumption that $\mathbf{M}$ is invertible and given explicitly by the positive frequency vector,

$$\tilde{\mathbf{p}} = \frac{\mathbf{M}^{-1}\mathbf{u}}{\langle \mathbf{u}, \mathbf{M}^{-1}\mathbf{u} \rangle}. \tag{10.3.45}$$

Suppose that such an equilibrium is stable. (This precludes the existence of phenotypic equilibria in the same allelic system; see Karlin and Lessard 1984.) According to (10.3.10), a new allele A_{n+1} will invade at a geometric rate if and only if

$$\frac{m_{n+1}(\tilde{\mathbf{p}})}{2\tilde{m}} + \frac{f_{n+1}(\tilde{\mathbf{p}})}{2\tilde{f}} > 1 \tag{10.3.46}$$

in the notation (10.3.7). In such a case, it can be shown (from Karlin and Lessard 1984) that in the augmented allelic system, from n to $n + 1$ alleles, either

(1) there exists a unique stable genotypic equilibrium $(\tilde{\tilde{\mathbf{p}}}, \tilde{\tilde{\mathbf{p}}})$ with $(\tilde{\tilde{m}}, \tilde{\tilde{f}})$ satisfying

$$\tilde{m} \times \tilde{f} < \tilde{\tilde{m}} \times \tilde{\tilde{f}} < \hat{m} \times \hat{f} \tag{10.3.47}$$

which does not coexist with any phenotypic equilibrium, or
(2) there exist phenotypic equilibria and every genotypic equilibrium $(\tilde{\tilde{\mathbf{p}}}, \tilde{\tilde{\mathbf{p}}})$ is unstable or completely separated from $(\tilde{\mathbf{p}}, \tilde{\mathbf{p}})$ by an equilibrium surface of phenotypic equilibria. This means that every continuous path of frequency vectors joining $\tilde{\tilde{\mathbf{p}}}$ and $\tilde{\mathbf{p}}$ meets at least one point $\mathbf{p}^*$ corresponding to a phenotypic equilibrium.

A tractable criterion for the existence of a phenotypic equilibrium surface separating two genotypic equilibria $(\tilde{\mathbf{p}}, \tilde{\mathbf{p}})$ and $(\tilde{\tilde{\mathbf{p}}}, \tilde{\tilde{\mathbf{p}}})$ is that

$$(\hat{m}, \hat{f}) = t(\tilde{m}, \tilde{f}) + (1 - t)(\tilde{\tilde{m}}, \tilde{\tilde{f}}) \tag{10.3.48}$$

for some $0 < t < 1$ (Lessard 1986).

Apart from isolated phenotypic equilibria in the form $(\mathbf{p}^*, \mathbf{p}^*)$, the condition (10.3.48) is necessary and sufficient for the existence of phenotypic equilibria. (See Lessard 1986 for more details and many variants.)

In many genetic models, the only genotypic equilibria are the fixation states. This is the case, for instance, in serial dominance models $(m_{ij} = m_{\min(i,j)})$, additive models $(m_{ij} = m_i + m_j)$ and multiplicative models $(m_{ij} = m_i m_j)$. [In all three cases it is assumed that $m_i \neq m_j$ for all $i \neq j$, and then the equilibrium condition (10.3.44) cannot be satisfied by a frequency vector having two or more positive components.]

The equilibrium configurations that are generally possible after the introduction of an invading allele at any genotypic equilibrium suggest that the only attainable equilibria following invasion exhibit a product $m \times f$ larger than it was at the invaded equilibrium. Convergence according to this theme was confirmed by simulations in general cases and proved analytically in the case where the invaded equilibrium is a fixation state of a single allele. Over successive equilibria, as new alleles are introduced one at a time, convergence to a phenotypic equilibrium surface on which $m \times f$ is globally maximized is generally expected.

The above conclusion is valid for a fitness set S reduced to any segment

$$S = \{t(m_1, f_1) + (1 - t)(m_2, f_2) : 0 \leq t \leq 1\} \tag{10.3.49}$$

except that the phenotypic equilibrium associated with $(\hat{m}, \hat{f})$ corresponds to a fixation state if $(\hat{m}, \hat{f}) = (m_1, f_1)$ or (m_2, f_2).

In the case of a fitness set reduced to a segment in the form

$$S = \{t(\hat{m}, \hat{f}) : 0 \leq t \leq 1\}, \tag{10.3.50}$$

which compels $\hat{m} f_{ij} = \hat{f} m_{ij}$ for all i, j, the product $m \times f$ even increases over successive generations (and then necessarily over successive equilibria) due to the Fundamental Theorem of Natural Selection (see, e.g., Crow and Kimura 1970; Karlin 1978; Ewens 1979; and references therein).

With a general fitness set S, the problem of the increase of $m \times f$ over successive equilibria is more delicate. At least in one case, this problem can easily be solved, namely, when $(\hat{m}, \hat{f})$ belongs to S and a dominant allele A_{n+1} with

$$m_{n+1,i} = \hat{m} \quad \text{and} \quad f_{n+1,i} = \hat{f} \quad \text{for} \quad i = 1, \ldots, n \tag{10.3.51}$$

is introduced into a population where alleles $A_1, \ldots, A_n$ are already represented. Then over two successive generations and in the notation of (10.3.1), we have

$$p'_{n+1} + q'_{n+1} = (p_{n+1} + q_{n+1})\left(\frac{\hat{m}}{2m} + \frac{\hat{f}}{2f}\right)$$

$$\geq p_{n+1} + q_{n+1}, \qquad (10.3.52)$$

with equality if and only if $(m, f) = (\hat{m}, \hat{f})$, due to Result 10.2. Therefore, any attainable equilibrium following the introduction of A_{n+1} must satisfy this optimality condition.

10.3.5. INITIAL INCREASE OF $m \times f$ FOLLOWING INVASION AT A SYMMETRIC EQUILIBRIUM

Consider the model (10.3.3) in the case of $(n + 1)$ alleles. Let $(\mathbf{p}^*, \mathbf{q}^*)$ be an equilibrium where alleles $A_1, \ldots, A_n$ are represented but not A_{n+1}, that is, $p^*_{n+1} = q^*_{n+1} = 0$, while all other components are positive. Writing $\mathbf{p} = \mathbf{p}^* + \boldsymbol{\xi}$ and $\mathbf{q} = \mathbf{q}^* + \boldsymbol{\eta}$ near $(\mathbf{p}^*, \mathbf{q}^*)$, we have

$$\begin{aligned} m &= \langle \mathbf{p}, \mathbf{Mq} \rangle \\ &= \langle \mathbf{p}^*, \mathbf{Mq}^* \rangle + \langle \boldsymbol{\xi}, \mathbf{Mq}^* \rangle + \langle \mathbf{p}^*, \mathbf{M}\boldsymbol{\eta} \rangle + \langle \boldsymbol{\xi}, \mathbf{M}\boldsymbol{\eta} \rangle \\ &\cong m^* + \langle \boldsymbol{\xi}, \mathbf{Mq}^* \rangle + \langle \boldsymbol{\eta}, \mathbf{Mp}^* \rangle, \end{aligned} \qquad (10.3.53)$$

and similarly for f, so that

$$m \times f \cong m^* \times f^*\left[1 + 2\left\langle \boldsymbol{\xi}, \frac{\mathbf{Mq}^*}{2m^*} + \frac{\mathbf{Fq}^*}{2f^*} \right\rangle + 2\left\langle \boldsymbol{\eta}, \frac{\mathbf{Mp}^*}{2m^*} + \frac{\mathbf{Fp}^*}{2f^*} \right\rangle\right]. \qquad (10.3.54)$$

Assuming that the equilibrium $(\mathbf{p}^*, \mathbf{q}^*)$ is *symmetric*, that is, $\mathbf{p}^* = \mathbf{q}^*$, a necessary condition for equilibrium is

$$\frac{(\mathbf{Mp}^*)_i}{2m^*} + \frac{(\mathbf{Fp}^*)_i}{2f^*} = 1 \qquad \text{for} \quad i = 1, \ldots, n, \qquad (10.3.55)$$

while the quantity

$$\frac{(\mathbf{Mp}^*)_{n+1}}{2m^*} + \frac{(\mathbf{Fp}^*)_{n+1}}{2f^*} = \rho \qquad (10.3.56)$$

represents the rate of increase or decrease in the frequency of A_{n+1} when rare. Since $\xi_{n+1} = -\sum_{i=1}^{n} \xi_i$ and $\eta_{n+1} = -\sum_{i=1}^{n} \eta_i$, we get

$$m \times f \cong m^* \times f^*[1 + 2(\rho - 1)(\xi_{n+1} + \eta_{n+1})]. \qquad (10.3.57)$$

Assuming $\rho \neq 1$, we conclude that $m \times f > m^* \times f^*$ as A_{n+1} is introduced if and only if $\rho > 1$. In such a case, the product $m \times f$ increases over two successive generations if and only if $(\xi_{n+1} + \eta_{n+1})$, that is $(p_{n+1} + q_{n+1})$, has the same property. This is the case after some generations as long as A_{n+1} is rare and the population state is near $(\mathbf{p}^*, \mathbf{p}^*)$ [see (10.3.13)].

In particular, if the equilibrium $(\mathbf{p}^*, \mathbf{p}^*)$ is stable when $\rho < 1$ and a nearby equilibrium is created by bifurcation as ρ crosses the value 1, then this equilibrium, which is stable when $\rho > 1$ and close to 1 (and to which there is convergence following departure from $(\mathbf{p}^*, \mathbf{p}^*)$ under the same conditions), will exhibit a product $m \times f$ larger than $m^* \times f^*$.

Let us highlight the main finding of this section.

RESULT 10.4. *Assuming nondegeneracies in linear approximations, a new allele A_{n+1} invades a symmetric equilibrium of* (10.3.3) *with n alleles represented if and only if it increases the product $m \times f$.*

The symmetry condition in Result 10.4 can be relaxed if the increase of $m \times f$ is understood locally in the long run since the approximation (10.3.57) with ρ given in (10.3.10) holds at least in the long run near any equilibrium $(\mathbf{p}^*, \mathbf{q}^*)$ where both $\mathbf{p}^*$ and $\mathbf{q}^*$ satisfy (10.3.55). This is the case, for example, when

$$\frac{m_{ij}}{2m^*} + \frac{f_{ij}}{2f^*} = 1 \qquad \text{for} \quad i, j = 1, \ldots, n. \qquad (10.3.58)$$

[See (10.3.60) for more details.]

10.3.6. Is the Product $m \times f$ Maximized at a Stable Equilibrium?

In this section, we examine the conditions for an equilibrium to maximize the product $m \times f$ within its own allelic system.

Consider an equilibrium $(\mathbf{p}^*, \mathbf{q}^*)$ for the n-allele model (10.3.3) with all positive components, and let (ξ, η) be small perturbations on these components. According to the first order approximation (10.3.54), the product $m \times f$ can be locally maximized at $(\mathbf{p}^*, \mathbf{q}^*)$ only if

$$\left\langle \xi, \frac{\mathbf{Mq}^*}{2m^*} + \frac{\mathbf{Fq}^*}{2f^*} \right\rangle \leq 0 \quad \text{and} \quad \left\langle \eta, \frac{\mathbf{Mp}^*}{2m^*} + \frac{\mathbf{Fp}^*}{2f^*} \right\rangle \leq 0 \quad (10.3.59)$$

for all $\xi \perp \mathbf{u}$ and $\eta \perp \mathbf{u}$ ($\perp \mathbf{u}$ for a vector perpendicular to the unit vector). We actually have equalities in (10.3.59) (take $-\xi$ and $-\eta$) and therefore

$$\frac{\mathbf{Mq}^*}{2m^*} + \frac{\mathbf{Fq}^*}{2f^*} = \frac{\mathbf{Mp}^*}{2m^*} + \frac{\mathbf{Fp}^*}{2f^*} = \mathbf{u}, \qquad (10.3.60)$$

with $m^* = \langle \mathbf{p}^*, \mathbf{M}\mathbf{q}^* \rangle$ and $f^* = \langle \mathbf{p}^*, \mathbf{F}\mathbf{q}^* \rangle$ as normalizing constants. Note that the condition (10.3.60) ensures only that $(\mathbf{p}^*, \mathbf{q}^*)$ is a critical point of $m \times f$.

We will examine two conditions:

(1)
$$\frac{\mathbf{M}}{2m^*} + \frac{\mathbf{F}}{2f^*} = \mathbf{U},$$

(2)
$$\mathbf{p}^* = \mathbf{q}^*.$$

In Condition (2), we have the equilibrium conditions

$$\mathbf{M}\mathbf{p}^* = m^*\mathbf{u} \quad \text{and} \quad \mathbf{F}\mathbf{p}^* = f^*\mathbf{u}. \tag{10.3.61}$$

Then a second-order approximation for $m \times f$ gives

$$m \times f \cong m^* \times f^* \left[1 + 2\left\langle \xi, \left(\frac{\mathbf{M}}{2m^*} + \frac{\mathbf{F}}{2f^*} \right)\eta \right\rangle \right]. \tag{10.3.62}$$

For $m^* \times f^*$ to be a local maximum (actually not a saddle point), we need

$$\left(\frac{\mathbf{M}}{2m^*} + \frac{\mathbf{F}}{2f^*} \right)\eta = c\mathbf{u} \tag{10.3.63}$$

for every $\eta \perp \mathbf{u}$ and some constant c. With the condition (10.3.60) in force, we actually have (10.3.63) for all η. Then every column of $\mathbf{M}/m^* + \mathbf{F}/f^*$ is a multiple of $\mathbf{u}$. This matrix being symmetric, the only possibility is to be a multiple of $\mathbf{U}$, actually $2\mathbf{U}$ because of the normalizing factors m^* and f^*. Then we are in the case of Condition (1). Of course, in this case, $m^* \times f^*$ is a global maximum on

$$D = \left\{ \sum_{i,j=1}^{n} p_i q_j(m_{ij}, f_{ij}) : p_i \geq 0, \sum_{i=1}^{n} p_i = 1, q_j \geq 0, \sum_{j=1}^{n} q_j = 1 \right\}$$

by appeal to Result 10.2. It is conjectured that Condition (1) is necessary and sufficient for this to happen. But it is possible to find stable symmetric equilibria $(\mathbf{p}^*, \mathbf{p}^*)$ that do not satisfy Condition (1) and then cannot maximize $m \times f$ on D. For instance, if $\mathbf{F} = \mathbf{U} - \mathbf{M}$, then Condition (1) becomes $(1 - 2m^*)\mathbf{M} = m^*(1 - 2m^*)\mathbf{U}$, which is precluded if $m^* \neq 1/2$ and $\mathbf{M} \neq m^*\mathbf{U}$. Such symmetric equilibria are known to exist and even to be the rule (see, e.g., Karlin and Lessard 1986).

RESULT 10.5. *A stable equilibrium $(\mathbf{p}^*, \mathbf{q}^*)$ of (10.3.3), even symmetric, may not be a local maximum of $m \times f$ with respect to perturbations on its positive components.*

10.3.7. REMARK ON THE OCCURRENCE OF PARTICULAR RELATIONSHIPS AT EQUILIBRIUM

When the entries of **M** and **F** are chosen at random in a general fitness set S (i.e., with a nonvoid interior or without any linear relationship), Conditions (1) and (2) of Section 10.3.6 are unlikely to happen at equilibrium apart from fixation states, for Condition (1) says that all (m_{ij}, f_{ij}) are on the same line, which is almost surely precluded with three points or more (i.e., two alleles or more if there is no dominance), while Condition (2) entails that a same combination of columns of **M** and **F** gives a multiple of **u** [see (10.3.61)], which is nongeneric for linearly unrelated **M** and **F**.

10.3.8. SIMULATIONS

In order to check whether or not the product $m \times f$ increases over successive equilibria and under what conditions, we ran simulations with eight different fitness sets, as illustrated in Figure 10.2. In Case 1 we chose (m_{ij}, f_{ij}) at random in the whole square $[0, 1] \times [0, 1]$, while in each other case we imposed a functional relationship $f_{ij} = f(m_{ij})$ to our choice. The

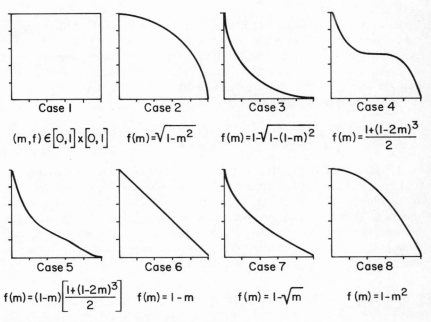

Figure 10.2. Fitness sets considered for simulations.

functions f considered are given in Figure 10.2. With reference to the sets of the first quadrant these functions bound above, two cases are convex (2 and 8), two concave (3 and 7), two are half-convex, half-concave (4 and 5), and one is linear (6). Cases 6, 7, and 8 belong to Charnov's class (10.2.7).

In our simulations, we started from a fixation state chosen at random, then a second allele was introduced and equations (10.3.1) iterated until an equilibrium was reached, then a third allele was introduced, and so on. For each case, we did this fifty times for five alleles with a new set of (m_{ij}, f_{ij}) each time.

The main conclusions are the following:

1. No symmetric equilibria other than fixation states were observed except in the linear Case (6).
2. In all cases, departure from a fixation state led to an equilibrium with higher $m \times f$.
3. In Cases 1, 2, 4, 6, the product $m \times f$ never decreased over successive equilibria.
4. In Cases 3, 5, 7, 8, the product $m \times f$ decreased a few times but only by a small amount each time from one equilibrium to the next.
5. In general, the increase of $m \times f$ over successive equilibria is more rapid and more effective (in approaching the optimal value) in "convex" cases (2, 8, and also 4) than in concave cases.
6. On the other hand, there are more alleles represented at equilibrium (higher polymorphism) and more equilibrium changes in "concave" cases (3, 5, 7).
7. The representation of more than four alleles at equilibrium practically never occurs except in the linear case (6).
8. In agreement with previous simulations in Case 6 (Karlin and Lessard 1983), an equilibrium with $m \times f = 1/4$ is attained very rapidly (following the introduction of the second or third allele), and then the introduction of any more alleles only slightly displaces the equilibrium achieved, maintaining $m \times f = 1/4$.

(See Tables 10.1, 10.2, 10.3 for more details).

To summarize, although the product $m \times f$ may decrease over two successive equilibria, such a decrease is rare and small. Note also that this occurred almost exclusively in "concave" cases (3, 5, 7). Even then, the increase tendency takes over rapidly and is definitely overwhelming in the long run.

Other observations can be made from Tables 10.1, 10.2, 10.3:

1. It is almost impossible to distinguish Case 2 from Case 1 on the basis of polymorphism. Also, the product $m \times f$ turns out to have

Table 10.1

Data on successive equilibria obtained by simulations.

Case	Fitness set Type	Number of equilibria reached according to the number of alleles represented (total = 200)[a]					Mean number of equilibrium changes (max = 4)[b]	Mean final number of alleles represented at equilibrium (max = 5)[b]
		1	2	3	4	5		
1	Square	67	105	26	2	0	1.5	2.0
2	Convex	69	107	24	0	0	1.5	2.0
3	Concave	21	143	33	3	0	2.1	2.3
4	Half convex, half concave	86	88	24	2	0	1.3	1.9
5[c]	Half weakly convex, half weakly concave	11	144	43	2	0	2.3	2.4
6	Linear	28	50	47	44	31	1.2	4.4
7	Weakly concave	9	145	46	0	0	2.1	2.4
8	Weakly convex	55	112	31	1	1	1.7	2.1

[a] Counting multiplicities as new alleles (up to five) are introduced, but excluding the starting fixation state. Keep in mind that an allele not represented at equilibrium is never completely eliminated once introduced into the system because of stopping rules for iterations in simulations. The whole procedure is repeated fifty times for each fitness set.

[b] After the introduction of the 5th allele.

[c] Case 5 is rather 80% concave.

Table 10.2

Data on the behavior of $m \times f$ over successive equilibria obtained by simulations.

Case	Fitness set Type	Initial value of $m \times f$ [a]	Final value of $m \times f$	Optimal value of $m \times f$ [b]	Point (m, f) where $m \times f$ is optimal	Number of decreases of $m \times f$ over two successive equilibria (max = 200)	Decreases of $m \times f$ over two successive equilibria
1	Square	0.249 (0.217)	0.485 (0.177)	1.000	$(1, 1)$	0	
2	Convex	0.351 (0.144)	0.475 (0.021)	0.500	$(1/\sqrt{2}, 1/\sqrt{2})$	0	
3	Concave	0.051 (0.030)	0.145 (0.041)	0.250	$(1/2, 1/2)$	2	0.00026 0.00434
4	Half convex, half concave	0.215 (0.102)	0.309 (0.021)	0.333	$(2/3, 1/2)$	0	
5	Half weakly convex, half weakly concave	0.083 (0.040)	0.178 (0.029)	0.250	$(1/2, 1/2)$	4	0.00052 0.00318 0.00006 0.00285
6	Linear	0.169 (0.080)	2.50 (0.001)	0.250	$(1/2, 1/2)$	0	
7	Weakly concave	0.102 (0.045)	0.182 (0.016)	0.250	$(1/2, 1/2)$	2	0.00015 0.00013
8	Weakly convex	0.266 (0.111)	0.372 (0.019)	0.385	$(1/\sqrt{3}, 2/3)$	1	0.00005

[a] Mean and standard deviation in fifty runs introducing five alleles successively up to equilibrium each run.
[b] Maximal value of $m \times f$ over the convex hull of fitness set.

Table 10.3a

An example of decrease of $m \times f$ over two successive equilibria.

Fitness set 3[a]	Male fitness matrix M					Female fitness matrix F				
	0.77349	0.01870	0.76593	0.98104	0.80716	0.02599	0.80751	0.02778	0.00018	0.01877
	0.01870	0.29510	0.97487	0.32393	0.90247	0.80751	0.29069	0.00032	0.26316	0.00477
	0.76593	0.97487	0.60850	0.32640	0.89234	0.02778	0.00032	0.07982	0.26090	0.00581
	0.98104	0.32393	0.32640	0.83393	0.50878	0.00018	0.26316	0.26090	0.01389	0.12896
	0.80716	0.90247	0.89234	0.50878	0.06351	0.01877	0.00477	0.00581	0.12896	0.64930

Equilibrium No.[b]	Frequencies in male gametes					Frequencies in female gametes					Product $m \times f$
	p_1	p_2	p_3	p_4	p_5	q_1	q_2	q_3	q_4	q_5	
1	1.00000	0.00000	0.00000	0.00000	0.00000	1.00000	0.00000	0.00000	0.00000	0.00000	0.02010
2	0.98231	0.01769	0.00000	0.00000	0.00000	0.51294	0.48706	0.00000	0.00000	0.00000	0.16428
3	0.50632	0.24383	0.24985	0.00000	0.00000	0.45468	0.54049	0.00483	0.00000	0.00000	0.15994[c]
4	0.50632	0.24383	0.24985	0.00000	0.00000	0.45468	0.54049	0.00483	0.00000	0.00000	0.15994
5	0.50632	0.24383	0.24985	0.00000	0.00000	0.45468	0.54049	0.00483	0.00000	0.00000	0.15994

[a] The male and female fitness matrices $M = \|m_{ij}\|$ and $F = \|f_{ij}\|$ are obtained by choosing (m_{ij}, f_{ij}) at random in fitness set 3.

[b] Alleles $A_1, \ldots, A_5$ are introduced one at a time in this order and lead successively (Eq. 10.3.1) to the equilibria $1, \ldots, 5$.

[c] Decrease of 0.00434 in the product of the male and female mean fitnesses at equilibrium.

Table 10.3b

Another example of decrease of $m \times f$ over two successive equilibria.

Fitness set 5[a]

Male fitness matrix M

0.83837	0.41327	0.76741	0.70193	0.11142
0.41327	0.78246	0.81671	0.33932	0.65645
0.76741	0.81671	0.46413	0.00278	0.87933
0.70193	0.33932	0.00278	0.78539	0.85834
0.11142	0.65645	0.87933	0.85834	0.06122

Female fitness matrix F

0.05577	0.29490	0.09851	0.13922	0.65282
0.29490	0.08916	0.06835	0.34130	0.16652
0.09851	0.06835	0.26803	0.98893	0.03399
0.13922	0.34130	0.98893	0.08735	0.04476
0.65283	0.16652	0.03399	0.04476	0.78662

Equilibrium No.[a]	Frequencies in male gametes					Frequencies in female gametes					Product $m \times f$
	p_1	p_2	p_3	p_4	p_5	q_1	q_2	q_3	q_4	q_5	
1	1.00000	0.00000	0.00000	0.00000	0.00000	1.00000	0.00000	0.00000	0.00000	0.00000	0.04676
2	0.45811	0.54189	0.00000	0.00000	0.00000	0.46138	0.53862	0.00000	0.00000	0.00000	0.11259
3	0.45811	054189	0.00000	0.00000	0.00000	0.46138	0.53862	0.00000	0.00000	0.00000	0.11259
4	0.00001	0.00000	0.00286	0.99712	0.00000	0.00000	0.00000	0.45221	0.54779	0.00000	0.21351
5	0.00000	0.00000	0.14787	0.69777	0.15436	0.00000	0.00000	0.48186	0.50884	0.00930	0.21033[b]

[a] See Table 10.3a for details.
[b] Decrease of 0.00318 in $m \times f$.

comparable means after the introduction of five alleles but with quite different variances (higher in Case 1, as could be expected).

2. Cases 5 and 7 are not very different from each other, but Case 5 leads to more equilibrium changes and more counterexamples of the increase of $m \times f$ over successive equilibria.

3. Among the "concave" cases (3, 5, 7), the weakly concave ones (5 and 7) favor a higher polymorphism and a product $m \times f$ closer to the optimal value with less variance.

4. The weakly convex case (8) favors more equilibrium changes, a higher polymorphism, and a product $m \times f$ closer to the optimal value than does the convex case (2).

5. Case 4, which combines strong convexity and strong concavity, promotes very little polymorphism but leads quickly to a product $m \times f$ close to the optimal value.

6. Most examples of decreasing $m \times f$ over two consecutive equilibria are found in "concave" cases (3, 5, 7).

10.4. Multilocus Resource Allocation Models with Frequency-Independent Fitnesses

A general formulation of two-sex multilocus models extending (10.3.3) can be found in Karlin and Liberman (1979). These authors were concerned with stability conditions at central Hardy-Weinberg equilibria in systems acted on by nonepistatic selection forces coupled to recombination events. We adhere to the more general formulation of the model to study initial increase properties of mutant alleles in multilocus systems in extension of Section 10.3.2.

Let $m(\mathbf{i}^{(0)}, \mathbf{i}^{(1)})$ and $f(\mathbf{i}^{(0)}, \mathbf{i}^{(1)})$ be constant male and female fitnesses associated with a genotype $(\mathbf{i}^{(0)}, \mathbf{i}^{(1)})$ of male and female parental gametes $\mathbf{i}^{(0)} = (i_1^{(0)}, \ldots, i_L^{(0)})$ and $\mathbf{i}^{(1)} = (i_1^{(1)}, \ldots, i_L^{(1)})$, respectively [see (10.2.1)]. Suppose that an individual of genotype $(\mathbf{i}^{(0)}, \mathbf{i}^{(1)})$ at the L autosomal loci produces a gamete $\mathbf{i}^{(\varepsilon)} = (i_1^{(\varepsilon_1)}, \ldots, i_L^{(\varepsilon_L)})$ with probability $R(\varepsilon)$, where $\varepsilon = (\varepsilon_1, \ldots, \varepsilon_L)$, $\varepsilon_l = 0$ or 1 for $l = 1, \ldots, L$. The recombination regime R satisfies

$$R(\varepsilon) = R(\mathbf{1} - \varepsilon) \quad \text{and} \quad \sum_\varepsilon R(\varepsilon) = 1 \qquad (10.4.1)$$

($\mathbf{1}$ for the L-dimensional unit vector).

(In dioecious populations, the recombination regimes in males and females may differ and be denoted, e.g., by R_m and R_f, respectively. This will not change our conclusions.) Moreover, it is assumed that the male and female fitnesses are independent of the position of the alleles, that is,

$$m(\mathbf{i}^{(\varepsilon)}, \mathbf{i}^{(1-\varepsilon)}) = m(\mathbf{i}^{(0)}, \mathbf{i}^{(1)}) \quad \text{and} \quad f(\mathbf{i}^{(\varepsilon)}, \mathbf{i}^{(1-\varepsilon)}) = f(\mathbf{i}^{(0)}, \mathbf{i}^{(1)}) \quad (10.4.2)$$

for all ε having $0 - 1$ components.

If $p(\mathbf{i}^{(0)})$ denotes the frequency of $\mathbf{i}^{(0)}$ gametes transmitted by males and $q(\mathbf{i}^{(1)})$ the frequency of $\mathbf{i}^{(1)}$ gametes transmitted by females, then the transformation equations generalizing (10.3.3) are

$$p'(\mathbf{i}^{(0)}) = \frac{\sum_{\varepsilon} R(\varepsilon) \sum_{\mathbf{i}^{(1)}} p(\mathbf{i}^{(\varepsilon)}) q(\mathbf{i}^{(1-\varepsilon)}) m(\mathbf{i}^{(\varepsilon)}, \mathbf{i}^{(1-\varepsilon)})}{m},$$

$$(10.4.3)$$

$$q'(\mathbf{i}^{(1)}) = \frac{\sum_{\varepsilon} R(\varepsilon) \sum_{\mathbf{i}^{(0)}} p(\mathbf{i}^{(1-\varepsilon)}) q(\mathbf{i}^{(\varepsilon)}) f(\mathbf{i}^{(1-\varepsilon)}, \mathbf{i}^{(\varepsilon)})}{f},$$

where the male and female mean fitnesses, m and f, are given by

$$m = \sum_{\mathbf{i}^{(0)}, \mathbf{i}^{(1)}} p(\mathbf{i}^{(0)}) q(\mathbf{i}^{(1)}) m(\mathbf{i}^{(0)}, \mathbf{i}^{(1)}),$$

$$(10.4.4)$$

$$f = \sum_{\mathbf{i}^{(0)}, \mathbf{i}^{(1)}} p(\mathbf{i}^{(0)}) q(\mathbf{i}^{(1)}) f(\mathbf{i}^{(0)}, \mathbf{i}^{(1)}).$$

Consider an equilibrium $(\mathbf{p}^*, \mathbf{q}^*)$ of (10.4.3) with male and female gamete frequencies $p^*(\mathbf{i}^{(0)})$ and $q^*(\mathbf{i}^{(1)})$ and male and female mean fitnesses m^* and f^*, respectively. Introduce a new allele in small frequency at any one of the L loci. Denote a male (female) gamete carrying the mutant allele by $\tilde{\mathbf{i}}^{(0)}(\tilde{\mathbf{i}}^{(1)})$ and any other male (female) gamete by $\mathbf{i}^{(0)}(\mathbf{i}^{(1)})$. Define the following mutant marginal gametic fitnesses at equilibrium:

$$m(\tilde{\mathbf{i}}^{(0)}; \mathbf{q}^*) = \sum_{\mathbf{i}^{(1)}} m(\tilde{\mathbf{i}}^{(0)}, \mathbf{i}^{(1)}) q^*(\mathbf{i}^{(1)}),$$

$$f(\tilde{\mathbf{i}}^{(0)}; \mathbf{q}^*) = \sum_{\mathbf{i}^{(1)}} f(\tilde{\mathbf{i}}^{(0)}, \mathbf{i}^{(1)}) q^*(\mathbf{i}^{(1)}),$$

$$(10.4.5)$$

$$m(\tilde{\mathbf{i}}^{(1)}; \mathbf{p}^*) = \sum_{\mathbf{i}^{(0)}} m(\mathbf{i}^{(0)}, \tilde{\mathbf{i}}^{(1)}) p^*(\mathbf{i}^{(0)}),$$

$$f(\tilde{\mathbf{i}}^{(1)}; \mathbf{p}^*) = \sum_{\mathbf{i}^{(0)}} f(\mathbf{i}^{(0)}, \tilde{\mathbf{i}}^{(1)}) p^*(\mathbf{i}^{(0)}).$$

Adding the equations (10.4.3) for all mutant gametes, rearranging the summations and using the properties (10.4.1) and (10.4.2), we get the linear approximation

$$\sum_{\tilde{\mathbf{i}}^{(0)}} p'(\tilde{\mathbf{i}}^{(0)}) + \sum_{\tilde{\mathbf{i}}^{(1)}} q'(\tilde{\mathbf{i}}^{(1)}) \cong \sum_{\tilde{\mathbf{i}}^{(0)}} p(\tilde{\mathbf{i}}^{(0)}) \left[\frac{m(\tilde{\mathbf{i}}^{(0)}; \mathbf{q}^*)}{2m^*} + \frac{f(\tilde{\mathbf{i}}^{(0)}; \mathbf{q}^*)}{2f^*} \right]$$

$$+ \sum_{\tilde{\mathbf{i}}^{(1)}} q(\tilde{\mathbf{i}}^{(1)}) \left[\frac{m(\tilde{\mathbf{i}}^{(1)}; \mathbf{p}^*)}{2m^*} + \frac{f(\tilde{\mathbf{i}}^{(1)}; \mathbf{p}^*)}{2f^*} \right] \quad (10.4.6)$$

near the equilibrium $(\mathbf{p}^*, \mathbf{q}^*)$. We conclude that the frequency of the mutant allele at conception, namely,

$$\frac{\sum_{\tilde{\mathbf{i}}^{(0)}} p(\tilde{\mathbf{i}}^{(0)}) + \sum_{\tilde{\mathbf{i}}^{(1)}} q(\tilde{\mathbf{i}}^{(1)})}{2} \tag{10.4.7}$$

will initially increase if and only if $\rho > 1$, where

$$\rho = \sum_{\tilde{\mathbf{i}}^{(0)}} \pi(\tilde{\mathbf{i}}^{(0)}) \left[\frac{m(\tilde{\mathbf{i}}^{(0)}; \mathbf{q}^*)}{2m^*} + \frac{f(\tilde{\mathbf{i}}^{(0)}; \mathbf{q}^*)}{2f^*} \right]$$

$$+ \sum_{\tilde{\mathbf{i}}^{(1)}} \sigma(\tilde{\mathbf{i}}^{(1)}) \left[\frac{m(\tilde{\mathbf{i}}^{(1)}; \mathbf{p}^*)}{2m^*} + \frac{f(\tilde{\mathbf{i}}^{(1)}; \mathbf{p}^*)}{2f^*} \right] \tag{10.4.8}$$

with $\pi(\tilde{\mathbf{i}}^{(0)})$ and $\sigma(\tilde{\mathbf{i}}^{(1)})$ for all $\tilde{\mathbf{i}}^{(0)}$ and $\tilde{\mathbf{i}}^{(1)}$ representing the stable distribution of male and female mutant gametes near equilibrium, satisfying

$$\sum_{\tilde{\mathbf{i}}^{(0)}} \pi(\tilde{\mathbf{i}}^{(0)}) + \sum_{\tilde{\mathbf{i}}^{(1)}} \sigma(\tilde{\mathbf{i}}^{(1)}) = 1. \tag{10.4.9}$$

These quantities exist because the frequencies of the mutant gametes are linearly related to a first approximation by a positive matrix when rare and the Perron-Frobenius theory applies. Equation (10.4.8) can be written in the form

$$\rho = \frac{\tilde{m}}{2m^*} + \frac{\tilde{f}}{2f^*} \tag{10.4.10}$$

where $\tilde{m}$ and $\tilde{f}$ represent the male and female marginal fitnesses of the mutant allele at equilibrium.

To summarize, we have the following general result.

RESULT 10.6. *In general two-sex multilocus frequency-independent selection models* [see (10.4.3)], *a mutant allele invades an equilibrium geometrically fast if and only if $\rho > 1$ where ρ represents the marginal reproductive value of the mutant allele at equilibrium as given in* (10.4.10).

10.5. General Case of Frequency-Dependent Fitness Functions

Let the genotype at L loci determine a resource allocation vector in a set X. Suppose that the corresponding male and female fitnesses depend on μ, the mean resource allocation vector in the population (see Section 10.2.4).

Denote the resource allocation vector associated with a genotype $(i^{(0)}, i^{(1)})$ by $x(i^{(0)}, i^{(1)})$ and corresponding male and female fitnesses by $m_\mu[x(i^{(0)}, i^{(1)})]$ and $f_\mu[x(i^{(0)}, i^{(1)})]$, respectively. Assume that $m_\mu(x)$ and $f_\mu(x)$ are *linear* in x [see (10.2.6)].

If p and q are the frequency distributions of male and female parental gametes, respectively, then

$$\mu = \sum_{i^{(0)}, i^{(1)}} p(i^{(0)})q(i^{(1)})x(i^{(0)}, i^{(1)}), \qquad (10.5.1)$$

under the assumption of random union of gametes.

An equilibrium (p^*, q^*) of (10.4.3) in the context at hand has $m_{\mu^*}(\mu^*)$ and $f_{\mu^*}(\mu^*)$ as male and female mean fitnesses, where μ^* is the mean resource allocation vector at equilibrium. Moreover, the male and female marginal fitnesses of a mutant allele are $m_{\mu^*}(\tilde{\mu})$ and $f_{\mu^*}(\tilde{\mu})$, respectively, where

$$\tilde{\mu} = \sum_{\tilde{i}^{(0)}, i^{(1)}} \pi(\tilde{i}^{(0)})q^*(i^{(1)})x(\tilde{i}^{(0)}, i^{(1)})$$

$$\sum_{i^{(0)}, i^{(1)}} \sigma(\tilde{i}^{(1)})p^*(i^{(0)})x(\tilde{i}^{(0)}, \tilde{i}^{(0)}) \qquad (10.5.2)$$

in the notation of (10.4.8). Defining

$$\rho_{\mu^*}(\tilde{\mu}) = \frac{m_{\mu^*}(\tilde{\mu})}{2m_{\mu^*}(\mu^*)} + \frac{f_{\mu^*}(\tilde{\mu})}{2f_{\mu^*}(\mu^*)}, \qquad (10.5.3)$$

the mutant allele will invade at a geometric rate if and only if

$$\rho_{\mu^*}(\tilde{\mu}) > 1. \qquad (10.5.4)$$

Observe that both μ^* and $\tilde{\mu}$ belong to

$$E(X) = \left\{ \sum_{i=1}^{r} \alpha_i x^{(i)} : x^{(i)} \in X, \alpha_i \geq 0, \sum_{i=1}^{r} \alpha_i = 1, r \geq 1 \right\}. \qquad (10.5.5)$$

Moreover, $\rho_{\mu^*}(\tilde{\mu})$ is linear with respect to $\tilde{\mu}$ and continuous with respect to μ^* (see the assumptions of Section 10.2.4).

In analogy with Result 10.3, we look for some $\hat{\mu}$ in $E(X)$ satisfying the conditions

(I) $\qquad\qquad\qquad\qquad \rho_{\hat{\mu}}(\mu) \leq 1$

and

(II) $\qquad\qquad\qquad\qquad \rho_\mu(\hat{\mu}) > 1$

for all μ in $E(X)$ near $\hat{\mu}$ but different from $\hat{\mu}$.

Since $\rho_{\hat{\mu}}(\hat{\mu}) = 1$, Conditions (I) and (II) assert, respectively, that $(\hat{\mu}, \hat{\mu})$ is a local maximum of $\rho_{\hat{\mu}}(\mu)$ and a strict local minimum of $\rho_{\mu}(\hat{\mu})$. The function $\rho_{\hat{\mu}}(\mu)$ being linear in μ, Condition (I) actually holds for all μ in $E(X)$.

Condition (I) guarantees that no alleles can invade geometrically fast an equilibrium with mean resource allocation $\hat{\mu}$, while Condition (II) ensures that a mutant allele with marginal resource allocation $\hat{\mu}$ invades geometrically fast every equilibrium with mean resource allocation different from but close to $\hat{\mu}$.

We claim that

$$\text{Condition (II)} \implies \text{Condition (I).} \tag{10.5.6}$$

Defining

$$\mu(\varepsilon) = (1 - \varepsilon)\hat{\mu} + \varepsilon\mu \tag{10.5.7}$$

for $\hat{\mu}$, μ in $E(X)$ and $\varepsilon > 0$ small enough, Condition (II) implies

$$\rho_{\mu(\varepsilon)}(\hat{\mu}) > 1, \tag{10.5.8}$$

or equivalently, using linearity properties,

$$\rho_{\mu(\varepsilon)}[\hat{\mu} - \mu(\varepsilon)] > 0, \tag{10.5.9}$$

which is possible only if

$$\rho_{\mu(\varepsilon)}(\hat{\mu} - \mu) > 0. \tag{10.5.10}$$

Letting $\varepsilon \to 0$, we have

$$\rho_{\hat{\mu}}(\hat{\mu} - \mu) \geq 0, \tag{10.5.11}$$

which is equivalent to Condition (I) and completes the proof of our claim.

In the linear case at hand, the gradient vector of $\rho_{\mu}(\tilde{\mu})$ with respect to $\tilde{\mu}$ is

$$\nabla\rho_{\mu} = \frac{\nabla m_{\mu}}{2m_{\mu}(\mu)} + \frac{\nabla f_{\mu}}{2f_{\mu}(\mu)} > 0 \tag{10.5.12}$$

independently of $\tilde{\mu}$ [see (10.2.6)]. In this context, Condition (II) is equivalent to

$$\langle \nabla\rho_{\mu}, \hat{\mu} - \mu \rangle \geq 0 \tag{10.5.13}$$

in a neighborhood of $\hat{\mu}$ in $E(X)$ with equality if and only if $\mu = \hat{\mu}$. This is possible only if $\hat{\mu}$ is on the boundary of the convex set $E(X)$, $\nabla\rho_{\hat{\mu}}$ is a vector normal to the boundary at $\hat{\mu}$, and $\nabla\rho\mu$ for $\mu \neq \hat{\mu}$ near $\hat{\mu}$ has a positive projection in the direction of $\hat{\mu}$ from μ. If $\hat{\mu}$ is in X, then $\nabla\rho_{\hat{\mu}}$ is a multiple of $\nabla g(\hat{\mu})$ [see (10.2.4)]. If $\hat{\mu}$ is not in X, then $\hat{\mu}$ belongs to a hyperplane perpendicular to $\nabla\rho_{\hat{\mu}}$ and bounding $E(X)$.

Case of Dioecious Populations

Similar conclusions can be drawn for dioecious populations if the resource allocation vectors in males and females, $\mathbf{x}(\mathbf{i}^{(0)}, \mathbf{i}^{(1)})$ and $\mathbf{y}(\mathbf{i}^{(0)}, \mathbf{i}^{(1)})$, associated with a genotype $(\mathbf{i}^{(0)}, \mathbf{i}^{(1)})$, are linearly related by a function h (see Section 10.2.5). Nonlinear models are still to be studied.

10.6. Summary and Discussion

10.6.1. PRODUCT CHARACTERIZATION IN GENERAL TWO-SEX FREQUENCY-INDEPENDENT SELECTION MODELS

It is tempting to conjecture that evolution in two-sex frequency-independent selection models is governed by a maximization principle that would extend the Fundamental Theorem of Natural Selection, namely, the increase of the mean fitness, to the case of sex-differentiated selection regimes. A candidate is the product of the male and female mean fitnesses, $m \times f$.

This is suggested by the optimality properties of the maximum value of $m \times f$. The maximum value of the product is taken over all convex combinations of male and female fitness pairs [the convex hull $E(S)$ of a fitness set S] corresponding to all possible distributions of genotypes within a given genetic framework for a Mendelian population. The maximum value is achieved at a unique point $(\hat{m}, \hat{f})$, and an equilibrium with $(m, f) = (\hat{m}, \hat{f})$ cannot be invaded geometrically fast by any mutant allele. On the other hand, a mutant allele with male and female marginal fitnesses $\tilde{m} = \hat{m}$ and $\tilde{f} = \hat{f}$, respectively, invades geometrically fast every equilibrium with $m \times f \neq \hat{m} \times \hat{f}$. This is true with any number of loci involved using the stable distribution of male and female mutant gametes near an equilibrium and the corresponding reproductive values according to the Shaw-Mohler formula in the computation of the mutant marginal fitnesses.

In the case of linearly related male and female fitnesses determined at a single locus, the product $m \times f$ should increase over successive equilibria as new alleles are introduced. This should also be the case in nonlinear models when the introduction of a new allele at a symmetric equilibrium (i.e., an equilibrium with same allelic frequencies in male and female gametes) is followed by convergence to a nearby equilibrium. However, even a stable symmetric equilibrium may not locally maximize $m \times f$.

Simulations for one-locus models have shown that a decrease of $m \times f$ is possible over successive equilibria. Such a decrease is generally small and rare and occurs mainly with concave fitness sets. Concave cases favor more polymorphism than convex cases but hamper the increase of $m \times f$.

All results have been obtained with the following assumptions: autosomal genetic determination without specific constraints in diploid populations, random mating (or random union of gametes), constant environment. We now discuss the effects of relaxing some of these basic assumptions.

Genetic Constraints. Implicitly it has been assumed that every genotype can be associated with any point in the fitness set S. But it may happen, for example, that heterozygotes have intrinsically greater fitnesses than homozygotes. In such a case, there should be higher polymorphism than what would be expected from our study.

Moreover, the occurrence of invading mutant alleles is not ensured on a purely genetic basis, and an equilibrium may be actually noninvadable for genetic reasons.

Partial Selfing. Charlesworth and Charlesworth (1981) have proposed the optimality criterion

$$\max\{m^{1-\alpha} \times f^{1+\alpha}\} \tag{10.6.1}$$

(see Charnov 1982, p. 230) in populations with a constant rate α of selfing. In the case of complete selfing ($\alpha = 1$), it can be proved that there will be global convergence to the equilibrium that maximizes f (see Karlin and Lessard 1986, Chap. 6, Appendix D).

LMC (Local Mate Competition) Models. For populations structured into colonies with local mating followed by dispersion, Charnov (1982, p. 68) has proposed the criterion

$$\max\left\{f \times m^{\frac{N-1}{N+1}}\right\}, \tag{10.6.2}$$

where N is the number of mated females per colony, while m and f represent the numbers of male and female offspring in the progeny. With the constraint $f = 1 - m$, the optimal value is

$$\hat{m} = \frac{N-1}{2N} \tag{10.6.3}$$

(Hamilton 1967). With a random N, the formula (10.6.3) is valid with N replaced by the expected number of mated females in nonempty colonies (Karlin and Lessard 1986).

Varying Environments. In geographically varying environments, the initial increase condition of a mutant allele depends on male and female migration rates between habitats and male and female fitnesses in each habitat. A simple product maximization principle seems to be precluded.

For temporally varying environments, Charnov (1986) proposes to maximize the quantity

$$E[\log(m \times f)] \tag{10.6.4}$$

(E for mathematical expectation).

Sex-Linked Models. In our simulations of X-linked one-locus selection models, the product $m \times f$, where m is the mean fitness in males and f the mean fitness in females, never decreased over successive equilibria. This is in agreement with previous theoretical results (Lessard 1987; see also Karlin 1972, and Cannings 1967).

In the case of Y-linked systems, there is global convergence to the maximum value of the mean male fitness m (see, e.g., Karlin and Lessard 1986, Chap. 6).

Haplo-Diploid Populations. In haplo-diploid (as well as diplo-diploid) models, where male and female fitnesses are determined by the mother's genotype, the optimality principle based on $m \times f$ still holds.

For haplo-diploid models with worker control, Charnov (1982, p. 298) suggests to maximize

$$m \times f^3. \tag{10.6.5}$$

Two-Sex Haploid Populations. In two-sex haploid populations, the frequency of an allele A_i increases if and only if its growth rate

$$\frac{m_i}{2m} + \frac{f_i}{2f} > 1 \tag{10.6.6}$$

where m_i and f_i are the male and female fitnesses of A_i, while m and f are the mean fitnesses in the whole population. With frequency-independent fitnesses, global convergence to the maximum value of $m \times f$ can be inferred from (10.6.6). (Compared with Lessard 1984 and Gregorius 1982.)

Heterostylous Plants. Resource allocation in heterostylous plants is a complex matter because of incompatibilities and fitness differences between morphs besides sex allocation patterns. We refer to Taylor (1984), Casper and Charnov (1982), Heuch (1979), and Lloyd (1977) for some theoretical studies.

10.6.2. Extension of ESS Theory to Two-Sex Multilocus Frequency-Dependent Selection Models

Evolutionarily Stable Strategies (ESS) were originally defined in a context of behavior patterns used in animal conflicts as a noninvadable strategy once adopted by all members of a population (Maynard Smith and Price 1973; see Maynard Smith 1982). A strategy is represented by a frequency vector (a nonnegative vector whose components sum up to 1) $\mathbf{x} = (x_1, \ldots, x_N)$ where x_i is the probability of adopting behavior i in a conflict with an opponent ($i = 1, \ldots, N$). In the simplest case, the opponent is chosen at random in the whole population, and the expected return to $\mathbf{x}$ when opposed to $\tilde{\mathbf{x}}$ is given in a bilinear form $\mathbf{x}^T \mathbf{A} \tilde{\mathbf{x}}$ for some nonnegative matrix $\mathbf{A}$. The quantity $\mathbf{x}^T \mathbf{A} \tilde{\mathbf{x}}$ is interpreted as a fitness and represents the growth rate of $\mathbf{x}$-strategists opposed to $\tilde{\mathbf{x}}$-strategists in a parthenogenetic population.

For $\hat{\mathbf{x}}$ to be noninvadable as a strategy of a monomorphic population, any alternative strategy $\tilde{\mathbf{x}}$ must be selected against at least when rare, that is,

$$\varepsilon \hat{\mathbf{x}} \mathbf{A} \tilde{\mathbf{x}} + (1 - \varepsilon)\hat{\mathbf{x}} \mathbf{A} \hat{\mathbf{x}} > \varepsilon \tilde{\mathbf{x}} \mathbf{A} \tilde{\mathbf{x}} + (1 - \varepsilon)\tilde{\mathbf{x}} \mathbf{A} \hat{\mathbf{x}} \qquad (10.6.7)$$

for every strategy $\tilde{\mathbf{x}} \neq \hat{\mathbf{x}}$ and every $\varepsilon > 0$ small enough. This is equivalent to either

$$\hat{\mathbf{x}} \mathbf{A} \hat{\mathbf{x}} > \tilde{\mathbf{x}} \mathbf{A} \hat{\mathbf{x}} \qquad (10.6.8a)$$

or

$$\hat{\mathbf{x}} \mathbf{A} \hat{\mathbf{x}} = \tilde{\mathbf{x}} \mathbf{A} \hat{\mathbf{x}} \quad \text{and} \quad \hat{\mathbf{x}} \mathbf{A} \tilde{\mathbf{x}} > \tilde{\mathbf{x}} \mathbf{A} \tilde{\mathbf{x}}. \qquad (10.6.8b)$$

(See, e.g., Maynard Smith 1974.) As noted in Hofbauer, Schuster, and Sigmund (1979), the conditions (10.6.8a) and (10.6.8b) are equivalent to

$$\hat{\mathbf{x}} \mathbf{A} \mathbf{x} > \mathbf{x} \mathbf{A} \mathbf{x} \qquad (10.6.9)$$

for every strategy $\mathbf{x} \neq \hat{\mathbf{x}}$ close to $\hat{\mathbf{x}}$. In (10.6.9), $\mathbf{x}$ represents a population strategy, more precisely, a combination of strategies $\hat{\mathbf{x}}$ and $\tilde{\mathbf{x}}$ in the form $\mathbf{x} = \varepsilon \tilde{\mathbf{x}} + (1 - \varepsilon)\hat{\mathbf{x}}$ for $\varepsilon > 0$ small.

In a more general context of frequency-dependent selection in polymorphic populations, let $f_\mu(\mathbf{x})$ denote the fitness of an $\mathbf{x}$-strategist in a population with mean individual strategy μ. Suppose that $f_\mu(\mathbf{x})$ is linear in $\mathbf{x}$ and continuous in μ. An ES (Evolutionarily Stable) state can be defined as a population strategy $\hat{\mu}$ such that

$$f_\mu(\hat{\mu}) > f_\mu(\mu) \qquad (10.6.10)$$

for every population strategy $\mu \neq \hat{\mu}$ close to $\hat{\mu}$. (Compare with Taylor and Jonker 1978 and Thomas 1984; see also Hines 1987 for a review of basic ESS theory.)

In general resource allocation models, we use as phenotypes nonnegative vectors **x** whose components may not necessarily sum up to 1. Moreover, the male and female fitness functions are distinguished and denoted by $m_\mu(\mathbf{x})$ and $f_\mu(\mathbf{x})$, respectively. Then an ES state is defined by Condition (II) of Section 10.5 [which implies Condition (I) by (10.5.6)]. Condition (I) bears on noninvadability of an ES state as a population equilibrium, while Condition (II) pertains to the invasion of alternative population equilibria by a subpopulation in the ES state.

The application of Conditions (I) and (II) to Mendelian populations is wide ranging. They can incorporate sex-dependent fitness functions as well as multilocus genetic determination [see (10.5.3)]. In a population genetic context, Conditions (I) and (II) refer to initial increase conditions of mutant alleles in comparing the mutant marginal reproductive value (the mean reproductive value over all genotypes carrying the mutant allele) to the reproductive value of all alleles present at equilibrium (equal to 1).

A mutant marginal reproductive value is in the form of a spectral radius, that is, the spectral radius of the matrix relating the frequencies of the male and female mutant gametes when rare. There is invasion when

$$\rho_\mu(\tilde{\mu}) > 1, \qquad (10.6.11)$$

where $\tilde{\mu}$ is the mutant marginal resource allocation vector and μ the mean resource allocation vector in the population at equilibrium. If (10.6.11) holds for $\tilde{\mu} = \hat{\mu}$ and every $\mu \neq \hat{\mu}$ close to $\hat{\mu}$, then $\hat{\mu}$ is an ES state.

An ES state is optimal with respect to initial increase properties. But this alone does not guarantee evolution to an ES state. Convergence to an ES state can be proved in some cases (e.g., for parthenogenetic populations or for some multiallele one-locus diploid populations; see Zeeman 1980, 1981, Cressman and Hines 1984, Lessard 1984, 1986, Thomas 1985b). In other cases, ES states may not exist and cycles can occur as a result of interactions (see, e.g., Zeeman 1980, Akin 1984). Cycles can also occur in frequency-independent selection regimes only because of multilocus genetic determination (Akin 1983; Hastings 1981). There may even be a contradiction between initial increase properties with respect to mutant alleles and stability with respect to existing alleles as a consequence of linkage (Lessard 1987).

For initial increase of mutant alleles, condition (10.6.11) is of general validity. For more complex frequency-dependent selection patterns (involving, e.g., group selection and kin selection), the means would be replaced by distributions. Equations in the form (10.6.11) can also be used to find ES genetic parameters (related e.g., to selfing rates, recombination rates, migration rates, segregation ratios, etc.).

Acknowledgment

Research was supported in part by the Natural Sciences and Engineering Research Council of Canada, Grant No. A8833.

References

Akin, E. 1982. Exponential families and game dynamics. *Can. J. Math.* 34: 374–405.

Akin, E. 1983. *Hopf Bifurcation in the Two-Locus Genetic Model.* Memoirs of the American Mathematical Society, No. 284. Providence, R.I.

Akin, E. 1984. Evolution: Game theory and economics. In *Population Biology*, pp. 37–61. Ed. S. A. Levin. Proceedings of Symposia in Applied Mathematics, vol. 30. American Mathematical Society, Providence, R.I.

Cannings, C. 1967. Equilibrium, convergence and stability at a sex-linked locus under natural selection. *Genetics* 56: 613–618.

Casper, B. B., and Charnov, E. L. 1982. Sex allocation in heterostylous plants. *J. Theor. Biol.* 96: 143–149.

Charlesworth, D., and Charlesworth, B. 1981. Allocation of resources to male and female functions in hermaphrodites. *Biol. J. Linn. Soc.* 14: 57–74.

Charnov, E. L. 1979. The genetical evolution of patterns of sexuality: Darwinian fitness. *Amer. Natur.* 113: 465–480.

Charnov, E. L. 1982. *The Theory of Sex Allocation.* Princeton University Press, Princeton, N.J.

Charnov, E. L. 1986. An optimisation principle for sex allocation in a temporally varying environment. *Heredity* 56: 119–121.

Charnov, E. L.; Maynard Smith, J.; and Bull, J. 1976. Why be an hermaphrodite? *Nature* 263: 125–126.

Cressman, R., and Hines, W.G.S. 1984. Evolutionary stable strategies of diploid populations with semi-dominant inheritance patterns. *J. Appl. Prob.* 21: 1–9.

Crow, J.-F., and Kimura, M. 1970. *An Introduction to Population Genetics Theory.* Harper and Row, New York.

Eshel, I. 1975. Selection on sex ratio and the evolution of sex determination. *Heredity* 34: 351–361.

Eshel, I. 1982. Evolutionary stable strategies and viability selection in Mendelian populations. *Theor. Pop. Biol.* 22: 204–217.

Eshel, I., and Feldman, M. W. 1982a. On evolutionary genetic stability of the sex ratio. *Theor. Pop. Biol.* 21: 430–439.

Eshel, I., and Feldman, M. W. 1982b. On the evolution of sex determination and the sex ratio in haplodiploid populations. *Theor. Pop. Biol.* 21: 440–450.

Eshel, I., and Feldman, M. W. 1984. Initial increase of new mutants and some continuity properties of ESS in two-locus systems. *Amer. Natur.* 124: 631–640.

Ewens, W. 1979. *Mathematical Population Genetics.* Springer-Verlag, Heidelberg.

Fisher, R. A. 1930. *The Genetical Theory of Natural Selection.* Oxford University Press, London. (Revised and enlarged edition, Dover, New York, 1958.)

Gantmacher, F. R. 1959. *The Theory of Matrices.* 2 vols. Chelsea, New York.

Gregorius, H. R. 1982. Selection in diplo-haplonts. *Theor. Pop. Biol.* 21: 289–300.

Hamilton, W. D. 1967. Extraordinary sex ratios. *Science* 156: 477–488.

Hastings, A. 1981. Stable cycling in discrete-time genetic models. *Proc. Natl. Acad. Sci. USA* 78: 7224–7225.

Heuch, I. 1979. Equilibrium populations of heterostylous plants. *Theor. Pop. Biol.* 15: 43–57.

Hines, W.G.S. 1980. Three characterizations of population strategy stability. *J. Appl. Prob.* 17: 333–340.

Hines, W.G.S. 1982. Strategy stability in complex randomly mating diploid populations. *J. Appl. Prob.* 19: 653–659.

Hines, W.G.S. 1987. *Evolutionarily stable strategies: A review of basic theory. Theor. Pop. Biol.* 31: 195 –272.

Hofbauer, J.; Schuster, P.; and Sigmund, K. 1979. A note on the evolutionary stable strategies and game dynamics. *J. Theor. Biol.* 81: 609–612.

Karlin, S. 1972. Some mathematical models of population genetics. *Amer. Math. Monthly* 79: 699–739.

Karlin, S. 1978. Theoretical aspects of multilocus selection balance, I. In *Studies in Mathematical Biology*, Part II: *Populations and Communities*, pp. 503–587. Ed. S. A. Levin. MAA Studies in Mathematics, vol. 16, Mathematical Association of America, Washington, D.C.

Karlin, S., and Lessard, S. 1983. On the optimal sex ratio. *Proc. Natl. Acad. Sci. USA* 80: 5931–5935.

Karlin, S., and Lessard, S. 1984. On the optimal sex ratio: A stability analysis based on a characterization for one-locus multiallele viability models. *J. Math. Biol.* 20: 15–38.

Karlin, S., and Lessard, S. 1986. *Theoretical Studies on Sex Ratio Evolution.* Princeton University Press, Princeton.

Karlin, S., and Liberman, U. 1979. Central equilibrium in multilocus systems, II. Bisexual generalized nonepistatic selection models. *Genetics* 91: 799–816.

Lessard, S. 1984. Evolutionary dynamics in frequency-dependent two-phenotype models. *Theor. Pop. Biol.* 25: 210–234.

Lessard, S. 1986. Evolutionary principles for general frequency-dependent two-phenotype models in sexual populations. *J. Theor. Biol.* 119: 329–344.

Lessard, S. 1987. The role of recombination and selection in the modifier theory of sex ratio distortion. *Theor. Pop. Biol.* 31: 339–358.

Lessard, S., and Karlin, S. 1982. A criterion for stability-instability at fixation states involving an eigenvalue one with applications in population genetics. *Theor. Pop. Biol.* 22: 108–126.

Lloyd, D. G. 1977. Genetic and phenotypic models of natural selection. *J. Theor. Biol.* 69: 543–560.

MacArthur, R. H. 1965. Ecological consequences of natural selection. In *Theoretical and Mathematical Biology*, pp. 388–397. Ed. T. H. Waterman and H. Morowitz. Blaisdell, New York.

Marsden, J. E., and McCracken, M. 1976. *The Hopf Bifurcation and Its Applications*. Applied Math. Sciences, vol. 19. Springer-Verlag, New York.

Maynard Smith, J. 1974. The theory of games and the evolution of animal conflict. *J. Theor. Biol.* 47: 209–221 (with an Appendix by J. Haigh).

Maynard Smith, J. 1981. Will sexual population evolve to an ESS? *Amer. Natur.* 117: 1015–1018.

Maynard Smith, J. 1982. *Evolution and the Theory of Games*. Cambridge University Press, Cambridge, Eng.

Maynard Smith, J., and Price, G. R. 1973. The logic of animal conflicts. *Nature* 246: 15–18.

Raper, J. K. 1983. General analysis of frequency-dependent sexual selection at a multi-allelic locus. *Theor. Pop. Biol.* 24: 192–211.

Shaw, R. F., and Mohler, J. D. 1953. The selective advantage of the sex ratio. *Amer. Natur.* 87: 337–342.

Speith, P. T. 1974. Theoretical considerations of unequal sex ratios. *Amer. Natur.* 108: 837–849.

Taylor, P. D. 1984. Evolutionarily stable reproductive allocations in heterostylous plants. *Evolution* 38: 408–416.

Taylor, P. D., and Bulmer, M. G. 1980. Local mate competition and the sex ratio. *J. Theor. Biol.* 86: 409–419.

Taylor, P. D., and Jonker, L. B. 1978. Evolutionarily stable strategies and game dynamics. *Math. Biosc.* 40: 145–156.

Thomas, B. 1984. Evolutionary stability: States and strategies. *Theor. Pop. Biol.* 26: 49–67.

Thomas, B. 1985a. Genetical ESS-models, II. Multi-strategy models and multiple alleles. *Theor. Pop. Biol.* 28: 33–49.

Thomas, B. 1985b. On evolutionary stable strategy sets. *J. Math. Biol.* 22: 105–115.

Uyenoyama, M. K., and Bengtsson, B. O. 1979. Towards a genetic theory for the evolution of the sex ratio. *Genetics* 93: 721–736.

Uyenoyama, M. K., and Bengtsson, B. O. 1981. Towards a genetic theory for the evolution of the sex ratio, II. Haplodiploid and diploid models with sibling and parental control of the brood sex ratio and brood size. *Theor. Pop. Biol.* 20: 57–79.

Uyenoyama, M. K., and Bengtsson, B. O. 1982. Towards a genetic theory for the evolution of the sex ratio, III. Parental and sibling control of brood investment ratio under partial sib-mating. *Theor. Pop. Biol.* 22: 43–68.

Zeeman, E. C. 1980. Population dynamics from game theory. In *Global Theory of Dynamical Systems*. pp. 471–497. Lecture Notes in Mathematics, vol. 819. Springer-Verlag, Berlin.

Zeeman, E. C. 1981. Dynamics of the evolution of animal conflicts. *J. Theor. Biol.* 89: 249–270.

CHAPTER ELEVEN

Sexual Selection Models
and the Evolution of Melanism
in Ladybirds

Peter O'Donald and
Michael E. N. Majerus

Introduction

Sexual selection by female preference is expected to give rise to a "rare male advantage" (O'Donald 1977, 1980). This arises because, at a low frequency, a preferred male phenotype will be preferred by a relatively greater proportion of females than at a high frequency. Stable polymorphisms will then be maintained, either when more than one male phenotype is the object of female preference, or when the sexual selection is balanced by natural selection (O'Donald 1973, 1980).

Karlin and O'Donald (1981) analyzed multiallelic models in which either (1) females exercised separate preferences for distinct phenotypes; or (2) females increasingly preferred males with greater development of the preferred character. In the latter models, there is a graded range of phenotypes, for example a graded range of melanic phenotypes. Some females prefer any of the melanics; others, more choosy, only prefer the more extreme or most extreme melanics. There is an increasing preference in favor of the more extreme phenotypes.

In models with separate preferences for each phenotype, all the alleles that determine the preferred phenotypes are maintained in the population to form a "protected polymorphism." In models with increased preference for more extreme phenotypes, only the most extreme phenotype is protected; the less extreme are eliminated. The two models thus make completely different predictions about the evolutionary outcome of sexual selection. Karlin and O'Donald (1981) called the models of increased preference for increased phenotypic expression "models with inclusion relationships," since different phenotypes were included in a phenotypic class subject to the same preference. For example, suppose alleles A_1, A_2, A_3 show a dominance relationship in the order

$$A_1 > A_2 > A_3 > A_4 \cdots A_n,$$

where A_1, A_2, and A_3 determine melanic phenotypes, while the remaining alleles, recessive to A_1, A_2, and A_3, might show dominance or codominance relationships to each other. Then we might have preferred classes as follows:

$$P_1 = \{A_1A_1, A_1A_2, A_1A_3, A_1A_4, \ldots, A_1A_n\}$$

$$P_{1,2} = \{A_1A_1, A_1A_2, \ldots, A_1A_n, A_2A_2, A_2A_3, \ldots, A_2A_n\}$$

$$P_{1,2,3} = \{A_1A_1, A_1A_2, \ldots, A_1A_n, A_2A_2, A_2A_3, \ldots, A_2A_n, A_3A_3,$$
$$A_3A_4, \ldots, A_3A_n\}.$$

$P_{1,2}$ and $P_{1,2,3}$ thus include the phenotypes determined by A_1 and A_2. If proportions α_1, α_2, and α_3 of the females prefer males in the classes P_1, $P_{1,2}$, and $P_{1,2,3}$, we have a greater preference for A_2 and the greatest preference for A_1. If the dominance ordering

$$A_1 > A_2 > A_3$$

reflects greater development of the male character in the phenotypes given by

$$A_1A_1, \ldots, A_1A_n$$

$$A_2A_2, \ldots, A_2A_n$$

$$A_3A_3, \ldots, A_3A_n,$$

we have the increased preference for the increased phenotypic expression of the male character. We may represent this situation by the notation

$$\{A_1\}, \begin{Bmatrix} A_1 \\ A_2 \end{Bmatrix}, \begin{Bmatrix} A_1 \\ A_2 \\ A_3 \end{Bmatrix}, \{A_4\}, \{A_5A_5\}, \{A_5A_6\}, \{A_6A_6\},$$

where the inclusion of an allele in any class signifies dominance over all alleles or genotypes in classes on the right of it; and the inclusion of genotypes signifies separable codominant phenotypes. In general, there may be a separate mating preference in favor of the phenotypes in each of the separately defined classes.

We shall call models with preference classes, such as

$$\{A_1\}, \begin{Bmatrix} A_1 \\ A_2 \end{Bmatrix}, \ldots, \qquad \text{corresponding to the preferred classes of phenotypes } P_1, P_{1,2}, \ldots$$

"inclusive preference models" to distinguish them from "separate preference models" in which, for example, each dominant or codominant phenotype may be preferred by a separate group of females.

The Two-spot Ladybird is polymorphic for a large number of melanic and nonmelanic phenotypes. The commoner phenotypes form an ordered dominant and codominant series (Majerus, O'Donald, and Weir 1982a; O'Donald et al. 1984), as Karlin and O'Donald postulated. The more melanic forms are dominant to the less melanic and nonmelanic forms, exactly as if in the inclusive preference model, the more extreme melanics are the more strongly preferred. This model then predicts that only the most extreme melanic should be maintained in the polymorphism. In the melanic series, in dominance and amount of melanism, we have *sublunata* (*L*) > *quadrimaculata* (*Q*) > *sexpustulata* (*X*). All these forms are dominant to *typica* (*T*), and extreme *annulata* (A_x), which are codominant producing heterozygous *annulata* (*A*). In a wild population at Keele, England, O'Donald et al. (1984) obtained the samples given in the following table.

Phenotypes	Females	Males
L	47	57
Q	2921	2907
X	414	446
T	4556	4422
$A + A_x$	2027	2083

The melanics form about 35% of the population, but there has been no tendency for *L* or *Q* to replace *X* as predicted by the inclusive preference model. Numbers of mating pairs and mating tests in population cages and mating chambers gave a consistent estimate of about 20% preference for melanics. All melanics appear to show a mating advantage in the wild population: about 20% of melanic males obtain mates in excess of what is expected under random mating. There is little or no assortment in these matings: females show a slight, but not significant tendency to mate with males of like phenotype. A general preference for any melanic will not by itself maintain a protected polymorphism of *L*, *Q*, and *X* melanic phenotypes. Genetic drift should eliminate two of the phenotypes since *L*, *Q*, and *X* would be strictly neutral with respect to each other, although at a mating advantage over nonmelanic phenotypes. The model, as described by the inclusive classes,

$$\{A_1\}, \{A_1, A_2\}, \{A_1, A_2, A_3\}, \ldots,$$

gives rise to the elimination of A_2 and A_3, since these phenotypes have a mating disadvantage relative to A_1. But where there is no extra preference for A_1 and A_2, the model

$$\{A_1, A_2, A_3\}, \ldots$$

entails the ultimate survival of any one and only one of these phenotypes. The alternative model of separate preferences for each phenotype

$$\{A_1\}, \{A_2\}, \{A_3\}, \ldots$$

of course entails a protected polymorphism of the corresponding phenotypic classes P_1, P_2, P_3, because each phenotype gains its own frequency-dependent mating advantage.

Majerus, O'Donald, and Weir (1982b) proved that the female ladybirds have a genetic preference for Q males: they selected females that mated with Q males, rapidly raising the proportion of matings with Q males. They showed that female preference had been selected because the high proportion of matings with Q males occurred when selected females chose males taken either from the selected stock or from the unselected original stock; but when males from the selected stock were chosen by females from the unselected stock, the proportion of matings with Q males was reduced to the original level. In other words, a female effect had been selected for increased propensity to mate with Q males. This experiment was later repeated and extended, selecting lines for both increased and reduced preference for Q males (O'Donald and Majerus 1985). At generation 10 in the "high" preference line, isofemale lines were started from single pair matings. These were intended to test the nature of the genetic variation in preference. Selection may have raised the level of a polygenically determined preference in most or all females, with some genetic variation around the mean level of preference. Alternatively, the preference may be determined by only a few genes or one. If so, the preference would be polymorphic, some females possessing preference genes and mating preferentially, others not possessing the genes and mating randomly. Provided eggs are laid and larvae reared in separate batches to avoid larval competition, each female can produce large numbers of progeny. Levels of preference can thus be measured in the progeny of each female. The preferences in the different isofemale lines strongly supported the hypothesis that a single preference gene determines the preference (Majerus et al. 1986). In some lines about 85–90% of females mated preferentially. In other lines, the level of preferential mating was about 60% or 40%, or zero. These results can be explained by postulating a single dominant gene, B. Females possessing this gene mate preferentially with probability about 0.85. Thus in the lines with 85% preferential mating, all females possessed the gene. In lines in which preferential mating was about 60%, 75% of females would possess the gene (being the product of matings $Bb \times Bb$). In lines with 40% preferential mating, the progeny would have been produced by matings $Bb \times bb$ and half would possess the preference gene. The remaining lines, showing random mating, would have been produced by matings $bb \times bb$, no females possessing the preference gene. This simple model accords closely with observations on the lines.

Although the female preference was selected as a preference for Q males, tests on the preferences in the isofemale lines showed that the females also mated preferentially with *sexpustulata* (X) males. The preference for X was roughly the same as the preference for Q in any particular line. (See Majerus et al. 1986). But sample sizes were small and differences in levels of preference for X and Q may have been appreciable but too small to have been detected at the sample sizes used in the experiment. Nevertheless, selection of a preference for Q produced roughly similar levels of increased preference for X. This result implies that the preferences are determined by the same gene: females with the preference gene do not discriminate between Q and X males. A general preference for the melanic phenotypes Q and X has been selected. The preferential mating of Q and X observed in natural populations is not produced by separate preferences for each phenotype exercised by different groups of females. Karlin and O'Donald's (1981) separate preference model is thus refuted by the results of the selection experiment. This model predicts that selecting the preference for Q will not raise the level of preference for X; selecting the preference for X will not raise the preference for Q. On the contrary, our results show that Q and X form an inclusive class of phenotypes both preferred by the same group of females. According to Lusis (1928), the allele s^m determines the four-spot melanic Q; s^p, recessive to s^m, determines the six-spot melanic X with two additional posterior red spots. The phenotypes Q and X are not completely distinct, however. Phenotypes classified as X at one week old may change to Q later as more melanin is laid down and the two posterior spots become obscured (Majerus, Kearns, and Ireland, in prep.). The common nonmelanic phenotypes *typica* (T) and *annulata* (A) are determined by alleles s^t and s^a, respectively. Lusis stated that s^t is dominant to s^a, but in British populations they are codominant (Majerus, Kearns, and Ireland, in prep.). The heterozygote $s^t s^a$ gives rise to the phenotype *annulata* (A), while the homozygote $s^a s^a$ gives rise to extreme *annulata* (A_x). The polymorphism is further complicated by the existence of two other rare nonmelanic phenotypes, *duodecempustulata* (D) and "spotty" determined by alleles s^d and s^{sp}. Both are codominant with s^t: the heterozygotes are often typical (T) but sometimes have extended black spots or additional satellite spots, placing them in the class *annulata*. *Annulata* is therefore a heterogeneous class of three different heterozygotes. But in the Keele population, s^d and s^{sp} are very rare; and, in respect of the commoner phenotypes, the inclusive preference model may be represented in Karlin and O'Donald's notation by the sets of genotypes,

$$\{s^m\}, \begin{Bmatrix} s^m \\ s^p \end{Bmatrix}, \{s^t s^t\}, \{s^t s^a\}, \{s^a s^a\},$$

where as before

$$\{s^m\} = \{s^m s^m, s^m s^p, s^m s^t, s^m s^a\},$$

and so on.

According to Karlin and O'Donald's analysis of this model, only the most extreme dominant should be maintained at stable polymorphic equilibrium: s^m should remain as a protected allele; s^p should be eliminated. The polymorphism should consist of phenotypes Q, A, T, A_x. Even if there were no additional preference for the most extreme melanic Q, we should still expect that either Q or X would be eliminated. In the dominant series of melanic phenotypes, we should expect to find only one melanic in any particular population. If there were some increase of preference for increasing melanism, only the most melanic form should be found. In natural populations with melanic ladybirds, however, the phenotypes L, Q, and X may all be found; L is usually very rare and may be absent from some populations, but Q and X may both be common. This observation refutes Karlin and O'Donald's inclusive preference model. Their separate preference model has already been refuted by the evidence of the genetics of preference. As we shall show in this chapter, a combination of the two models may explain this polymorphism.

The other common phenotypes *typica* (T), *annulata* (A), and extreme *annulata* (A_x) are codominant. There is strong evidence of assortative mating for *annulata* (Majerus, O'Donald, and Weir 1982a; O'Donald et al. 1984). Any separate preference, assortative or nonassortative, for a heterozygote would be sufficient to maintain a polymorphism. Evidence from the natural population at Keele (where our original stocks for the selection of preference were taken) shows quite strong assortative mating for the heterozygote A. This would certainly maintain a polymorphism of the codominant alleles s^t and s^a. In general, it is difficult to see how a preference for a heterozygote might evolve. Computer simulations show that genetic preferences for heterozygotes are usually eliminated (O'Donald 1980). The preference for A_x has not been tested, so the preference for

A may also include preference for A_x. In terms of preference, A and A_x may be dominant to T, although codominant in terms of expression of melanic pattern: O'Donald et al.'s model may still be valid for this case. Moreover, the preference for A has not yet been estimated for individuals of known genotype, for *annulata* phenotypes are determined by the alleles s^a, s^d, and s^{sp} when heterozygous with s^t. The extent of sexual selection and assortative mating for the different *annulata* heterozygotes, for $s^a s^a$, and for the rare homozygotes *duodecempustulata* (D) ($s^d s^d$) and spotty ($s^{sp} s^{sp}$), has not yet been ascertained. Explanation of how s^d and s^{sp} are maintained in British populations must be left for future investigations.

The object of our present study is to explain the stability of polymorphism of the melanic phenotypes Q and X. Karlin and O'Donald's inclusive preference model can be extended to give stable polymorphisms of the alleles in the ordered dominance series. Our experiments have certainly shown that Q and X are both preferred by a common group of females that have a genetically determined preference. The inclusive preference model postulates an additional, separate preference for the more extreme Q phenotype. Suppose there were also a separate preference for X. This additional preference might permit the maintenance of a protected polymorphism for both phenotypes.

Of course, the polymorphism might be maintained by natural selection differing between Q and X males. But this would entail some frequency dependence in the relative advantage of Q and X, without which a polymorphism could not be maintained. Melanic ladybirds have been shown to be at a selective disadvantage related to levels of atmospheric pollution: as pollution has declined, melanics have declined in frequency (Creed 1971). This disadvantage in natural selection is certainly balanced by the mating preference for melanics. As O'Donald et al. (1984) showed, the polymorphism is maintained by sexual selection and assortative mating. The natural selection merely shifts the point of equilibrium. A preference in favor of both Q and X with constant natural selection against them will not provide the frequency-dependent advantage of Q relative to X necessary to maintain the polymorphism. Separate preferences for both Q and X would also be required to provide the necessary frequency dependence.

Models of Sexual Selection for Ladybirds

Karlin (1978) gave a general formulation of models of sexual selection in which females mate preferentially but without assortment. Any model of fixed or frequency-dependent expression of preference has the same general form. In fact, Karlin's formulation has a wider application to any model with frequency-dependent fitness or mating. Consider a number of alleles

$$A_1, A_2, \ldots, A_n.$$

Let

$$u_{ij} = \text{freq}(A_iA_j) \text{ in the female population}$$

$$p_i = \text{freq}(A_i) = u_{ii} + \frac{1}{2}\sum_{\substack{r=1\\r\neq i}}^{n} u_{ir}.$$

Suppose some of the alleles show a dominance ordering

$$A_1 > A_2 > A_3 > \cdots A_{n-1}, A_n,$$

giving classes of phenotypes

$$P_1 = \{A_1A_1, A_1A_2, \ldots, A_1A_n\}$$
$$P_2 = \{A_2A_2, A_2A_3, \ldots, A_2A_n\},$$

while other alleles show codominance giving separate phenotypic classes

$$P_{n-1,n-1} = \{A_{n-1}A_{n-1}\}$$
$$P_{n-1,n} = \{A_{n-1}A_n\}$$
$$P_{n,n} = \{A_nA_n\}.$$

Before mate choice, male genotypes have the same frequencies as female genotypes. After mate choice, male genotypes have frequencies $f_{ij} = \text{freq}(A_iA_j)$ among males chosen to mate. This formulation obviously applies to males after any form of selection has acted upon them. In general, f_{ij} will be some frequency-dependent function of the frequencies u_{ij} $(i, j = 1, \ldots, n)$. Among the mated males the gametic frequency is

$$f_i = f_{ii} + \frac{1}{2}\sum_{\substack{r=1\\r\neq i}}^{n} f_{ir}.$$

When there is no assortment and the females' choice of mate is not determined by the females' own genotypes, then the zygotic frequencies in the next generation are given by the following recursion equations (Karlin 1978):

$$u'_{ij} = \tfrac{1}{2}[p_if_j + p_jf_i] \qquad (i, j = 1, \ldots, n),$$

thus giving the recursion equation for gene frequencies

$$p_i' = \tfrac{1}{2}[p_i + f_i],$$

and hence at equilibrium

$$p_i = f_i.$$

SEPARATE PREFERENCE MODEL

In the complete dominance ordering

$$A_1 > A_2 > A_3 > \cdots > A_n,$$

giving phenotypic classes of the males

$$P_1, P_2, \ldots, P_n$$

each with separate mating preferences of

$$\alpha_1, \alpha_2, \ldots, \alpha_n$$

and total proportion of preferential matings

$$\theta = \sum_{i=1}^{n} \alpha_i,$$

it can be shown that the phenotypic equilibrium frequencies are exactly proportional to the preferential mating rates

$$\pi_i^* = \alpha_i/\theta \qquad (i = 1, \ldots, n)$$

with allelic frequencies

$$p_i^* = \sqrt{\sum_{r=i}^{n} \alpha_r/\theta} - \sqrt{\sum_{r=i+1}^{n} \alpha_r/\theta}$$

$$i = 1, \ldots, n.$$

Karlin and O'Donald gave the corresponding solution for the general model with a series of both dominant and codominant alleles. In each case, there is a single stable equilibrium point. Preferences for all dominant phenotypes and phenotypes containing each codominant allele produce a protected polymorphism of all alleles. Since our genetic experiments clearly refute the separate preference model, detailed results of the general dominance/codominance model are not given.

Inclusive Preference Model

Karlin and O'Donald (1981) analyzed the model with phenotypic classes

$$P_1 = \{A_1A_1, A_1A_2, A_1A_3\} = \{A_1\} = \text{dominant class of } A_1$$

$$P_{1,2} = \{A_1A_1, A_1A_2, A_1A_3, A_2A_2, A_2A_3\}$$

$$= \begin{Bmatrix} A_1 \\ A_2 \end{Bmatrix} = \text{inclusive dominant class of } A_1 \text{ and } A_2$$

$$P_3 = \{A_3A_3\},$$

with preferential mating rates α, β, and γ for these three classes of males. Since A_1 has a consistent advantage over A_2, A_2 is always eliminated, A_1 and A_3 remaining at the phenotypic equilibria

$$\pi_1^* = \text{freq}\{P_1\} = \frac{\alpha + \beta}{\alpha + \beta + \gamma}$$

$$\pi_2^* = 0 = \text{freq}\{P_2\}$$

$$\pi_3^* = \text{freq}\{P_3\} = \frac{\gamma}{\alpha + \beta + \gamma}.$$

More complex models show the same consistent feature; the allele in the inclusive class that is not the object of a separate preference is always eliminated. Thus if P_1 is the more extreme melanic phenotype compared to the less extreme melanic P_2, the less extreme, with less preferential mating advantage, is eliminated. Since this does not occur in ladybird populations, an additional preference must exist to favor P_2 as a separate class. A general model for the Two-spot Ladybird would be as follows:

$$\{s^m\}, \{s^p\}, \begin{Bmatrix} s^m \\ s^p \end{Bmatrix}, \{s^ts^t\}, \{s^ts^a\}, \{s^as^a\},$$

where each of these phenotypic groupings may be the object of female preference. Let the phenotypes, frequencies and preferential mating rates be

	Phenotypes				
	Q	X	T	A	A_x
frequencies	π_1	π_2	π_3	π_4	π_5
separate preferences	α_1	α_2	γ_1	γ_2	γ_3
inclusive preference	β				

The alleles s^m, s^p, s^t, and s^a have gene frequencies p_1, p_2, p_3, and p_4. This could be stated more clearly as follows:

phenotypic preferred classes

	$\{Q\}$	$\{X\}$	$\{M = Q + X\}$	$\{A\}$	$\{T\}$	$\{A_x\}$
preferences	α_1	α_2	β	γ_1	γ_2	γ_3

Genotypes then mate at the following frequencies:

Genotypes	Females	Males
$s^m s^m$	u_{11}	$\alpha_1 u_{11}/\pi_1 + \beta u_{11}/(\pi_1 + \pi_2) + u_{11}(1 - \theta)$
$s^m s^p$	u_{12}	$\alpha_1 u_{12}/\pi_1 + \beta u_{12}/(\pi_1 + \pi_2) + u_{12}(1 - \theta)$
$s^m s^t$	u_{13}	$\alpha_1 u_{13}/\pi_1 + \beta u_{13}/(\pi_1 + \pi_2) + u_{13}(1 - \theta)$
$s^m s^a$	u_{14}	$\alpha_1 u_{14}/\pi_1 + \beta u_{14}/(\pi_1 + \pi_2) + u_{14}(1 - \theta)$
$s^p s^p$	u_{22}	$\alpha_2 u_{22}/\pi_2 + \beta u_{22}/(\pi_1 + \pi_2) + u_{22}(1 - \theta)$
$s^p s^t$	u_{23}	$\alpha_2 u_{23}/\pi_2 + \beta u_{23}/(\pi_1 + \pi_2) + u_{23}(1 - \theta)$
$s^p s^a$	u_{24}	$\alpha_2 u_{24}/\pi_2 + \beta u_{24}/(\pi_1 + \pi_2) + u_{24}(1 - \theta)$
$s^t s^t$	u_{33}	$\gamma_1 + u_{23}(1 - \theta)$
$s^t s^a$	u_{34}	$\gamma_2 + u_{34}(1 - \theta)$
$s^a s^a$	u_{44}	$\gamma_3 + u_{44}(1 - \theta)$,

where $\theta = \alpha_1 + \alpha_2 + \beta + \gamma_1 + \gamma_2 + \gamma_3$

$$\pi_1 = u_{11} + u_{12} + u_{13} + u_{14}$$

$$\pi_2 = u_{22} + u_{23} + u_{24}$$

$$\pi_3 = u_{33}$$

$$\pi_4 = u_{34}$$

$$\pi_5 = u_{44}$$

In females, gametic frequencies in generation $t + 1$ are simply given by the zygotic frequencies in generation t:

$$p_1 = u_{11} + \tfrac{1}{2}u_{12} + \tfrac{1}{2}u_{13} + \tfrac{1}{2}u_{14}$$

$$p_2 = u_{22} + \tfrac{1}{2}u_{12} + \tfrac{1}{2}u_{23} + \tfrac{1}{2}u_{24}$$

$$p_3 = u_{33} + \tfrac{1}{2}u_{13} + \tfrac{1}{2}u_{23} + \tfrac{1}{2}u_{34}$$

$$p_4 = u_{44} + \tfrac{1}{2}u_{14} + \tfrac{1}{2}u_{24} + \tfrac{1}{2}u_{34},$$

while in males the gametic frequencies are given by

$$f_1 = \alpha_1 p_1/\pi_1 + \beta p_1/(\pi_1 + \pi_2) + p_1(1 - \theta)$$

$$f_2 = \alpha_2 q_2/\pi_2 - \alpha_1 q_2/\pi_1$$
$$+ \alpha_1 p_2/\pi_1 + \beta p_2/(\pi_1 + \pi_2) + p_2(1 - \theta)$$

$$f_3 = \gamma_1 + \tfrac{1}{2}\gamma_2 + p_3(1 - \theta)$$
$$+ \tfrac{1}{2}u_{13}[\alpha_1/\pi_1 + \beta/(\pi_1 + \pi_2)]$$
$$+ \tfrac{1}{2}u_{23}[\alpha_2/\pi_2 + \beta/(\pi_1 + \pi_2)]$$

$$f_4 = \gamma_3 + \tfrac{1}{2}\gamma_2 + p_4(1 - \theta)$$
$$+ \tfrac{1}{2}u_{14}[\alpha_1/\pi_1 + \beta/(\pi_1 + \pi_2)]$$
$$+ \tfrac{1}{2}u_{24}[\alpha_2/\pi_2 + \beta/(\pi_1 + \pi_2)],$$

where

$$q_2 = u_{22} + \tfrac{1}{2}u_{23} + \tfrac{1}{2}u_{24}.$$

Putting $p_1 = f_1$, $p_2 = f_2$ for equilibrium, it follows immediately that

$$\alpha_1/\pi_1 + \beta/(\pi_1 + \pi_2) - \theta = 0$$
$$\alpha_2/\pi_2 - \alpha_1/\pi_1 = 0,$$

so that

$$\pi_2 = \pi_1 \alpha_2/\alpha_1$$

and

$$\pi_1^* = \frac{\alpha_1(\alpha_1 + \alpha_2 + \beta)}{(\alpha_1 + \alpha_2)\theta}.$$

Since also $\alpha_2/\pi_2 + \beta/(\pi_1 + \pi_2) - \theta = 0$, it follows that

$$\pi_2^* = \frac{\alpha_2(\alpha_1 + \alpha_2 + \beta)}{(\alpha_1 + \alpha_2)\theta}.$$

Substituting θ for $\alpha_1/\pi_1 + \beta/(\pi_1 + \pi_2)$ in f_3 and θ for $\alpha_2/\pi_2 + \beta/(\pi_1 + \pi_2)$ in f_4, we obtain at equilibrium

$$p_3 = \gamma_1 + \tfrac{1}{2}\gamma_2 + p_3(1 - \theta) + \tfrac{1}{2}\theta(u_{13} + u_{23})$$
$$p_4 = \gamma_3 + \tfrac{1}{2}\gamma_2 + p_4(1 - \theta) + \tfrac{1}{2}\theta(u_{14} + u_{24})$$

or

$$\gamma_1 + \tfrac{1}{2}\gamma_2 - \theta(u_{33} + \tfrac{1}{2}u_{34}) = 0$$
$$\gamma_3 + \tfrac{1}{2}\gamma_2 - \theta(u_{44} + \tfrac{1}{2}u_{34}) = 0$$

Karlin (1978) showed that, for all such models at equilibrium, the genotypes arise in Hardy-Weinberg ratios. Therefore,

$$u_{33} = p_3^2, \quad u_{34} = 2p_3 p_4, \quad u_{44} = p_4^2,$$

and hence

$$p_3(p_3 + p_4) = (\gamma_1 + \tfrac{1}{2}\gamma_2)/\theta$$

$$p_4(p_3 + p_4) = (\gamma_3 + \tfrac{1}{2}\gamma_2)/\theta.$$

Since

$$p_3 + p_4 = 1 - p_1 - p_2$$

$$= \sqrt{1 - \pi_1 - \pi_2}$$

$$= \sqrt{\frac{\gamma_1 + \gamma_2 + \gamma_3}{\theta}},$$

therefore

$$p_3^* = \left(\frac{\gamma_1 + \tfrac{1}{2}\gamma_2}{\gamma_1 + \gamma_2 + \gamma_3}\right)\sqrt{\left(\frac{\gamma_1 + \gamma_2 + \gamma_3}{\theta}\right)}$$

$$p_4^* = \left(\frac{\gamma_3 + \tfrac{1}{2}\gamma_2}{\gamma_1 + \gamma_2 + \gamma_3}\right)\sqrt{\left(\frac{\gamma_1 + \gamma_2 + \gamma_3}{\theta}\right)}.$$

The full set of equilibrium phenotypic frequencies are therefore

$$\pi_1^* = \frac{\alpha_1(\alpha_1 + \alpha_2 + \beta)}{(\alpha_1 + \alpha_2)\theta} \qquad \pi_2^* = \frac{\alpha_2(\alpha_1 + \alpha_2 + \beta)}{(\alpha_1 + \alpha_2)\theta}$$

$$\pi_3^* = \frac{(\gamma_1 + \tfrac{1}{2}\gamma_2)^2}{\theta(\gamma_1 + \gamma_2 + \gamma_3)} \qquad \pi_4^* = \frac{2(\gamma_1 + \tfrac{1}{2}\gamma_2)(\gamma_3 + \tfrac{1}{2}\gamma_2)}{\theta(\gamma_1 + \gamma_2 + \gamma_3)}$$

$$\pi_5^* = \frac{(\gamma_3 + \tfrac{1}{2}\gamma_2)^2}{\theta(\gamma_1 + \gamma_2 + \gamma_3)},$$

with gene frequencies

$$p_1^* = 1 - \sqrt{1 - \frac{\alpha_1(\alpha_1 + \alpha_2 + \beta)}{\theta(\alpha_1 + \alpha_2)}}$$

$$p_2^* = \sqrt{1 - \frac{\alpha_1(\alpha_1 + \alpha_2 + \beta)}{\theta(\alpha_1 + \alpha_2)}} - \sqrt{\frac{\gamma_1 + \gamma_2 + \gamma_3}{\theta}}$$

$$p_3^* = \left(\frac{\gamma_1 + \tfrac{1}{2}\gamma_2}{\gamma_1 + \gamma_2 + \gamma_3}\right)\sqrt{\frac{\gamma_1 + \gamma_2 + \gamma_3}{\theta}}$$

$$p_4^* = \left(\frac{\gamma_3 + \tfrac{1}{2}\gamma_2}{\gamma_1 + \gamma_2 + \gamma_3}\right)\sqrt{\frac{\gamma_1 + \gamma_2 + \gamma_3}{\theta}}.$$

This equilibrium is easily shown to be globally stable using the gene frequency recursion equations. Fixation states are unstable, showing stability of the single internal polymorphic equilibrium derived above.

Extending the inclusive preference model to allow for separate preferences for both Q and X has a dramatic effect on the evolutionary outcome of the selection. The phenotype X is no longer eliminated: it is maintained at stable equilibrium with Q. The relative frequencies of Q and X are simply determined by the ratio of the separate preferences in their favor. These preferences may be very small compared to the preference for Q and X as an inclusive class of preferred phenotypes. But provided some females prefer $\{Q\}$ rather than $\{Q, X\}$ and others prefer $\{X\}$, a stable polymorphism will be maintained, even though the preference for the inclusive class $\{Q, X\}$ may be much the largest component. In Karlin and O'Donald's original inclusive preference model, the extra preference was assumed to exist only for the more extreme Q phenotype in the dominance series: some females are sufficiently stimulated by either melanic Q or X to mate preferentially with either; others, more choosy, require the greater stimulus of the more extreme melanic Q. The model analyzed here implies that, although most females prefer either Q or X, some are more choosy for Q and others are more choosy for X. In the general genetic preference for melanics, either Q or X, there must in fact be a range of "choosiness" among the females. Our model assumes that some, at one end of the range, have the stronger preference for Q; others, at the other end of the range, have the stronger preference for X, thus producing some females that exercise a preference for Q, others that exercise a preference for X. Strength of preference must in fact vary between females; the variation might itself be partly genetic: the expression of the gene for the preference for Q and X might be subject to modifier genes. The important point is that in the genetic preference for the melanics Q and X, an extra choosiness by a few females for Q and by a few others for X will be sufficient to maintain the stable polymorphism of both melanics.

Experiments on Expression of Preference in Ladybirds

In previous experiments it has been shown that females have a genetic preference, probably determined by a single dominant gene, for the melanic phenotypes Q and X (Majerus, O'Donald, and Weir 1982b; Majerus et al. 1986). No preference has been observed for the *typica* (T) phenotype. Selection of females that mated with T males merely eliminated the preference for Q but produced no preference for T. Heterozygous *annulata* (A) appears to mate assortatively in a natural population, but experimental tests on A phenotypes of known genotype have not yet been undertaken. In natural populations, A phenotypes are probably heterozygotes of at least three different alleles, although the corresponding homozygotes may also be involved in assortative mating.

The experiments carried out hitherto have not been designed to determine what male characters are the objects of female choice. Do the females choose on the basis of melanism, or do they choose by some pleiotropic effect of the gene for melanism—perhaps a pheromonal effect? If females choose by melanism, then we should expect that preference would be expressed in proportion to the degree of melanism of the phenotype. According to the corresponding inclusive preference model, this would produce polymorphism with only the most extreme melanic maintained in the population. All other melanics will be eliminated unless there is some additional preference for each of them, as for example when there is some additional preference for both Q and X separately.

In the natural population at Keele from which our experimental stocks have been derived, Q and X occur at frequencies

$$\pi_1 = 0.29 \quad \text{(of } Q\text{)}$$

$$\pi_2 = 0.043 \quad \text{(of } X\text{)}.$$

We know that the preference for Q and X is about $\beta = 0.2$ in the natural population. Estimates consistent with this value have also been obtained in tests on samples in population cages and mating chambers (O'Donald et al. 1984). These experimental estimates were derived from tests using only Q and T males. An overall average of $\beta = 0.24$ was obtained from a large number of such tests (O'Donald and Majerus 1984). This figure will of course include any excess preference for Q not shared by X. Thus in the general inclusive preference model of the previous section it represents the preference $\alpha_1 + \beta$. Strictly, we should put

$$\alpha_1 + \beta = 0.24$$

if the model is valid for the Keele population. The ratio of the frequencies of Q and X is then proportional to the separate preferences α_1 and α_2 for Q and X, respectively. Thus

$$\pi_1^* = 0.29 = \frac{\alpha_1(\alpha_1 + \alpha_2 + \beta)}{\theta(\alpha_1 + \alpha_2)}$$

$$\pi_2^* = 0.043 = \frac{\alpha_2(\alpha_1 + \alpha_2 + \beta)}{\theta(\alpha_1 + \alpha_2)}.$$

Hence

$$0.043/0.29 = \alpha_2/\alpha_1 = 0.15.$$

If the total preference for Q, $\alpha_1 + \beta$, is very similar to the total preference for X, $\alpha_2 + \beta$, it follows that since $\alpha_2 = 0.15\alpha_1$, α_1 must itself be a small quantity in order not to be statistically significant in comparison with β.

EXPERIMENTAL DESIGN

Females from two of the high preference isofemale lines and two of the low preference isofemale lines were used to test the level of preference for a range of melanic phenotypes. The high preference lines were Y19 and Z37; the low preference lines were Y4 and Z33. Lines Y19 and Z37 gave estimated preferences of about 80% (Majerus et al. 1986); Y4 and Z33 preferences of about 4%, not differing significantly from zero. Y4 and Y19 consist solely of T phenotypes (having been produced by $T \times T$ matings). Z33 consists only of Q phenotypes. Z37 consists of both Q and T. In the mating preference tests, females used were as follows:

$$Q \female\female \text{ from Z37 and Z33}$$

$$T \female\female \text{ from Y19 and Y4.}$$

Each female was offered two types of male, one male from series A and the other male from series B, as shown in the table.

Series A males	Series B males
typica (*T*) from Keele	*quadrimaculata* (*Q*), from Keele
quadrimaculata (*Q*) from Keele	*sexpustulata* (*X*) from Keele
	melanic *annulata* from Cambs.
	extreme *annulata* from Keele
	duodecempustulata from Cambs.
	spotty from Cambs.
	bar *annulata* from Keele
	typica from Keele

Figure 11.1 shows the pattern of melanic pigment in these forms, in the order of the degree of melanism. Melanic *annulata* from Cambridgeshire is produced by the *annulata* homozygote $s^a s^a$ in conjunction with a modifier for increased melanism: it has a phenotype very similar to the melanic *sexpustulata*, but is in effect a modified nonmelanic genotype. Extreme *annulata* is produced by the homozygote $s^a s^a$. Spotty and *duodecempustulata* are homozygous for s^{sp} and s^a. Bar *annulata* phenotypes were produced from crosses of $s^t s^t \times s^a s^a$ and are thus known to be $s^t s^a$ hetero-

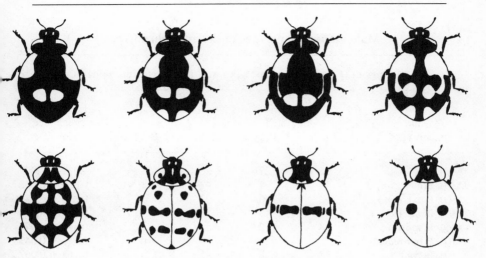

Figure 11.1. Phenotypes of the 2-spot ladybird used in the mating tests. The figure shows the following eight phenotypes: top row, left to right, *quadrimaculata, sexpustulata,* melanic *annulata,* extreme *annulata*; bottom row, left to right, *duodecempustulata,* spotty, bar *annulata, typica.*

zygotes. The order of the genotypes in the list reflects the level of melanism, with *quadrimaculata* the most melanic and *typica* the least.

Four series of mating tests were carried out. The males in series B were tested both against *typica* A males and against *quadrimaculata* A males. The tests, were carried out with both high and low preference females. The results, shown in Table 11.1, are very striking. High preference females strongly prefer Q and X males, which are the normal melanic forms determined by alleles s^m and s^p. Other males, even melanic *annulata,* which closely resembles X, but which is determined by the genotype $s^a s^a$ for the nonmelanic extreme *annulata* in combination with a modifier, is the object of a preference no greater than any of the other nonmelanic phenotypes—no preference at all. There is no significant difference in preference when nonmelanics, including melanic *annulata* and extreme *annulata,* are tested against *typica.* The only two significant effects are (1) in the comparison of Q and X against all other phenotypes when chosen by the high preference females, and (2) in a difference between lines of low preference females when choosing between T and other males. Table 11.2 shows the full analysis of χ^2 for all four sets of mating tests. The analysis has four components: these are the separate components of the analysis of the $2 \times 2 \times 7$ contingency tables in which there are two lines of females (e.g., lines Y19 and Z37 with high preference), A and B males, and seven

Table 11.1

Male mating frequencies of different phenotypes.

Matings of Q and other phenotypes with high preference females

Male phenotypes	Y19♀♀ A♂♂	Y19♀♀ B♂♂	Z37♀♀ A♂♂	Z37♀♀ B♂♂	Estimate of preference for A
A(Q), B(X)	25	27	28	22	0.039 ± 0.099
A(Q), B(mel. *ann.*)	43	4	42	1	0.889 ± 0.048
A(Q), B(ext. *ann.*)	52	6	37	3	0.816 ± 0.058
A(Q), B(*duo.*)	41	3	46	4	0.851 ± 0.054
A(Q), B(spotty)	48	2	53	5	0.870 ± 0.047
A(Q), B(bar *ann.*)	37	4	50	0	0.912 ± 0.043
A(Q), B(*typ.*)	37	8	45	3	0.763 ± 0.067

Matings of Q and other phenotypes with low preference females

Male phenotypes	Y4♀♀ A♂♂	Y4♀♀ B♂♂	Z33♀♀ A♂♂	Z33♀♀ B♂♂	Estimate of preference for A
A(Q), B(X)	30	23	19	35	-0.084 ± 0.096
A(Q), B(mel. *ann.*)	28	20	26	21	0.137 ± 0.102
A(Q), B(ext. *ann.*)	22	28	23	27	-0.100 ± 0.099
A(Q), B(*duo.*)	27	21	23	27	0.020 ± 0.101
A(Q), B(spotty)	17	29	23	23	-0.130 ± 0.103
A(Q), B(bar *ann.*)	30	22	21	30	-0.010 ± 0.099
A(Q), B(typ.)	25	25	33	20	0.126 ± 0.098

NOTE: In this table we have phenotypes: Q ... *quadrimaculata* (4-spot melanic); X ... *sexpustulata* (6-spot melanic); mel. *ann.* ... melanic *annulata*; ext. *ann.* ... extreme *annulata*; *duo.* ... *duodecempustulata*; spotty; bar *ann.* ... bar *annulata*; T ... *typica*.

continued

phenotypes within the B series of males. The components, as shown in Table 11.2, are as follows:

1. Between phenotypes of males. This component tests whether females differ in their preferences for the different phenotypes.
2. Between female lines. This tests whether there are differences in the numbers of mating females in the two female lines. It depends mainly

Table 11.1 (continued)

Matings of T and other phenotypes with high preference females

Male phenotypes	Y19♀♀ A♂♂	B♂♂	Z37♀♀ A♂♂	B♂♂	Estimate of preference for B
A(T), B(Paris Q)	6	40	8	40	0.684 ± 0.075
A(T), B(X)	6	42	3	45	0.812 ± 0.059
A(T), B(mel. *ann.*)	23	18	25	30	0.000 ± 0.102
A(T), B(ext. *ann.*)	23	30	22	24	0.091 ± 0.100
A(T), B(*duo.*)	28	20	26	28	−0.059 ± 0.099
A(T), B(spotty)	31	24	21	27	−0.010 ± 0.099
A(T), B(bar *ann.*)	27	23	22	27	0.010 ± 0.100

Matings of T and other phenotypes with low preference females

Male phenotypes	Y4♀♀ A♂♂	B♂♂	Z33♀♀ A♂♂	B♂♂	Estimate of preference for B
A(T), B(Paris Q)	28	23	23	30	0.019 ± 0.098
A(T), B(X)	21	25	18	29	0.161 ± 0.102
A(T), B(mel. *ann.*)	23	25	23	30	0.089 ± 0.099
A(T), B(ext. *ann.*)	25	18	35	18	−0.250 ± 0.099
A(T), B(*duo.*)	33	17	25	30	−0.105 ± 0.097
A(T), B(spotty)	30	20	23	30	−0.029 ± 0.098
A(T), B(bar *ann.*)	28	23	23	25	−0.030 ± 0.100

on the total numbers of females used for testing and is not relevant to the experiment.

3. Between female lines and totals of males. This tests the significance of the 2 × 2 table of female lines and total of males over all phenotypes. It tests whether there may have been an overall difference in preference between the two female lines.
4. Residual interaction. This tests whether females in the two lines have different preferences for particular phenotypes.

The analysis shows what is immediately obvious from Table 11.1. The females from the high preference lines (selected, remember, for a preference for Q males) show their high preference for both the melanics Q and X,

Table 11.2
Analysis of χ^2 of data of male mating frequencies.

Matings of Q and other phenotypes with high preference females

Component of variation	Value of χ^2	Dfs	Value of P
1. Between phenotypes of males:			
Between melanics and nonmelanics	121.114	1	—
Within melanics and nonmelanics	2.849	5	0.723
2. Between female lines (Y19 and Z37)	5.480	6	0.484
3. Between female lines and totals of males	3.331	1	0.068
4. Residual interaction	4.696	6	0.583
Total	137.470	19	

Matings of Q and other phenotypes with low preference females

Component of variation	Value of χ^2	Dfs	Value of P
1. Between phenotypes of males	6.770	6	0.343
2. Between female lines (Y4 and Z33)	0.135	6	1.000
3. Between female lines and totals of males	0.967	1	0.325
4. Residual interaction	11.133	6	0.084
Total	19.005	19	

Matings of T and other phenotypes with high preference females

Component of variation	Value of χ^2	Dfs	Value of P
1. Between phenotypes of males:			
Between melanics and nonmelanics	79.759	1	—
Within melanics and nonmelanics	2.051	5	0.842
2. Between female lines (Y19 and Z37)	3.378	6	0.760
3. Between female lines and totals of males	2.227	1	0.136
4. Residual interaction	2.965	6	0.813
Total	90.380	19	

Matings of T and other phenotypes with low preference females

Component of variation	Value of χ^2	Dfs	Value of P
1. Between phenotypes of males	10.274	6	0.114
2. Between female lines (Y4 and Z33)	1.001	6	0.986
3. Between female lines and totals of males	5.057	1	0.025
4. Residual interaction	5.535	6	0.477
Total	21.867	19	

NOTE: In this table "melanics" means Q and X phenotypes.

but show no preference for other phenotypes, not even for melanic *annulata*, which is phenotypically similar to X (*sexpustulata*). Thus we have proved that melanic pigmentation is not the character by which the female choice is exercised. There is no increasing preference for increasing levels of melanism. Evidently the alleles s^m (for Q) and s^p (for X) have some other pleiotropic effect, which perhaps may be pheromonal, that determines the female preference. This effect is not produced by the other alleles, s^t, s^a, and spotty whether combined with modifiers for melanism or not.

The two low preference lines of females, Y4 and Z33, appear to differ slightly in the frequencies of their matings with the T and other males. The totals of matings in Table 11.1 (iv) are as follows:

Y4♀♀		Z33♀♀	
A♂♂	B♂♂	A♂♂	B♂♂
188	151	170	192

The Y4 females mated more often with the A(T) males: the Z33 females mated more often with the B series of males. This gives rise to the significant value

$$\chi_1^2 = 5.057$$

in Table 11.2 (iv). But there is no difference between these lines when choosing between Q and other males.

Conclusion

As we have seen, models with preferences that increase with melanism all predict that only the most extreme (i.e., Q) melanic will ultimately survive. This prediction, derived originally from Karlin and O'Donald's (1981) inclusive preference model, was refuted by the general observation that Q and X and other rarer melanics are usually found in populations polymorphic for melanic phenotypes. We have now refuted the model at the level of the mechanism of choice: females do not choose by the melanism. However, the extended inclusive preference model, analyzed in the previous section of this paper, may still apply to the pleiotropic effect of the alleles s^m and s^p on which the choice, we must now presume, is exercised. All we require is that females should show a spectrum of choice in favor of the melanics, from a greater preference for Q to a greater preference for X. Experimentally, we must test the relative preferences for Q and X over a range of frequencies. This will provide estimates of the separate preferences for Q and X. We know the relative magnitudes of these preferences from the equilibrium frequencies of Q and X. As shown in the previous section,

$$\pi_1^*/\pi_2^* = \alpha_1/\alpha_2,$$

giving

$$\alpha_2 = 0.15\alpha_1.$$

Since the preference for X can only be 15% of the preference for Q, we must include very low frequencies of X against Q in order to detect the frequency-dependent mating advantage of X produced by the low preference α_2. So far, we have only carried out tests of Q and X at the same frequencies, This procedure cannot detect the separate preferences for Q and X postulated in the model. The mating tests required to corroborate or refute the generalized inclusive preference model will be carried out in future experiments.

For the present, we regard the inclusive preference model, developed in collaboration with Sam Karlin, as the most plausible explanation of the polymorphism of melanic ladybirds. We are pleased to acknowledge that Sam's general multiallelic models of sexual selection (Karlin 1978) were a major inspiration of our work on ladybirds, providing a basis for the analysis and estimation of the selection of the phenotypes.

References

Creed, E. R. 1971. Industrial melanism in the Two-spot Ladybird and smoke abatement. *Evolution* 25: 290–293.

Karlin, S. 1978. Comparisons of positive assortative mating and sexual selection models. *Theor. Pop. Biol.* 14: 281–312.

Karlin, S., and O'Donald, P. 1981. Sexual selection at a multiallelic locus with complete or partial dominance. *Heredity* 47: 209–220.

Lusis, J. J. 1928. On inheritance of colour and pattern in ladybeetles *Adalia bipunctata* and *Adalia decempunctata. Izv. Byuro. Genet. Leningrad* 6: 89–163.

Majerus, M.E.N.; O'Donald, P.; Kearns, P.W.E.; and Ireland, H. 1986. Genetics and evolution of female choice. *Nature* 321: 164–167.

Majerus, M.E.N.; O'Donald, P.; and Weir, J. 1982a. Evidence for preferential mating in *Adalia bipunctata. Heredity* 49: 37–49.

Majerus, M.E.N.; O'Donald, P.; and Weir, J. 1982b. Female mating preference is genetic. *Nature* 300: 521–523.

O'Donald, P. 1973. Models of sexual and natural selection in polygynous species. *Heredity* 31: 145–156.

O'Donald, P. 1977. The mating advantage of rare males in models of sexual selection. *Nature* 267: 151–154.

O'Donald, P. 1980. *Genetic Models of Sexual Selection.* Cambridge University Press, Cambridge, Eng.

O'Donald, P., and Majerus, M.E.N. 1984. Polymorphism of melanic ladybirds maintained by frequency-dependent sexual selection. *Biol. J. Linn. Soc.* 23: 101–111.

O'Donald, P., and Majerus, M.E.N. 1985. Sexual selection and the evolution of preferential mating in ladybirds, I. Selection for high and low lines of female preference. *Heredity* 55: 401–412.

O'Donald, P.; Derrick, M.; Majerus, M.E.N.; and Weir, J. 1984. Population genetic theory of the assortative mating, sexual selection and natural selection of the Two-Spot Ladybird *Adalia bipunctata. Heredity* 52: 43–61.

CHAPTER TWELVE

The Evolution of Marine

Life Cycles

Jonathan Roughgarden

Biological Background

KINDS OF LIFE CYCLES

Although the marine environment harbors animal species with a great diversity of life cycles, some generalizations have long been known. By 1900 two cycles had been identified for the species that have their adult phase in a benthic habitat (i.e., the adult lives on, or in, a rocky or muddy substrate). Either the entire life cycle is spent in the benthic habitat, or the life cycle begins with a morphologically distinctive larval phase that feeds in the water column (a pelagic larva) before metamorphosis into the benthic adult phase. Furthermore, in the entirely benthic life cycle, either the adult gives birth to juveniles that have developed within the parent, or the adult lays eggs in protective egg capsules that it attaches to the substrate. Such eggs then hatch into juveniles with the same morphology as the adult. By either developmental mechanism, a larval phase is bypassed.

A pelagic larval phase implies long-range dispersal because a larva typically feeds for weeks to months in the coastal waters, and is carried scores of kilometers during that time. Also, a pelagic larval phase implies that an individual experiences two types of environments during its life—it must grow and survive in the conditions of both the water column and the benthic habitat. Conversely, the absence of a pelagic larval phase entails limited dispersal, and a life lived in only one environmental circumstance.

This distinction between a purely benthic life cycle and a mixed pelagic-benthic life cycle initiated attempts to identify and classify still more types of life cycles. Here is a summary that encompasses much of the variety, beginning with a completely pelagic life cycle, extending through various types of mixed pelagic-benthic cycles, and ending with completely benthic life cycles.

1. Completely pelagic life cycle—a great many species, from several phyla, that reside in the water column for their entire life cycle.

2. Pelagic life cycle except for benthic spawning—species, such as salmon, that reside in the water column their entire life except for a brief period devoted almost exclusively to reproduction. Also, this category may include hydrozoans that exist primarily in a pelagic medusid phase with a reduced sessile hydroid phase.

3. Tiny feeding larvae with benthic adults—a large proportion of invertebrates. The cycle begins with a tiny larva that feeds for weeks or months in the water column until a certain size necessary for metamorphosis is attained. These larvae are called "planktotrophic" because they consume plankton, usually algae, as food. While most such larvae are near the surface, and are termed "pelagic," still other species have tiny feeding larvae that are nearly sedentary on the ocean floor, and are called "demersal." After metamorphosis to the adult morphology, the juveniles also feed, but in the benthic habitat, until they attain a size at which reproduction begins.

4. Large nonfeeding larvae, or no larvae at all, with benthic adults—species either with live birth (viviparity) or with large eggs deposited in protective egg capsules that hatch into small replicas of an adult (direct development). Also, some species produce a large larva that settles within minutes of release and metamorphoses into the adult morphology, as with the tadpole larva of colonial ascidians. Finally, some species produce a large relatively long-lived larva provisioned with enough yolk so that feeding is not necessary prior to metamorphosis. Such a larva is called "lecithotrophic," because it feeds upon its yolk during development. Thus this category lumps various patterns of development and embryogenesis that share the feature of having all growth (gain in weight) occur in the benthic habitat.

5. Reproduction through budding by benthic adults—species, often colonial, as, for example, most corals, bryozoans, many sea anemones, tunicates, and sponges that reproduce through binary fission or by budding off fully functional adults from specialized polyps. However, few such species reproduce only by budding; colonial ascidians usually brood embryos and release a large tadpole larva, while corals release a tiny short-lived planula larva.

This classification offers an overview of the variety of life cycles that have to be explained; but classifying the diversity remains a difficult prerequisite to formulating the evolutionary issues. Perhaps several distinct types of life cycles have been incorrectly lumped into the classification above. An especially vexing problem is that larvae classified as nonfeeding may, in fact, be feeding on bacteria or on organic substances dissolved in sea water. Nonetheless, a spectrum is evident that extends from completely pelagic, through mixed cycles with a nonreproductive planktonic phase and a reproductive benthic phase, to completely benthic.

Evolutionary Issues

During the early 1900s ecological correlations were suggested for the life cycles of some species, but the present-day discussion was initiated by the great Danish marine biologist, Gunnar Thorson (1936, 1946, 1950). Here are the main issues:

1. Most invertebrates with some benthic part to their life cycle have a pelagic larva (Mileikovsky 1971), and many, perhaps a majority, of these are planktotrophic. Therefore, is there some advantage to a pelagic larva, and to a planktotrophic larva in particular, that might account for the prevalence of these life cycles? Thorson mentioned two further points:

a. The planktotrophic life cycle inherently involves a huge "wastage" of larvae; each adult produces tens of thousands of larvae, nearly all of which are either eaten while in the water column, or carried away from shore by oceanic currents.

b. The population dynamics of species with pelagic larvae are very erratic because of their susceptibility to large unpredictable changes in the offshore currents relative to otherwise similar species having direct-developing benthic larvae, or no larvae at all.

Thus, in understanding why planktotrophic larvae are so common among species, one may also explain why so many marine species suffer large larval wastage and erratic changes in population size.

2. Does the sequence of life cycles ranging from purely pelagic, through mixed cycles, to purely benthic represent distinct classes of strategies or does the sequence represent points uniformly distributed along a continuum of possible strategies?

Thorson observed that the size of the egg together with any accompanying nutritive and protective material may serve as a morphological index of a planktotrophic or direct-development life cycle. Tiny eggs, barely visible to the naked eye, are typical of life cycles with a planktotrophic larval phase, while large eggs (e.g., 1 mm or more in diameter) are typical of life cycles where the embryos develop in benthic egg capsules. Moreover, the egg size shows little variation within a species. Therefore, the distribution of egg sizes from a collection of species can indicate the distribution of life-cycle types among those species. Figure 12.1 shows the distribution of egg diameters for a group of snails (Prosobranchs) from the vicinity of Denmark. A clear tendency to bimodality is apparent. While

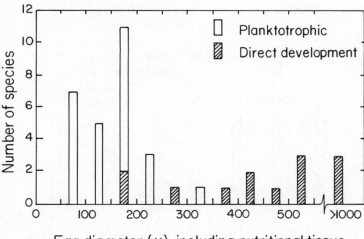

Figure 12.1. Histogram of egg diameter for Prosobranch snails, plotted from data in Thorson (1946). The solid bars (or parts thereof) refer to species with direct development, and the open bars to species with planktotrophic development. For species with direct development, any accompanying nutritive material is included by computing the diameter of an effective sphere consisting of both the egg and its supplementary material.

most species have tiny eggs and a planktotrophic larval phase, a subset has large eggs and direct development. These data underlie the proposition that the life-cycle categories represent distinct classes of strategies.

Nonetheless, contemporary investigators with other groups of animals usually do not confirm this observation of bimodality in egg size.

a. In some groups, like acorn barnacles, direct development is nearly absent (Strathmann 1977). Such groups consist almost entirely of the planktotrophic-larva benthic-adult life cycle, inviting the suspicion that a "phylogenetic constraint" retards the evolution of alternative life cycles.

b. In groups having a small percentage of direct-developing species, the distribution of egg sizes is not bimodal but positively skewed (cf., Perron and Carrier 1981; Grant 1983).

Still, the distribution of marine life cycles is neither uniform nor normal, and perhaps can be thought of as concentrated at foci representing distinct classes of life cycles.

3. Thorson discovered a latitudinal gradient in the percentage of species having a planktotrophic life cycle in contrast to a direct-development life cycle. Consider a set of comparable species, say, all those in the same genus, or family. Thorson observed that the fraction of direct-development species in such a set increases monotonically from 0 to nearly 1 as one

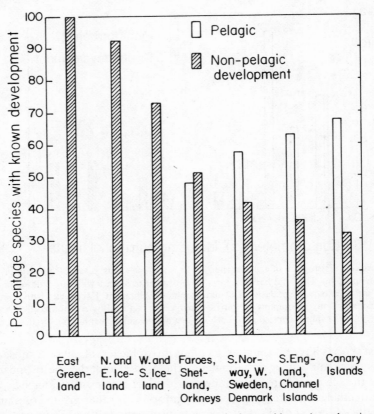

Figure 12.2. Distribution of planktotrophic and nonplanktotrophic species at locations of differing latitude in the North Atlantic (adapted from Thorson 1950).

traverses a path from the equator to the North Pole. As Figure 12.2 shows for a group of snails, toward the equator all species have pelagic larvae while above the arctic circle most have direct-development life cycles. This latitudinal gradient has been strongly corroborated with further data on snails and for some other groups (Mileikovsky 1971). More recently, however, Pearse (pers. comm.) has pointed out that this generalization may not be as valid as often believed; it evidently does not apply to sea urchins and may not be as valid in gradients from the equator to the South Pole as it is for gradients from the equator to the North Pole.

Furthermore, Thorson also noted that lecithotropic larvae predominate in the tropics, suggesting a latitudinal gradient whereby the fraction of lecithotropic species declines on a path from the equator to the poles.

4. Thorson did not discuss fish or plants, and discussions of marine life cycles are normally focused only on invertebrates with a benthic adult phase.

The life cycle of marine fish seems to accord qualitatively with that of invertebrates. Fish that are benthic as adults, for example, those with territories or home ranges, usually have pelagic larvae. Ecologists studying coral reef fish report a pattern of population dynamics for the benthic population (the stock) governed by larval settlement similar in principle to that of invertebrates with pelagic larvae (Victor 1983, 1986).

Like benthic marine animals, many seaweeds have biphasic live cycles. But in contrast, the phases involve an alternation of ploidy, and both phases are typically benthic. A life cycle in which one of the phases of a plant releases a tiny spore to the water column that then grows there to a medium-sized structure, such as a ball or sheet, before attaching to benthic substrate for further growth and for reproduction, is rare or nonexistent for seaweeds. Yet this is the most common life cycle for benthic marine animals.

THE QUESTIONS POSED

Thus, a theory for the evolution of life cycles in marine animals should eventually answer all of the following questions:

1. Why are marine life cycles patterned in discrete types of strategies— ranging from purely pelagic, through pelagic-larva with a benthic-adult, to purely benthic?
2. Why do many species with a benthic adult stage possess a planktotrophic pelagic larva?
3. Why do most species groups possess only a subset of the possible types of life cycles?
4. Why do plant life cycles differ from animal life cycles?
5. Why may the percentage of species that have direct development increase, and that have lecithotrophic larvae decrease, from the equator to the poles?

Theoretical Background

LARVAL FEEDING

Vance (1973) offered the first theoretical contribution to the evolution of marine life cycles; the focus was on larval feeding. He posed the following strategy question: How much material should an adult place into each egg to maximize the number of offspring that eventually attain a certain size? At this desired size the larva undergoes metamorphosis to the adult phase. The breeding adult is understood to have a finite pool of material to place into eggs. Its options range from making many small eggs to a few large eggs. The small eggs give rise to larvae that must spend a long time feeding in the water column, while larger eggs produce larvae that spend less time in the plankton; in the extreme, eggs large enough to metamorphose immediately are regarded as bypassing the larval phase altogether, as in direct development.

To formulate this question Vance (1973) assumed a linear growth law in the larval phase

$$s(t) = g_l t + s_n, \tag{12.1}$$

where s is the size (weight) of the larva at time t after release, s_n its size when it was initially released, and g_l is the larval growth rate (in units of weight per time). If the size needed to metamorphose is s_c, then the duration of larval life, t_l, as a function of the initial larval size, s_n, needed to attain that size is

$$t_l(s_n) = \frac{(s_c - s_n)}{g_l}. \tag{12.2}$$

Also, the adult has some quantity of material to put into eggs, say Q. The number of eggs that can be produced is therefore Q/s_n. The number of offspring from the adult that eventually reach the size for metamorphosis is then the number of eggs produced, times the probability of survival through the larval phase. If the instantaneous mortality rate for larvae in the water column is v, then the number of offspring left by the adult that reach the size, s_c, is

$$N(s_n) = \left(\frac{Q}{s_n}\right) e^{-v(s_c - s_n)/g_l}. \tag{12.3}$$

This function is either monotonically decreasing or U-shaped on the interval $[0, s_c]$ depending on whether its minimum is greater than or less than s_c. This result was taken to indicate that selection either unequivocally favors lowering s_n to as low as possible [when $N(s_n)$ is monotonically decreasing], or favors two distinct strategies, both $s_n \to 0$ and $s_n \to s_c$ [when $N(s_n)$ is U-shaped]. Thus either all taxa should converge on tiny planktotrophic larva, or the taxa should split into two distinct classes, one with tiny planktotrophic larva and the other with a direct-development life cycle that bypasses the pelagic larval stage altogether.

Vance's (1973) strategy argument offered a point of departure for over a decade of further discussion (see review in Strathmann 1985). Among the modifications studied are sigmoid larval growth with size-dependent larval mortality (Christiansen and Fenchel 1979), a cost of egg encapsulation that increases with egg size (Perron and Carrier 1981), alternative adult fecundity functions (Caswell 1981), and composite formulations that combine all of these additional ingredients (Grant 1983). Nonetheless, no analysis to date rests on a population model for the entire life cycle; all the extensions of Vance's original contribution still pertain only to a piece of the life cycle.

LARVAL DISPERSAL

Another major function of a pelagic larval phase is to effect dispersal. After Vance's model was published, Crisp (1974, 1976), Doyle (1975), Frank (1981), and Palmer and Strathmann (1981) explored the evolutionary significance of the dispersal resulting from a pelagic larval phase. The common message from these papers can be expressed as a comparison between the arithmetic and geometric means of all the local-population growth rates that occur in the geographic region occupied by the species. In essence, a metapopulation (species population) with pelagic larvae grows by an arithmetic average of the various local-population growth rates, while a metapopulation with a direct-development life cycle grows by a geometric average of the local-population growth rates. Since the arithmetic average is greater than the geometric average, the life cycle with pelagic larvae is advantageous.

Consider, for example, two generations with both types of life cycles. Suppose four sites are initially started with 100 animals apiece, and that the population geometric growth factors (R) in these sites are $1/2$, 2, $1/2$, and 2, respectively. At time-1 the populations at the sites then are 50, 200, 50, and 200. In a direct-development species, these local populations remain at the sites. Let the next generation bring a new set of R's, say, 2, $1/2$, 2, and $1/2$. So, the populations at the sites at time-2 are 100, 100, 100, and 100 again. Compare now a species with pelagic larvae. All is the same after time-1. But to begin time-2, we combine the animals from the four sites and redistribute them. So time-2 begins with a population of 125 in each of the four sites. Then after time-2, the abundances are 250, 62.5, 250, and 62.5. So, the total for the metapopulation is 625. After two intervals, the metapopulation with pelagic larvae has increased by a factor of $625/400 = 25/16$, which is the square of the arithmetic average of the R's. In contrast, the nondispersing species with direct-development simply has increased by a factor of 1, which is the square of the geometric average of the R's.

This essential point is pursued in various ways by all the papers mentioned above. In particular, Frank (1981) notes, in effect, that both stages should not be mobile. If the larvae distribute at random, then the adults should stay put, for otherwise the adults would experience a geometric average of the mortality rates among the sites, thereby removing any advantage of the arithmetic averaging that has resulted from the larval dispersal.

The models investigating the importance of dispersal demonstrate, I think, that dispersal is not the major consideration in the evolution of the life cycle consisting of pelagic larvae and benthic adults. The advantage to dispersal discussed above, while genuine, is fully attained with only a little bit of dispersal. All that is needed is enough dispersal so the larvae "see" an independent sample of environmental conditions. The scale of dispersal has to exceed the scale of any spatial autocorrelation in environmental conditions. Similarly, Palmer and Strathmann (1981) conclude that "feeding larvae of benthic invertebrates with their concomitant long planktonic period, receive little if any advantage from increased scale of dispersal, and consequently . . . the advantage to planktotrophy . . . must lie in . . . the ability to produce a greater number of smaller eggs" (p. 308). In short, the function of the larval feeding stage *is* to feed; this feeding is not incidental snacking while on the road to a new location.

Moreover, if dispersal, rather than feeding, were the paramount consideration underlying a pelagic larval phase, one would expect the dispersal sometimes to be carried out by the adult while the larva is sedentary. In terrestrial insects, for example, the typical life cycle involves larvae (grubs or caterpillars) feeding intensively in a small area and winged adults who disperse both for mating and in searching for egg-laying sites. This life cycle is the reverse of that most common in the marine environment, but is just as effective in achieving the advantages of environmental averaging through dispersal.

Two Further Considerations

Todd and Doyle (1981) have suggested that in special circumstances the duration of the larval phase is such that the larvae arrive at a desired habitat at a critical time. The example comes from a very seasonal site in the North Atlantic. Also, Strathmann (1978) has emphasized that evolution of a lecithotrophic larva, or a direct-development life cycle, both of which involve the loss of larval feeding structures, is irreversible. These structures, once lost, are apparently not recoverable.

The Evolution of Coloniality

Several phyla have sessile adults that are especially prone to coloniality, including corals, ascidians, sponges, bryozoans, and coelenterates. A literature on the evolution of coloniality has been developing recently, especially through the efforts of Jackson and coworkers (cf. Jackson 1977, 1986; Jackson and Winston 1982; Palumbi and Jackson 1983; Coates and Jackson 1985), much of which has focused on the advantages to coloniality in competition for space, and on the relationship between coloniality and whether the mode of reproduction is asexual or sexual. This literature is independent of the earlier literature that focused on feeding and dispersal reviewed above. It would be interesting, therefore, if a formulation could be found that could theoretically integrate the life histories of phyla where solitary life forms predominate with the life histories of phyla that are typically colonial.

The Theoretical Task

What remains to be developed then, is theory for the evolution of a planktotrophic larval phase that takes into account the entire life cycle, and that admits the potential of tradeoffs among different phases of the entire life cycle. A formulation to accomplish this task is sketched in the remainder of this chapter. Also, this new formulation allows a wider range of questions to be addressed than was previously possible.

A Population Model for a Marine Life Cycle

Population Dynamics

To obtain a population-dynamic model for the full life cycle, we adapt a formulation developed primarily for rocky intertidal barnacles (Roughgarden, Iwasa, and Baxter, 1985; Roughgarden and Iwasa 1986; Iwasa and Roughgarden 1985), but that seems simple enough to serve as a metaphor for many other organisms as well. The environment is imagined to consist of a benthic substrate of total area A, together with an offshore water column where a "larval pool" can exist. The $N(x, t)dx$ is the number of adult organisms on the benthic substrate whose age is between x and $x + dx$ at time t. Similarly, $L(x, t)$ is the number of larval organisms whose age is between x and $x + dx$ at time t. Age is measured separately for larvae and adults; for larvae it is measured from their time of release into the water column, and for adults from their time of settlement onto the benthic substrate. The population dynamics of this population then satisfy

$$\frac{\partial N(x, t)}{\partial t} + \frac{\partial N(x, t)}{\partial x} = -\mu(x)N(x, t)$$

$$N(0, t) = F(t) \int_0^t c(x)L(x, t)\,dx$$

$$A \equiv F(t) + \int_0^t a(x)N(x, t)\,dx \tag{12.4}$$

$$\frac{\partial L(x, t)}{\partial t} + \frac{\partial L(x, t)}{\partial x} = -[v(x) + F(t)c(x)]L(x, t)$$

$$L(0, x) = \int_0^t b(x)L(x, t)\,dx + \int_0^t m(x)N(x, t)\,dx.$$

The first equation pertains to the aging and mortality of organisms in the benthic habitat, where the instantaneous age-dependent mortality rate is $\mu(x)$. The second equation describes the settlement into the benthic habitat; this occurs with mass-action kinetics onto the vacant space, $F(t)$, on the substrate. The $c(x)$ is a rate coefficient for settlement by larvae of age x, and the integral is taken over all ages of larvae in the larval pool. The third equation stipulates that the total area of the substrate is conserved. The integral here is the occupied space on the substrate, where $a(x)$ is the age-dependent basal area for a sessile adult, or the home range size for a sedentary adult, and the integral is taken over all ages of adults in the benthic habitat. The fourth equation describes the aging and mortality of organisms in the larval pool, with mortality rate $v(x)$. The last equation describes the formation of larvae. In principle, new larvae could be produced by older organisms in the larval pool, with a birth function $b(x)$, and also by the organisms from the benthic habitat, with a corresponding birth function of $m(x)$. Few species are known that actually do reproduce simultaneously in both stages (some coelenterates reproduce both as sessile polyps and as pelagic medusae), but the reason why selection does not seem to favor this situation in general cannot be discussed in a model that does not allow the possibility to begin with. Typically, the model above will be used where either $m(x)$ is identically zero, to indicate a purely pelagic life cycle, or where $b(x)$ is identically zero, to indicate any of the life cycles in which the reproductive phase is benthic.

The dynamics for this model have not been fully solved. If both $m(x)$ and $c(x)$ are zero, while $b(x)$ is positive, then the larval population is simply the exponentially growing age-structured population of classical demographic theory. In the other extreme, both $c(x)$ and $m(x)$ positive, and $b(x)$ zero, the population often approaches a stable steady state as a result of the space limitation in the benthic habitat, and some special cases where this steady state is stable have been detailed in Roughgarden and Iwasa (1986).

THE STEADY STATES

The steady state, whether stable or not, is easily exhibited. Actually, there are two steady states, one representing extinction $N(x, t) \equiv 0$, $L(x, t) \equiv 0$, and one representing the interior steady state, $\hat{F}$, obtained as the root of

$$\hat{F} = \frac{1 - \bar{b}}{\bar{c}\bar{m}}, \tag{12.5a}$$

where

$$\bar{b} = \int_0^\infty b(x)e^{-\int_0^x [v(u) + \hat{F}c(u)]\,du}\,dx$$

$$\bar{c} = \int_0^\infty c(x)e^{-\int_0^x [v(u) + \hat{F}c(u)]\,du}\,dx$$

$$\bar{m} = \int_0^\infty m(x)e^{-\int_0^x \mu(u)\,du}\,dx \tag{12.6}$$

$$\bar{a} = \int_0^\infty a(x)e^{-\int_0^x \mu(u)\,du}\,dx.$$

The $\hat{F}$ indicates the degree of vacant space in the benthic habitat at the steady state, and the remaining variables at steady state are found successively from

$$A \equiv \hat{F} + \bar{a}\hat{N}$$

$$\hat{N} = \bar{c}\hat{F}\hat{L}$$

$$N(x) = \hat{N}e^{-\int_0^x \mu(u)\,du} \tag{12.5b}$$

$$L(x) = \hat{L}e^{-\int_0^x [v(u) + \hat{F}c(u)]\,du}.$$

In the following, we simply assume this steady state is stable (as in the cases detailed in Roughgarden and Iwasa 1986), and investigate the evolutionary fate of such a population.

THE EVOLUTIONARY CRITERIA

Contemporary life-history theory offers two approaches to determining the strategy to be considered as "optimum," that is, as resulting somehow from the evolutionary process. The first approach is to consider alleles that "mold" the life history presently found in the population. Such alleles, when expressed in a heterozygote, confer a life history altered from that currently fixed in the population. The type of alteration that increases the fastest, when rare, is the most likely alteration (abbreviated here as MLA) and represents the direction in which evolution is most likely to "go" relative to its current state. Also, an allele with pleiotropic effects can be studied. The rate of increase of an allele making a "unit change" to some part of the life cycle can be viewed as the value of that change. Then the fate of an allele that affects more than one place in the life cycle can be predicted by adding the values of the effects at each of the places affected. This approach was pioneered by Hamilton (1966) who proposed a genetical theory for the evolution of senescence, with a major focus on pleiotropic alleles that eroded survivorship at one age while increasing it at another.

More recently, Maynard Smith and Price (1973) proposed a conceptually different criterion whose predictions, nonetheless, are often the same as those obtained by considering allelic alterations. An evolutionarily stable strategy (ESS) is a strategy such that, if *it* is fixed in the population, produces the lowest rate of increase for any other strategy that might be introduced. Such a strategy is thus resistant to invasion, and represents a state where evolutionary stasis is likely to persist, until, that is, conditions change enough to start a new game. Also, by such a definition an ESS is not guaranteed to remain a pure state. The rate of increase for any alternative strategy must actually be negative to guarantee that the ESS strategy will remain in the population as the only strategy; otherwise a polymorphism will presumably result.

A Threshold for the Evolution of Benthic Life Cycles

We begin with the ESS criterion for an optimal life cycle, and later discuss its synonymy with an MLA criterion. To identify an ESS strategy we must show that it leads to the lowest rate of increase for any alternative type. Suppose then that an alternative type appears, as identified by a "2" in subscript; the original type is left unsubscripted. The dynamics for the new type, given that it is rare and that the original type exists in steady state, is given by

$$\frac{\partial N_2(x, t)}{\partial t} + \frac{\partial N_2(x, t)}{\partial x} = -\mu_2(x)N_2(x, t)$$

$$N_2(0, t) = \hat{F} \int_0^t c_2(x)L_2(x, t)\,dx$$

$$\frac{\partial L_2(x, t)}{\partial t} + \frac{\partial L_2(x, t)}{\partial x} = -[v_2(x) + \hat{F}c_2(x)]L_2(x, t)$$

$$L_2(0, t) = \int_0^t b_2(x)L_2(x, t)\,dx + \int_0^t m_2(x)N_2(x, t)\,dx.$$

(12.7)

The exponential rate of increase for the introduced type is found by seeking a separable trial solution, $N_2(x, t) = \tilde{N}_2(x)\hat{N}_2(t)$, and similarly for $L_2(x, t)$. This asymptotic solution is then found to be

$$\tilde{N}_2(x) = e^{-r_2 x - \int_0^x \mu_2(u)\, du}$$
$$\tilde{L}_2(x) = e^{-r_2 x - \int_0^x [\nu_2(u) + \hat{F}c_2(u)]\, du}$$
$$\hat{N}_2(t) = c_N e^{r_2 t} \tag{12.8}$$
$$\hat{L}_2(t) = c_L e^{r_2 t},$$

and the exponential rate of increase for the alternative type, r_2, satisfies

$$1 = \bar{b}_2(r_2, \hat{F}) + \hat{F}\bar{c}_2(r_2, \hat{F})\bar{m}_2(r_2), \tag{12.9}$$

where

$$\bar{b}_2(r_2, \hat{F}) = \int_0^\infty b_2(x) e^{-r_2 x - \int_0^x [\nu_2(u) + \hat{F}c_2(u)]\, du}\, dx$$
$$\bar{c}_2(r_2, \hat{F}) = \int_0^\infty c_2(x) e^{-r_2 x - \int_0^x [\nu_2(u) + \hat{F}c_2(u)]\, du}\, dx \tag{12.10}$$
$$\bar{m}_2(r_2) = \int_0^\infty m_2(x) e^{-r_2 x - \int_0^x \mu_2(u)\, du}\, dx.$$

The original type is an ESS if it minimizes the r_2 determined as the root of (12.9), for any choice of invading phenotype. The properties of the original type in the population enter only through $\hat{F}$. Therefore, whatever traits constitute an ESS do so by virtue of their effect on $\hat{F}$. Specifically, if decreasing $\hat{F}$ lowers r_2, then an ESS is a strategy that minimizes $\hat{F}$, provided the derivative of r_2 with respect to $\hat{F}$ is positive, and conversely if the derivative is negative. So, to find the effect of increasing or decreasing $\hat{F}$ on r_2, we differentiate (12.9) taking r_2 as an implicit function of $\hat{F}$. The result is

$$\frac{dr_2}{d\hat{F}} = \frac{\bar{c}_2 \bar{m}_2 - \overline{c_2 b_2} - \hat{F}\overline{c_2 1}\bar{m}_2}{Z}, \tag{12.11}$$

where $\bar{c}_2$ and $\bar{m}_2$ are defined in (12.10), and

$$\overline{c_2 1} = \int_0^\infty \left(\int_0^x c_2(u)\, du \right) e^{-r_2 x - \int_0^x [\nu_2(u) + \hat{F}c_2(u)]\, du}\, dx$$
$$\overline{c_2 b_2} = \int_0^\infty \left(\int_0^x c_2(u)\, du \right) b_2(x) e^{-r_2 x - \int_0^x [\nu_2(u) + \hat{F}c_2(u)]\, du}\, dx, \tag{12.12}$$

and Z is a positive quantity. Thus, if the numerator in (12.11) is positive, then an ESS is a life cycle that minimizes $\hat{F}$. In this situation some life cycle with a benthic component (either completely benthic or benthic only in the adult phase) is an ESS. But if the numerator of (12.12) is negative, then a life cycle with some benthic component is not an ESS, as the population presumably evolves to a purely pelagic state, that is, *occupancy* of benthic space is minimized because $\hat{F}$ is maximized. In more detail, the expression $\bar{c}_2\bar{m}_2 - \overline{c_2b_2}$ indicates the comparative reproductive success inherent to the benthic and pelagic habitats, respectively. If this is positive, the benthic habitat offers some inherent advantage. However, the abundance that can be realized in that habitat must also be sufficiently high (i.e., $\hat{F}$ sufficiently low) for the numerator in (12.12) to be positive, and accordingly for some benthic life cycle to be an ESS. Therefore, a threshold exists—expressed as a minimum abundance that must be achieved in the benthic habitat. If the realized abundance is not higher than this threshold, even if the benthic habitat is inherently better for reproduction, then the species should be purely pelagic.

Properties of an Optimal Benthic Life Cycle

Now let us suppose that the threshold condition for a benthic life cycle derived above is satisfied and explore what an optimal life cycle with a benthic component should look like. The mathematical criterion will be to find the life cycle that leads to a minimum $\hat{F}$. This cannot be done in the abstract, however, because the problem has been posed in terms of particular traits, such as the size of the larval stage at its time of release, the duration of the larval phase, the size of the larvae at the time of metamorphosis, the size of the adult at which it begins to reproduce, and so forth. To explore an optimal life history in terms of such traits, we need special submodels that summarize the interrelations among the traits—the constraints and the tradeoffs. Other submodels may be used if those introduced here prove too unrealistic.

THE SUBMODELS

1. The mortality rate in the benthic habitat is taken as independent of age,

$$\mu(x) = \mu. \tag{12.13}$$

2. The mortality rate in the larval habitat is also taken as independent of age,

$$\nu(x) = \nu. \tag{12.14}$$

3. The settlement coefficient as a function of age is taken as a step function,

$$c(x) = 0 \quad (0 \le x < t_l)$$
$$= c \quad (x \ge t_l).$$

(12.15)

The t_l is the age at which the larvae have attained the size, s_c, necessary for metamorphosis.

4. Growth in the larval habitat is assumed to follow a power law until the size of metamorphosis is reached, after which growth is assumed to stop:

$$s(t) = [g_l(t + s_n^{1/k_l}/g_l)]^{k_l} \quad (0 \le t \le t_l).$$

(12.16)

Here t is the time since release, t_l is the duration of the larval phase, s is the larval weight at time t, and s_n is the larval weight when initially released. The g_l is the growth rate in the larval habitat, and k_l is the power constant for larval growth—this reflects the allometry inherent in the body plan of the animal. The parametrization in (12.16) allows the initial larval size to be viewed as giving the larva a "head start" in its growth along the curve, $(g_l t)^{k_l}$. If the growth is linear, $k_l = 1$, then this assumption reduces to that made by Vance (1973).

5. Similarly, growth in the adult habitat is also assumed to follow a power law between the time of settlement and the start of reproduction, after which all net assimilation is assumed to be placed into reproductive material:

$$s(t) = [g_a(t + s_c^{1/k_a}/g_a)]^{k_a} \quad (t_l \le t \le t_a).$$

(12.17)

Here, t is the time since metamorphosis, t_a is the duration of the pre-reproductive period in the benthic habitat, s is the adult weight at time t, and s_c is the weight of the organism after metamorphosis. Also, g_a is the growth rate in the benthic habitat, and k_a is the power constant for growth in the benthic habitat. Again the parametrization in (12.17) allows the initial size, s_c, to be viewed as providing a head start in growth. Growth in the benthic habitat follows the curve, $(g_a t)^{k_a}$, representing a possibly quite different allometry from the larval phase, because the body plans of larval and adult phases may be strikingly different.

6. The fecundity of an adult consists of two components, the age-dependent reproductive activity ("effort"), and the allocation of that activity into producing eggs of some size.

a. The reproductive activity is assumed to follow a step function. It is zero until t_a, the age at which exclusive growth ceases, and remains a positive constant thereafter. Furthermore, the quantity of reproductive material that the adult can place into eggs is assumed to depend on the size it has attained when it switches from growth to reproduction. This quantity of reproductive material is assumed to be a power function of the adult's total weight at the time of the switch.

b. The number of eggs is assumed to be inversely proportional to the egg size, that is, the number of eggs times the weight per egg must equal the quantity of reproductive material in the adult. This assumption follows Vance (1973).

Putting these components together yields a maternity function in the benthic habitat,

$$m(x) = 0 \qquad (0 \le x < t_a)$$
$$= m \frac{s_a^{k_m}}{s_n} \qquad (x \ge t_a). \tag{12.18}$$

The next step will be to substitute these submodels into the formula for $\hat{F}$.

FINDING THE OPTIMAL LIFE HISTORY

Inspecting (12.5a) reveals that minimizing $\hat{F}$ is equivalent to maximizing the quantity,

$$\bar{c}\bar{m} = \int_0^\infty c(x)e^{-\int_0^x [v(u) + \hat{F}c(u)]\,du}\,dx \int_0^\infty m(x)e^{-\int_0^x \mu(u)du}\,dx. \tag{12.19}$$

However, $\hat{F}$ is implicitly a function of $\bar{c}\bar{m}$ itself, so a complete analysis is difficult. Nonetheless, the condition for a life cycle with a benthic component to be an ESS, is, as noted above, only satisfied if $\hat{F}$ is sufficiently small. Therefore, it seems appropriate to investigate optimal life histories in the limit as $\hat{F}$ approaches zero, or, in biological terms, that the larval mortality rate, $v(x)$, is much higher than the settlement rate into the benthic habitat, $\hat{F}c(x)$. In this limit, the optimal life history can be found from maximizing the expression, W, defined as

$$W = \int_0^\infty c(x)e^{-\int_0^x v(u)\,du}\,dx \int_0^\infty m(x)e^{-\int_0^x \mu(u)\,du}\,dx. \tag{12.20}$$

So, all that remains is to substitute the submodels into (12.20) to obtain the expression whose maximization yields the optimal life history. From (12.16) we have the duration of the larval phase in terms of the size at which it begins and ends as

$$t_l = \frac{(s_c^{1/k_l} - s_n^{1/k_l})}{g_l}, \tag{12.21}$$

and, similarly, from (12.17) we have the duration of the prereproductive episode in the benthic habitat in terms of the size at metamorphosis and size at the start of reproduction as

$$t_a = \frac{(s_a^{1/k_a} - s_c^{1/k_c})}{g_a}.$$

(12.22)

Then upon substituting (12.13) and (12.14) for the mortality rates, (12.15) for the settlement coefficient, and (12.18) for the benthic maternity function into (12.20), and using (12.21) and (12.22) for the duration of the larval and prereproductive benthic phases, and then doing the integral, we obtain a scalar-valued fitness expression defined on the entire life cycle,

$$W(s_n, s_c, s_a) = \left(\frac{mc}{\mu v}\right)\left(\frac{s_a^{k_m}}{s_n}\right) e^{-\frac{\mu s_a^{1/k_a}}{g_a} + \frac{\mu s_c^{1/k_a}}{g_a} - \frac{v s_c^{1/k_l}}{g_l} + \frac{v s_n^{1/k_l}}{g_l}}.$$

(12.23)

The optimal life cycle is then defined as a vector, $\hat{S} = (\hat{s}_n, \hat{s}_c, \hat{s}_a)$ that maximizes W, with the constraint that $0 \leq \hat{s}_n \leq \hat{s}_c \leq \hat{s}_a$. The components of the vector are the size of the larva when released by the adult, the size of the larva at metamorphosis, and the size of the adult at the beginning of reproduction. These sizes can be inserted into (12.21) and (12.22) to determine the length of the larval phase, and the prereproductive phase spent in the benthic habitat.

PIECES OF THE OPTIMAL LIFE CYCLE

To characterize the optimal life cycle defined by maximizing W in (12.23), the partial derivatives of W with respect to each of the variables should be examined. We consider these variables one at a time.

s_n: W, with respect to the size of the larva upon its release, s_n, has the form

$$W \longrightarrow \left(\frac{1}{s_n}\right) e^{\frac{v s_n^{1/k_l}}{g_l}},$$

(12.24)

which is, in general, $\bigcup$-shaped, with minimum at

$$\check{s}_n = \left(\frac{k_l g_l}{v}\right)^{k_l}.$$

(12.25)

This relation between W and s_n is qualitatively the same as that derived by Vance (1973) and is here embedded in the full life cycle.

s_c: W, with respect to the size of the larva at metamorphosis, s_c, has the form

$$W \longrightarrow e^{\frac{\mu s_c^{1/k_a}}{g_a} - \frac{v s_c^{1/k_l}}{g_l}},$$

(12.26)

which is $\bigcap$-shaped if $k_a > k_l$, and $\bigcup$-shaped if $k_a < k_l$. The optimum size of metamorphosis is then

$$\hat{s}_c = \left(\frac{\dfrac{k_l g_l}{v}}{\dfrac{k_a g_a}{\mu}}\right)^{\frac{k_a k_l}{k_a - k_l}} \qquad (k_a > k_l). \qquad (12.27)$$

Alternatively, a minimum occurs at

$$\check{s}_c = \left(\frac{\dfrac{k_a g_a}{\mu}}{\dfrac{k_l g_l}{v}}\right)^{\frac{k_a k_l}{k_l - k_a}} \qquad (k_l > k_a). \qquad (12.28)$$

s_a: W, with respect to the size at which the adult switches from growth to reproduction, s_a, has the form

$$W \longrightarrow s_a^{k_m} e^{-\frac{\mu s_a^{1/k_a}}{g_a}}, \qquad (12.29)$$

which always has a maximum at

$$\hat{s}_a = k_m^{k_a} \left(\frac{k_a g_a}{\mu}\right)^{k_a}. \qquad (12.30)$$

Putting the Pieces Together

These pieces can fit together only in certain ways, with the result that five classes of optimal life cycles are possible in this formulation. The first two types represent the planktotrophic-larva benthic-adult cycle and the direct-development cycle as discussed in earlier studies. The remaining three are more difficult to interpret, and yet may reflect the diversity of life cycles in nature better than formulations that allow only two types of strategies.

$\hat{s}_1 = (0, \hat{s}_c, \hat{s}_a)$. In this type of solution, the initial larval size is as close to 0 as possible; the larva grows in the plankton until it metamorphoses at size $\hat{s}_c$; and then grows in the benthic habitat until size $\hat{s}_a$, at which it ceases growth and directs its energies into reproduction. This cycle represents the sequence {begin as tiny larva, planktotrophic larval growth, benthic prereproductive growth, benthic reproduction}.

$\hat{s}_2 = (\hat{s}_c, \hat{s}_c, \hat{s}_a)$. Here the larval phase is bypassed because the larva is released at the size, $\hat{s}_c$, needed for metamorphosis to begin with. The juvenile then grows in the benthic habitat to the size $\hat{s}_a$ at which it begins reproduction. This formulation does not distinguish among viviparity, direct development, and lecithotrophy, all of which are lumped together in this strategy type. The cycle thus represents the sequence {begin as large larva or small juvenile, benthic prereproductive growth, benthic reproduction}.

$\hat{s}_3 = (0, \hat{s}_a, \hat{s}_a)$. This life cycle has an extended planktotrophic larval phase; it begins with a larval size near 0 and lasts until the size at which reproduction can occur, and then reproduction begins immediately upon settlement into the benthic habitat. This could represent a species that returns to a benthic habitat only to reproduce, but that is otherwise pelagic. The cycle represents the sequence {begin as tiny larva, planktotrophic larval growth, benthic reproduction}.

$\hat{s}_4 = (\hat{s}_a, \hat{s}_a, \hat{s}_a)$. This solution means that a reproductive adult gives birth to another reproductive adult. In effect, this occurs through budding. A great many invertebrates are colonial, and possibly this solution represents such organisms, although a short-lived larva is often produced as well. (If the reproduction is primarily through budding, perhaps a little energetic investment in short-lived larva can be excused as a small effort at dispersal.) Thus this cycle represents the (degenerate) sequence {begin as adult, benthic reproduction}.

$\hat{s}_5 = (0, 0, \hat{s}_a)$. This solution is the most difficult to interpret. It indicates skipping the planktonic phase, yet entering the benthic habitat at a size near zero and growing there to adult size. I am not aware of any possible example of this strategy type. The cycle represents the sequence {begin as tiny benthic larva or tiny juvenile, benthic prereproductive growth, benthic reproduction}.

To determine which of these strategies is an ESS in a particular situation, a key can be used. We begin by noting that W is always $\cap$-shaped with respect to s_a, and there is therefore always an optimum adult size for beginning reproduction, $\hat{s}_a$, given by (12.30). Next we move to the size at metamorphosis, s_c, and a comparison is now needed between the power coefficient for growth in the benthic habitat and in the larval habitat. If the benthic habitat provides faster growth (left branch in Figure 12.3), then W is $\cap$-shaped with respect to s_c. In this case, there is a single optimum size for metamorphosis, $\hat{s}_c$, given by (12.27), and the question becomes whether it satisfies the constraint that $\hat{s}_c \leq \hat{s}_a$. Alternatively, if the larval habitat provides faster growth (right branch in Figure 12.3), then W is $\cup$-shaped with respect to s_c, and this implies that an optimum size for

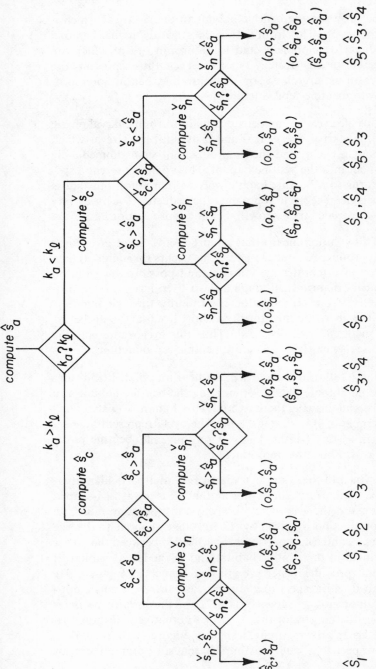

Figure 12.3. Mathematical conditions for life cycles with a benthic component to be evolutionarily optimal. To use the graph, first compute the optimal size for the adult at its time of reproduction $\hat{s}_a$, using Eq. (12.30). Next, if the power coefficient in the allometric growth law for the adult phase is higher than that for the larval phase, $k_a > k_l$, then take the left branch; otherwise, take the right branch. Now, following the left branch as an example, calculate the optimal size for metamorphosis from the larval phase to the adult phase, $\hat{s}_c$, using Eq. (12.27). If this is less than the optimal size for reproduction, continue with the left branch; otherwise go to the right for strategies in which a benthic prereproductive growth phase is bypassed. Finally, to continue with the left branch as an example, compute the worst size for the larva at its time of release, $\check{s}_n$, from Eq. (12.25). If this is greater than the optimal size at metamorphosis, then the best larval size at time of release is zero; otherwise, two sizes are optimal (local maxima): both zero and $\check{s}_c$. An optimal size at time of release of zero indicates a planktotrophic larval phase, while an optimal size at time of release of $\check{s}_c$ indicates that a planktotrophic larval phase has been bypassed. Other possibilities, at other end points of the graph, include the strategies of benthic spawning, as in

metamorphosis is either $\hat{s}_c \rightarrow \hat{s}_n$, or $\hat{s}_c \rightarrow \hat{s}_a$, or both. Finally, W is always $\cup$-shaped with respect to s_n, and so the optimum initial larval size is either $\hat{s} \rightarrow 0$, or $\hat{s}_n \rightarrow \hat{s}_c$, or both. Figure 12.3 presents a key in the form of a decision tree that can be traversed in any particular situation to find what the ESS life cycles are. I am uncomfortable, however, with the entire right branch in Figure 12.3 ($k_a < k_l$) because it applies when the larval habitat is, in a sense, better than the adult habitat. This condition could then favor a purely pelagic life cycle instead of those listed at the terminus of the right branch, all of which include a benthic adult phase. Thus, I feel the left branch represents a more reliable set of theoretical predictions.

Figure 12.3 can, in principle, be combined with data on the distribution over various potential habitats of the rates of growth, birth, and death for various body plans (considered as properties inherent to a high biological taxon, such as a family or class). Figure 12.3, together with such data, permits the frequency distribution of life-cycle types to be predicted. This may even be practical to a limited degree. Here, however, it suffices to note some general features of Figure 12.3.

1. Almost all the ESS life cycles begin with a tiny planktonic larva, although it may or may not be long-lived and planktotrophic. The conditions for the direct-development strategy ($\hat{S}_2$) to be favored are restrictive. Indeed, this cycle is, at best, only one of two simultaneously stable ESS life cycles, the other ESS cycle having a tiny larva. The full condition for direct-development to be one of the possibilities (the condition for the second branch from the left in Figure 12.3) is

$$1 < \frac{E_l}{E_a^{k_a/k_l}} < k_m^{\frac{(k_a - k_l)}{k_l}}, \qquad (12.31)$$

where

$$k_a > k_l$$

$$E_h = \frac{k_h g_h}{v_h} \qquad (h = a, l; \; v_a = \mu; \; v_l = v). \qquad (12.32)$$

This condition imposes strong constraints on the allometry of growth for both larval and adult morphologies, on the allometric relation between adult size and the quantity of reproductive material available for eggs, and on the ratio of the parameters describing the environmental conditions in both larval and adult habitats. So it does not seem surprising that many higher taxa totally lack species with direct development.

2. Figure 12.3 represents the first theoretical study of the evolution of marine life histories that predicts when budding and colonial life cycles should evolve.

Equivalence with an Allelic Alteration Approach

Here it is noted briefly that the ESS life cycles displayed in Figure 12.3 coincide with the optimal strategies predicted according to a most likely alteration (MLA) criterion. As above, the instantaneous settlement rate $\hat{F}c_2(x)$ is taken as small relative to the instantaneous larval mortality rate $v_2(x)$, and the pelagic reproduction, $b_2(x)$, is assumed zero. Then, from (12.9), the MLA strategy is found by maximizing r_2, defined as the root of

$$W_2(r_2) = \int_0^\infty c_2(x)e^{-(v_2+r_2)x}\,dx \int_0^\infty m_2(x)e^{-(\mu_2+r_2)x}\,dx = \frac{1}{\hat{F}}. \quad (12.33)$$

If $c_2(x)$, $m_2(x)$, $v_2(x)$, and $\mu_2(x)$ are taken as functions of some parameter, s, based on a submodel, and if r_2 is viewed as an implicit function of this s, then the MLA value of s is found by differentiating $W_2(r_2)$ in (12.33) implicitly with respect to s. After solving for dr_2/ds, one would obtain an expression of the form

$$\frac{dr_2}{ds} = -\left(\frac{\dfrac{\partial W}{\partial s}}{\dfrac{\partial W}{\partial r_2}}\right). \quad (12.34)$$

However, $\partial W/\partial r_2$ is negative, so maximizing r_2 with respect to s is equivalent to maximizing W with respect to s. Therefore, an ESS life cycle as presented in Figure 12.3 coincides with a MLA life cycle.

Life Cycles for Marine Plants

As a final note, the importance of the detailed assumptions embodied in the submodels can be illustrated by showing that plants should not evolve a life cycle analogous to that of noncolonial animals. The reason traces to a fundamentally different allometry between the growth of plants and noncolonial animals. Plants, and perhaps some colonial animals, essentially produce structures that compound, so growth is exponential and not according to a power law. Thus, the size of a plant during a hypothetical pelagic period is given by

$$s(t) = e^{g_1 t}s_n, \quad (12.35)$$

and during a subsequent benthic period by

$$s(t) = e^{g_a t}s_c. \quad (12.36)$$

Therefore, the durations of the "larval" and "adult" growth periods, as functions of the sizes at the end of these periods, are given by

$$t_l = \frac{\ln(s_c) - \ln(s_n)}{g_l}$$

$$t_a = \frac{\ln(s_a) - \ln(s_c)}{g_a}. \tag{12.37}$$

Upon using these submodels for growth, and keeping the other submodels used for Figure 12.3, we obtain from (12.20) a fitness criterion

$$W(s_n, s_c, s_a) = \left(\frac{mc}{\mu v}\right) s_n^{\frac{v}{g_l} - 1} s_c^{\frac{\mu}{g_a} - \frac{v}{g_l}} s_a^{k_m - \frac{\mu}{g_a}}. \tag{12.38}$$

This criterion is monotonic in all three variables, and so no biphasic life cycle is optimal.

Discussion

This chapter extends previous theoretical investigations of the evolution of marine life cycles by developing a formulation that pertains to the entire life cycle. Previous studies have examined only pieces, usually the beginning, of the life cycle. New conclusions include the discovery of a threshold that must be met for the evolution of a life cycle with a benthic component; populations that do not attain a minimum abundance in the benthic habitat should carry out a purely pelagic life cycle even if the benthic habitat is better for growth and reproduction. Given that this threshold abundance is achieved, the characteristics of optimal life cycles with a benthic component are derived based on a collection of submodels, including growth according to a general allometric power law in both larval and adult phases. Five classes of benthic life cycles are shown to be evolutionarily optimal under various conditions. The classes include cycles with planktotrophic larvae; direct development to a juvenile phase; benthic spawning; and budding, as in colonial forms, where a reproductive adult gives "birth" to a fully reproductive adult. The conditions for these to be optimal depend on the relative allometric growth rates inherent in the body plans of the larval and adult morphologies, and on the environmental conditions (including mortality and growth rates) in both the larval and adult habitats. It is also shown that the allometry of plant growth leads to a different optimality criterion than that obtained for the growth of animals, with the result that a life cycle with both planktonic and benthic phases is not optimal for plants.

THE MAJOR QUESTIONS REEXAMINED

How well does evolutionary theory, consisting of the results of earlier studies spliced together with those derived here, presently account for the major questions posed in the introduction to this chapter? Here then is a status report.

1. Why do marine life cycles occur in discrete classes, and not in a uniformly or normally distributed continuum? Optimal life cycles, according to evolutionary fitness criteria, belong to six qualitative classes (1 purely pelagic, 5 with benthic adults), and more refined models would probably lead to some further classes. And while these classes are distinct qualitatively, a particular life cycle in one class may be quantitatively close to a life cycle in another class, so that the classes should seem to intergrade somewhat.

2. Why do a great many species with benthic adults possess a planktotrophic larva? As Figure 12.3 shows, life cycles with planktotrophic larva figure in most of the optimal life cycles. It appears that under most circumstances the advantage in fecundity to having many tiny larva exceeds the disadvantage of forcing those larvae to survive for a long time while feeding in the water column. Furthermore, other studies, as cited earlier in this chapter, have shown that larval dispersal provides a genotype with an arithmetic average of the environmental conditions throughout the species range, which is generally better than the geometric average seen by a nondispersing genotype. This advantage to dispersal further enhances the evolutionary pressure favoring a planktotrophic larva. This advantage may especially contribute to explaining the evolution of a nonfeeding pelagic larval phase, as in lecithotrophic larvae, where a feeding-only theory cannot distinguish lecithotrophic larvae from direct development or viviparity.

3. Why do most groups possess only a subset of the possible life cycles? According to the formulation developed here, allometric relationships during growth play a major role in determining what classes of life cycles are optimal. The allometry of growth is a basic property of the characteristic body plan used to define a higher biological taxon such as a family or class. Hence higher taxa should automatically sort into separate life-cycle categories. One might consider the allometry of a taxon's body plan as an expression of its inherent "phylogenetic constraints."

4. Why do plants not possess a life cycle like that of many invertebrates in having a small spore that grows to intermediate size in the water column

before settling into a benthic habitat where it continues its growth until reproduction? The allometry of growth for a photosynthesizing structure is fundamentally different from that of an animal. Photosynthetic structures can compound themselves, and this allometry favors all-or-none strategies instead of the mixed strategies exhibited by animals.

5a. Why does the percentage of species with direct development increase from the equator to the poles? The answer is complicated because evolutionary pressures throughout the entire life cycle may change with latitude as well as any pressure on the larval phase itself. The basic requirement is found from comparing $\check{s}_n$ of (12.25) to $\hat{s}_c$ of (12.27), and $\hat{s}_c$ in turn, to $\hat{s}_a$ of (12.30). The result is that the theory can explain the latitudinal gradient, provided environmental conditions in the larval habitat deteriorate as fast or faster than those in the adult habitat as one proceeds along a transect from the equator to the poles.

   i. If only the larval habitat is deteriorating along a transect from the equator to the poles, then more species should have direct development toward the poles, *and* the size at metamorphosis should also decrease along the transect while the size at reproduction should remain unchanged. [The $\check{s}_n$ decreases in (12.25) as g_l/v decreases, and $\hat{s}_c$ in (12.27) also decreases with g_l/v, while $\hat{s}_a$ of (12.30) remains unchanged.]

   ii. If both the larval and adult habitats are deteriorating along a transect from the equator to the poles (i.e., both larval and adult habitats covary somewhat in quality), then again there should be more species with direct development toward the poles, although now the size at metamorphosis should remain relatively unchanged along the transect, while the size at reproduction should become smaller. [As before, the $\check{s}_n$ decreases in (12.25) as g_l/v decreases, but now $\hat{s}_c$ in (12.27) is not affected much because both g_l/v and g_a/μ are decreasing and their quotient may remain nearly the same; however, $\hat{s}_a$ of (12.30) declines as g_a/μ declines.]

If both larval and adult habitats deteriorate to about the same degree with increasing latitude, then an explanation is offered for an observed relationship between adult size and the tendency to have a direct-development life cycle. Typically, within a taxonomic group having both direct-development species and planktotrophic species, direct development occurs in the species with small adult sizes, while a planktotrophic larva is produced in species whose adults are large (Strathmann 1985). If both larval and adult habitat quality simultaneously decline from the tropics to the poles (as in case ii. above) then the presence of a direct-development life cycle should correlate with a smaller adult size. This correlation should thus reflect covarying evolution on separate traits, and not be the expression of an inherent morphological constraint by which the traits are automatically coupled.

5b. Why does the percentage of species with lecithotrophic larvae decline from the equator to the poles? One possible explanation is an extension of the latitudinal gradient in the relative quality of the benthic and larval habitats just discussed. Perhaps in the tropics the water column is a safer place to be than the rocky substrate. If so, producing large eggs in an attached egg mass may expose the eggs to nearly certain mortality, while allowing large eggs to float about as the embryos develop may lead to less egg mortality. Indeed, there is much evidence for extremely high predation pressure on hard substrates from grazing fish in the tropics (Menge and Lubchenco 1981), and this has been invoked to explain the high prevalence of toxic sponges in the tropics (Bakus 1969), the paucity of fleshy seaweeds, especially laminarians (Gaines and Lubchenco 1982), and the restriction of many seaweeds to sites inaccessible to grazing fish (e.g., Brock 1979). Indeed, an abundant tropical population of fleshy seaweeds is actually pelagic—sargasso weed of the Sargasso Sea, which is derived from a benthic stock (Parr 1939). Alternatively, the pressure favoring lecithotrophic larvae in the tropics may be primarily dispersal. If so, a larger scale of dispersal in the tropics would require a higher spatial correlation in habitat conditions there, so that longer dispersal would be needed to achieve an independent sample of the possible conditions. Finally, one must remain cautious in interpreting this latitudinal gradient because the larvae thought to be lecithotrophic may, in the future, be found to be feeding on dissolved organic material, and not to be lecithotrophic at all.

Thus, the advance offered by this chapter is the ability now to address theoretically a richer set of basic evolutionary questions concerning marine life cycles than was previously possible. It is premature, however, to expect this theory to do more than point the way for more specific formulations. All this will need some "chewing over" from a purely theoretical standpoint before its empirical significance should be taken too seriously. We especially need to know how robust this formulation is to refinement or substitution of the submodels that have been used.

Some Final Conjectures

To conclude, it is interesting to note parallels between the life cycles of marine and terrestrial animals, since they come from habitats so different, and often from phyla so different, that comparisons are rarely attempted.

Cody (1971) documented the major biogeographic pattern of life histories known for birds. The clutch size (number of eggs in the nest) increases along any transect from the equator to the poles. Subsequent research implicates a corresponding latitudinal gradient in the level of density-independent mortality as the cause. Situations of high mortality, as are increasingly realized near the poles, lead to the evolution of a suite of traits—including early reproduction, high clutch size, and rapid senescence. This latitudinal gradient in the life cycles of a class of terrestrial vertebrates parallels the latitudinal gradient in the occurence of planktotrophic larva vs. direct-development life cycles described by Thorson (1936) for snails. Both biogeographic patterns share much the same cause: a latitudinal gradient in environmental quality.

The extinction of the dinosaurs by the end of the Cretaceous has been followed by the emergence of mammals and birds, both of which are endothermic and typically have life histories involving extensive parental care. In mammals, both the placental and marsupial lines independently evolved different mechanisms that allow a large offspring to be released into the adult habitat. Similarly, most birds have life histories where a small number of eggs (usually less than five) are incubated in a nest, and tended until fledged into the adult habitat at nearly the adult size. These are the life cycles also characteristic of marine invertebrates in high temperate to polar latitudes, where viviparity and direct development are common. In contrast, reptiles typically have a life cycle closer to that of most marine invertebrates, especially those from warmer seas, where copious eggs are laid and, upon hatching, yield juveniles that fend for themselves until they attain reproductive size. Even more comparable to most marine invertebrates are amphibians—for these have a multistage life cycle with morphologically distinct larval and adult phases occupying different habitats.

While the cause of dinosaur extinction remains controversial, the extinction is only half of the problem—the other half is to explain why dinosaurs were replaced by birds and mammals rather than by some other dinosaurs. One class of hypothesis to explain dinosaur extinction invokes a cooling trend known to have occurred since the late Cretaceous (Savin, Douglas, and Stehli 1975), much as cooling has been demonstrated to have caused mass extinction among marine bivalves in the late Pliocene (Stanley 1986); the cooler environment is supposed to be unfavorable for cold-blooded forms like reptiles and amphibians. In fact, warm body temperatures are often attained by amphibians and reptiles in field conditions (e.g., Roughgarden et al. 1981). Yet, the life cycles of the forms that replaced the dinosaurs are those expected in cold climates, according to the biogeographic pattern of marine invertebrates, which lends support to the climate-deterioration hypothesis. In short, the present-day latitudinal pattern of marine invertebrate life cycles may mirror the temporal pattern of change since the Cretaceous for the life cycles of dominant terrestrial vertebrates.

Thus, the extinction of dinosaurs may well have been primarily caused by a cooling trend since the Cretaceous because dinosaurs have been replaced by forms with obvious adaptations to cold conditions, such as endothermy with insulating feathers and fur. And as just noted, both birds and mammals also possess the kind of life cycles characteristic of marine invertebrates in arctic conditions.

Acknowledgments

I wish, above all, to thank Samuel Karlin for his role in furthering theoretical studies in biology, and for his support and interaction as a colleague and friend at Stanford. His contributions set a new standard in theoretical population genetics, a standard that has proven equally definitive and inspiring in the development of theoretical ecology.

Also, I thank John McLaughlin for analyzing the data for Figure 12.1, and I thank him, together with S. Blower and S. Gaines, for discussion during the course of this research. Moreover, I thank Yoh Iwasa, John Pearse, Richard Strathmann, and Shripod Tuljapakur for reviewing the manuscript. Any errors that remain are, of course, solely my responsibility. Finally, I gratefully aknowledge major support from the Department of Energy (DE-FGO-3-85ER60362), with supplemental funding from the National Science Foundation (OCE-8514755).

References

Bakus, G. L. 1969. Energetics and feeding in shallow marine waters. *Int. Rev. Gen. Expt. Zool.* 4: 275–369.
Brock, R. E. 1979. An experimental study on the effects of grazing by parrotfishes and role of refuges in benthic community structure. *Mar. Biol.* 51: 381–388.
Caswell, H. 1981. The evolution of "mixed" life histories in marine invertebrates and elsewhere. *Amer. Natur.* 117: 529–536.
Christiansen, F. B., and Fenchel, T. M. 1979. Evolution of marine invertebrate reproductive patterns. *Theor. Popul. Biol.* 16: 267–282.
Coates, A. G., and Jackson, J.B.C. 1985. Morphological themes in the evolution of clonal and aclonal marine invertebrates. In *Population Biology and Evolution of Clonal Organisms*, pp. 67–106. Ed. J.B.C. Jackson, L. W. Buss, and R. E. Cook. Yale University Press, New Haven, Conn.
Cody, M. L. 1971. Ecological aspects of reproduction. *Avian Biology* 1: 461–512.
Crisp, D. J. 1974. Energy relations of marine invertebrate larvae. *Thal. Jugoslavica* 10: 103–120.
Crisp, D. J. 1976. The role of the pelagic larva. In *Perspectives in Experimental Zoology*, vol. 1, pp. 145–155. Ed. P. S. Davies. Pergamon Press.

Doyle, R. W. 1975. Settlement of planktonic larvae: A theory of habitat selection in varying environments. *Amer. Natur.* 109: 113–126.

Frank, P. W. 1981. A condition for a sessile strategy. *Amer. Natur.* 118: 288–290.

Gaines, S. D., and Lubchenco, J. 1982. A unified approach to marine plant-herbivore interactions, II. Biogeography. *Ann. Rev. Ecol. Syst.* 13: 111–138.

Grant, A. 1983. On the evolution of brood protection in marine benthic invertebrates. *Amer. Natur.* 122: 549–555.

Hamilton, W. D. 1966. The moulding of senescence by natural selection. *J. Theor. Biol.* 12: 12–45.

Iwasa, Y., and Roughgarden, J. D. 1985. Evolution in a metapopulation with space-limited subpopulations. *IMA J. Math. Appl. Med. Biol.* 2: 93–107.

Jackson, J.B.C. 1977. Competition on marine hard substrata: The adaptive significance of solitary and colonial strategies. *Amer. Natur.* 111: 743–767.

Jackson, J.B.C. 1986a. Modes of dispersal of clonal benthic invertebrates: Consequences for species' distributions and genetic structure of local populations. *Bull. Mar. Sci.* 39: 588–606.

Jackson, J.B.C. 1986b. Life cycles and evolution of clonal (modular) animals. *Phil. Trans. R. Soc. Lond. B* 313: 7–22.

Jackson, J.B.C., and Winston, J. E. 1982. Ecology of cryptic coral reef communities, I. Distribution and abundance of major groups of encrusting organisms. *J. Exp. Mar. Biol. Ecol.* 57: 135–147.

Maynard Smith, J., and Price, G. R. 1973. The logic of animal conflict. *Nature* 248: 15–18.

Menge, B. A., and Lubchenco, J. 1981. Community organization in temperate and tropical rocky intertidal habitats: Prey refuges in relation to consumer pressure gradients. *Ecol. Monogr.* 51: 429–450.

Mileikovsky, S. A. 1971. Types of larval development in marine bottom invertebrates, their distribution and ecological significance: A re-evaluation. *Mar. Biol.* 10: 193–213.

Palmer, A. R., and Strathmann, R. R. 1981. Scale of dispersal in varying environments and its implications for life histories of marine invertebrates. *Oecologia (Berl.)* 48: 308–318.

Palumbi, S. R., and Jackson, J.B.C. 1983. Aging in modular organisms: Ecology of zooid senescence in *Steginoporella* sp. (Bryozoa; Cheilostomata). *Biol. Bull.* 164: 267–278.

Parr, A. E. 1939. Quantitative observations on pelagic Sargassum vegetation of the western North Atlantic. *Bull. Bingham Oceanogr. Coll.* 6: 1–94.

Perron, F. E., and Carrier, R. H. 1981. Egg size distributions among closely related marine invertebrate species: Are they bimodal or unimodal? *Amer. Natur.* 118: 749–755.

Roughgarden, J. D.; Porter, W.; and Heckel, D. 1981. Resource partitioning of space and its relationship to body temperature in *Anolis* lizard populations. *Oecologia* 50: 256–264.

Roughgarden, J. D.; Iwasa, Y.; and Baxter, C. 1985. Demographic theory for an open marine population with space-limited recruitment. *Ecology* 66: 54–67.

Roughgarden, J. D., and Iwasa, Y. 1986. Dynamics of a metapopulation with space-limited subpopulations. *Theor. Pop. Biol.* 29: 235–261.

Savin, S. M.; Douglas, R. G.; and Stehli, F. G. 1975. Tertiary marine paleotemperatures. *Geol. Soc. Amer. Bull.* 86: 1499–1510.

Stanley, S. M. 1986. Anatomy of a regional mass extinction: Plio-Pleistocene decimation of the western Atlantic bivalve fauna. *Palaios* 1: 17–36.

Strathmann, R. R. 1977. Egg size, larval development, and juvenile size in benthic marine invertebrates. *Amer. Natur.* 111: 373–376.

Strathmann, R. R. 1978. The evolution and loss of feeding larval stages of marine invertebrates. *Evolution* 32: 894–906.

Strathmann, R. R. 1985. Feeding and nonfeeding larval development and life-history evolution in marine invertebrates. *Ann. Rev. Ecol. Syst.* 16: 339–361.

Thorson, G. 1936. The larval development, growth and metabolism of Arctic marine bottom invertebrates compared with those of other seas. *Meddr. Gronland* 100: 1–155.

Thorson, G. 1946. Reproduction and larval development of Danish marine bottom invertebrates, with special reference to the planktonic larvae in the Sound (Oresund). *Meddr. Kommn. Danm. Fisk. -og Havunders. (Ser. Plankton)* 4: 1–529.

Thorson, G. 1950. Reproductive and larval ecology of marine bottom invertebrates. *Biol. Rev.* 25: 1–45.

Todd, C. D., and Doyle, R. W. 1981. Reproductive strategies of marine benthic invertebrates: A settlement-timing hypothesis. *Mar. Ecol. Prog. Ser.* 4: 75–83.

Vance, R. R. 1973. On reproductive strategies in marine benthic invertebrates. *Amer. Natur.* 107: 339–352.

Victor, B. C. 1983. Recruitment and population dynamics of a coral reef fish. *Science* 219: 419–420.

Victor, B. C. 1986. Larval settlement and juvenile mortality in a recruitment-limited coral reef fish population. *Ecol. Monogr.* 56: 145–160.

CHAPTER THIRTEEN

An Evolutionary Model for

Highly Repeated Interspersed

DNA Sequences

Norman L. Kaplan and
Richard R. Hudson

Introduction

Within the genomes of mammalian species there are many distinct families of repetitive DNA elements. The lengths of the repeated DNA sequences vary from a few nucleotides to several thousand, and their copy number from several hundred to several million. For some families the repeats occur tandemly, while for others the individual copies are dispersed throughout the genome. In this paper we are interested in studying evolutionary models for highly repeated short interspersed families (SINEs as denoted by Singer 1982). Examples of such families are the Alu family in humans, the B1 family in mice, the Monomer family in Galagos, the rodent B2 family, and the rat identifier family (Deininger and Daniels 1986).

Until the last few years few data were available to help explain the evolutionary dynamics of SINE families. Recent discoveries, however, make it possible to draw some conclusions about their possible origins and how they might have spread through the population. New evidence suggests that most if not all SINE families have originated from a functional RNA gene via a retroposition mechanism (Deininger and Daniels 1986). Retroposition is the process whereby RNA molecules are reverse transcribed into DNA and then inserted into the genome. The human Alu family, which was the first SINE family for which the ancestral gene was identified, is believed to be descended from the 7SL RNA gene (Ullu and Tschudi 1984). This conclusion is based in part on the fact that the 7SL RNA gene is well conserved in all species from Drosophila to human, and that in Drosophila the Alu sequence exists only as part of the 7SL RNA gene, of which there are just two copies. More recent data suggest that several other mammalian SINE families are derived from transfer RNA genes (Daniels and Deininger 1985).

It is not presently known whether most members of a family or only a small subset actively generate transcripts that can be reverse transcribed and inserted at new genomic locations. For Alu-like SINE families, RNA polymerase III transcripts are the primary RNA intermediates and the promoter is located within the transcribed gene. Hence, newly inserted copies have the potential to generate more transcript. There is, however, some evidence that gene flanking regions may be involved in the regulation of RNA polymerase III transcription (Ullu and Weiner 1985), and so it may be that there are only a few parent genes.

Some recent theoretical work also suggests that only a small subset of a SINE family can actively transcribe. If most copies of a family can transpose, then it is not unreasonable to suppose that the population dynamics of a SINE family are similar to those of a transposable element family. In a recent paper, Brookfield (1985) adopted this approach for studying the evolutionary relationship between copies at different genomic locations. More specifically, he proposed to estimate the nucleotide mutation rate for a SINE family by dividing the base pair divergence for a pair of randomly chosen copies by an estimate of $E(T)$, the expected time to the most recent common ancestor of the two copies. His work focused on predicting the equilibrium value of $E(T)$. Using a finite population model proposed by Langley, Brookfield, and Kaplan (1983) for the evolution of a transposable element family, he showed that at equilibrium $E(T)$ is approximately equal to $\Lambda/(2\mu)$ generations, where Λ is the average number of copies per genome at equilibrium and μ is the probability each generation that a copy is excised from the genome. For the human Alu family, Λ is approximately 300,000, and θ, which equals $4N_e\mu$, is most likely small (Brookfield 1985). The quantity N_e is the effective population size. If one assumes as Brookfield did, that $N_e \simeq 10^4$, then an overestimate of $1/(2\mu)$ is 2×10^4. A conservatively low estimate of $E(T)$ is therefore, 6×10^9 generations. This estimate is much too large, and so Brookfield concluded that the predictions for $E(T)$ based on the model of Langley et al. most likely do not apply.

In this paper we propose an alternative model for the evolution of a SINE family. Rather than assuming that most, if not all, copies are capable of transposing, we assume the other extreme, which is that there is one parent gene that creates new copies each generation. This assumption may be more appropriate if copies are processed pseudogenes. In the next section we formally define the model and develop a formula for $E(T)$ at equilibrium. We also investigate several other properties of the model.

Our theoretical results are used to analyze recent sequence data for seven human Alu copies and seven chimpanzee Alu copies, and we obtain estimates of the parameters of the model, including the excision rate of Alu copies, from these data. We also consider whether the model and the estimates are consistent with the following observations: (1) the α-globin region of chimpanzee contains Alu copies in each of the seven locations that contain copies in the human α-globin region (Sawada et al. 1985); (2) the divergence of flanking direct repeats of Alu copies is approximately 12% (Schmid and Shen 1985); and (3) a sample of six Alu copies from galago shows species-specific differences from human Alu copies (Daniels et al. 1983).

Theory

The model we consider here is similar in many respects to the one studied by Langley, Brookfield, and Kaplan (1983). We assume that the host population is a finite random mating Mendelian population of 2N haploids. Within the genome are presumably a large number of locations at which copies can insert, and so we assume that a new copy always inserts at a location that is not occupied in any of the individuals in the population. Any location at which there is a copy in at least one individual is called a site. At this point we make no assumptions about the rate of recombination between sites. In each generation, any copy has a probability μ of being excised from the host genome. Wright-Fisher sampling occurs in each generation, and so we describe the population dynamics of any site by a Wright-Fisher process with deletion (Ewens 1979).

The remaining assumption of the model concerns the dynamics of creating new sites. While Langley et al. assumed that the rate of transposition was copy-number dependent, we assume that the number of copies inserted each generation in new locations in each genome has a Poisson distribution with mean β. We also assume that there is only one parental gene that generates the new copies and that there is free recombination between the parental gene and each of the other sites in the genome. The parental genes of human Alu and mouse B1 families are presumably derived from and closely resemble the 7SL RNA gene. Similarly, the parental genes of mouse B2a and B2b families are presumably derived from Ser-tRNA, and the rat identifier (ID) family parental gene is derived from Ala-tRNA (Deininger and Daniels 1986).

The arguments of Langley et al. can be applied directly to the parental gene model to show that at equilibrium the expected number of sites in the population with frequencies between x and $x + dx$ is given by

$$f(x) = \Lambda\theta x^{-1}(1 - x)^{\theta - 1}\, dx, \qquad (13.1)$$

where $\theta = 4N\mu$ and $\Lambda = \beta/\mu$, which equals the expected number of copies per genome at equilibrium. Any predictions based on the sample frequency spectra are thus the same for the parental gene model and the transposition model of Langley et al. Differences between the models do occur, as we shall show, with regard to their predictions about the distribution of T, the number of generations back to the most recent common ancestor of two randomly chosen copies at different sites.

The distribution of T is important because of its relation to the distribution of S, the number of sites that differ between two randomly chosen copies. If each generation the number of mutations occuring in each copy in each genome has a Poisson distribution with mean μ_m, then it follows from arguments similar to those given by Watterson (1975) that for an infinite sites model at equilibrium,

$$P(S = j) = \int_0^\infty e^{-2\mu_m t} \frac{(\mu_m t)^j}{j!} F(dt), \tag{13.2}$$

where $F(t) = P(T \le t)$. Hence, at equilibrium the distribution of S is a compound Poisson where the compounding distribution is that of $2\mu_m T$. The moments of S are easily obtained from (13.2), for example,

$$E(S) = 2\mu_m E(T) \tag{13.3a}$$

and

$$\mathrm{Var}(S) = 2\mu_m E(T) + 4\mu_m^2 \, \mathrm{Var}(T). \tag{13.4a}$$

It is common to measure T in units of $2N$ generations. In this case (13.3a) and (13.4a) become

$$E(S) = \theta_m E(T) \tag{13.3b}$$

and

$$\mathrm{Var}(S) = \theta_m E(T) + \theta_m^2 \, \mathrm{Var}(T), \tag{13.4b}$$

where θ_m equals $4Nu_m$. If one samples copies from more than two different sites, then the number of variable sites in the sample still has a compound Poisson distribution, except that the compounding distribution is that of $\mu_m T_{\mathrm{tot}}$ where T_{tot} is the sum of the lengths of all the branches (measured in generations) of the ancestral tree of the sampled copies (Watterson 1975).

Hudson and Kaplan (1986a) recently showed for the model of Langley et al. that the distribution of T (measured in units of $2N$ generations) is approximately the same as the sum of two exponential random variables with means $\Lambda(1 + \theta)/\theta$ and $1/(1 + \theta)$, respectively. Brookfield's conclusion that the equilibrium value of $E(T)$ (measured in generations) is approximately equal to $\Lambda/(2\mu)$ if θ is small is immediate from this result. We now compute the distribution of T for the parental gene model.

We suppose that the population was originally devoid of copies. At some time in the ancient past, an allele of the parent gene with the ability to generate new copies appeared and quickly spread through the population. Hence, we assume without loss of generality that initially the number of copies in each genome has a Poisson distribution with mean β. Since it is assumed that the excision process is independent of the insertion process, it is not difficult to show that for Wright-Fisher sampling the distribution of the number of copies in a random genome from the population t generations later ($t \geq 1$) is also Poisson, but with mean equal to

$$\Lambda_t = \beta\left(\sum_{j=0}^{t-1}(1-\mu)^j\right).$$

This result does not depend on the amount of linkage between sites because of the additive property of the Poisson distribution. As t becomes large, Λ_t approaches β/μ, which equals the average number of copies per genome at equilibrium.

In any generation, copies can be categorized in two ways: copies that are at locations that were occupied in at least one genome in the previous generation and copies that are newly inserted. If X and Y denote the numbers of these two types of copies in a randomly chosen genome, then

$$\eta = E\left(\frac{X}{X+Y}\right)$$

is the probability that a randomly chosen copy is at a location in the genome that was occupied in at least one genome in the previous generation. Since X and Y are independent, Y has a Poisson distribution with mean β and $X + Y$ has a Poisson distribution with mean β/μ, η at equilibrium equals $1 - \mu$. This follows from the result that if Z and W are Poisson variables with means μ and v, respectively, then the conditional distribution of Z given $Z + W$ is binomial with parameters $Z + W$ and $\mu/(\mu + v)$. The same kind of argument in conjunction with the strong Markov property of the sampling process shows that the probability that a randomly chosen copy is at a location that is occupied in at least one genome in each of the j previous generations equals $(1 - \mu)^j$. Let $T(1)$ denote the ancestral generation in which a copy of the parental gene was first inserted in the population at the sampled site in the genome. The above discussion implies that $T(1)$ (measured in units of $2N$ generations) has a negative exponential distribution with parameter $\theta/2$.

Suppose we sampled k copies at different sites. One can show just as before that the probability that each of the k copies is at a site which was occupied in at least one genome in the previous generation equals $(1 - \mu)^k$. Furthermore, it does not matter whether all k copies are from different genomes or from the same. If $T(k)$ denotes the most recent ancestral generation in which a copy of the parent gene was first inserted at one of the k sampled sites in the genome, then $T(k)$ has an exponential distribution with parameter $k\theta/2$ when measured in units of $2N$ generations.

We are now in a position to determine the distribution of T. The distribution of $T(2)$, the most recent ancestral generation (measured in units of $2N$ generations) in which a copy was first inserted in one or the other of the two sampled sites, was shown to be negative exponential with parameter θ. Since we are at equilibrium, the distribution of the additional number of generations $T(1)$, until a copy is first inserted at the remaining sampled location, is negative exponential with parameter $\theta/2$. In the $[T(2) + T(1)]$th ancestral generation the ancestors of the two sampled copies are two random parental genes. Hence, an additional number of generations $T(0)$ is required until the two parental genes have a common ancestor. For Wright-Fisher sampling, the distribution of $T(0)$ (measured in units of $2N$ generations) is exponential with parameter 1 (Watterson 1975). We have thus shown that

$$T = T(2) + T(1) + T(0),\qquad(13.5)$$

where $T(2)$, $T(1)$, and $T(0)$ are independent exponential random variables with parameters θ, $\theta/2$, and 1, respectively. It should be noted that the distributions of $T(2)$ and $T(1)$ only depend on the insertion and deletion processes and not on the sampling process. Only the distribution of $T(0)$ depends on the sampling model.

It follows directly from (13.5) that

$$E(T) = \frac{3}{\theta} + 1\qquad(13.6)$$

and

$$\text{Var}(T) = \frac{5}{\theta^2} + 1.\qquad(13.7)$$

The formula for $E(T)$ in (13.6) is quite different from Brookfield's in that Λ does not enter into the calculation. For example, if one uses the values that Brookfield did, then the estimate of $E(T)$ in generations for the parental gene model is of the order $3/(2\mu) \simeq 1.5 \times 10^5$, which is several orders of magnitude smaller than Brookfield's result. One should note, however, that if $\mu \simeq 10^{-5}$ and $\Lambda \simeq 10^6$, then $\beta \simeq 10$. Hence, new sites are being created at a rate which is six orders or magnitude higher than the rate of excision.

If one were to sample k copies at different sites, then the distribution of T_{tot} is much more complicated. However, if θ is small, then it is not difficult to show that

$$T_{\text{tot}} \simeq kT(k) + kT(k-1) + (k-1)T(k-2) + \cdots + 2T(1),$$

where the $\{T(j), 1 \leq j \leq k\}$ are independent exponential random variables with parameters $\{j\theta/2, 1 \leq j \leq k\}$, respectively. The mean and variance of T_{tot} are approximately

$$E(T_{\text{tot}}) \simeq \frac{2}{\theta}\left(k + \sum_{j=1}^{k-1} \frac{1}{j}\right) \tag{13.8}$$

and

$$\text{Var}(T_{\text{tot}}) \simeq \frac{4}{\theta^2}\left(1 + \sum_{j=1}^{k-1}\left(\frac{j+1}{j}\right)^2\right). \tag{13.9}$$

The only samples we have considered so far are those where all copies are at different sites. Suppose, instead, we sample a copy from a random genome and sequence it, sample another genome, and assuming that it has a copy at the sampled site, sequence that copy too. The number of variable sites still has a compound Poisson distribution, but the distribution of T is conditional on there being copies at the same site in both genomes (Hudson and Kaplan 1986b). To determine the conditional distribution of T, we first need to calculate the probability that the second genome has a copy at the sampled site. This probability, which we denote by Q, is easy to compute in light of our previous results. We have already noted that the distribution of $T(0)$, the time (measured in units of $2N$ generations) to the most recent common ancestor of the two genomes, is exponential with parameter 1. Given the assumption of insertion only into unoccupied sites, the only way that a copy can be present in two genomes at the same location is for the copy to be at that location in every ancestor since the most recent common one. Hence,

$$\begin{aligned} Q &= E((1-\mu)^{2T(0)}) \\ &\simeq E(e^{-\theta T(0)}) \\ &= \frac{1}{1+\theta}. \end{aligned} \tag{13.10}$$

We can now calculate the conditional distribution of T. Indeed,

$P[T \in (t, t+dt)|\text{both genomes have a copy at the sampled site}]$
$$\begin{aligned} &= (1+\theta)e^{-\theta t}P[T(0) \in (t, t+dt)] \\ &= (1+\theta)e^{-(1+\theta)t}\,dt. \end{aligned}$$

Thus the conditional distribution of T is exponential with parameter $1 + \theta$. We note in passing that if we were to sample n additional genomes instead of just one, then we can show that the probability that exactly k of these genomes would have a copy at the sampled site equals $[(k + 1)/(n + 1)]E(\alpha_{k+1})$, where α_{k+1} equals the number of alleles of frequency $k + 1$ in a sample of $n + 1$ genes for an infinite alleles model (Ewens 1979).

Until now we have considered only samples from one population. An alternative sampling scheme is to sample, say, k copies at different sites in one species and then sample an additional k copies at the same sites in a diffferent species. These kinds of data were recently published for seven copies of human and chimpanzee Alu (Sawada et al. 1985), and so it is of interest to determine the distribution of the time to the most recent common ancestor of copies at the same site in different species.

To begin the analysis, we suppose that we sample a copy at the same site in two species that diverged $2Nt_0$ generations ago. Since copies are always inserted at new locations in the genome, it must be that when the two species diverged, there was at the sampled location an ancestral copy of each of the sampled copies. Hence, at the time of the split, we have effectively sampled two random copies at the same site. The equilibrium distribution of T^*, the additional time until the two ancestral copies have a common ancestor, has already been shown to be exponential with parameter $1 + \theta$. Thus T satisfies

$$T = t_0 + T^*.$$

It follows that

$$E(T) = t_0 + \frac{1}{1 + \theta} \tag{13.11}$$

and

$$\text{Var}(T) = \frac{1}{(1 + \theta)^2}. \tag{13.12}$$

We next consider the case where k copies are sampled at different sites in one species and k copies at the same sites in the other species. Since all copies must have ancestors at the time of the split, we can suppose that at the time of the split we have sampled two copies at k different sites. If θ is small, then the times to the coalescences at the k sites are negligible. Hence, the distribution of the sum of the lengths of the branches of the ancestral tree of the $2k$ copies is approximately the same as $2kt_0 + T_{\text{tot}}$, where T_{tot} is given in (13.8).

To complete this section, we consider the following problem. Suppose we sample k copies at different sites in one species. For convenience we label the sites 1 to k. We next sample k genomes from a related species, label them from 1 to k and determine whether in genome i there is a copy at site i ($1 \leq i \leq k$). The problem is to compute the probability, $R(k)$, that for all i, the ith genome has a copy at site i. This probability is of interest for the human and chimpanzee Alu data, which is analyzed in the next section.

Under the assumptions of the model, the time (measured in unit of $2N$ generations) to the recent common ancestor of the DNA at site i equals $t_0 + T_i(0)$, where $T_i(0)$ has an exponential distribution with parameter 1. Hence, conditional on $T_i(0)$, the probability that the ith genome has a copy at site i equals

$$(1 - 2\mu)^{2N[t_0 + T_i(0)]} \simeq e^{-\theta[t_0 + T_i(0)]}.$$

Since the excision process is independent for each site,

$$R(k) = e^{-k\theta t_0} E(e^{-\theta \sum_{i=1}^{k} T_i(0)}).$$

For Wright-Fisher sampling, the joint distribution of the $\{T_i(0)\}$ is complex and depends on what assumptions are made about the rate of recombination between sites, for example, if the sites are unlinked then the $\{T_i(0)\}$ are independent. It follows from Jensen's inequality and the fact that $E[T_i(0)] = 1$ that $R(k)$ always satisfies

$$R(k) \geq e^{-k\theta(t_0 + 1)}. \tag{13.13}$$

Applications to ALU Sequences of Humans, Chimpanzees, and Galago

In this section we focus on the sequence data of Sawada et al. (1985). The data consist of the sequences of seven Alu copies from the α-globin region of the human genome and seven Alu copies found in the same sites in the chimpanzee genome. For our calculations we use only the 265 aligned sites that are present in all fourteen of the Alu copies. Using the theory of the previous section, we estimate $\theta_m(t_0 + 1)$, θ_m/θ, and the lower bound on $R(7)$ in equation (13.13). Since there is an independent estimate of t_0, we can also estimate μ_m and μ. We then consider whether our model and the estimates of the parameters are consistent with the observed level of divergence of the direct repeats flanking Alu copies of humans and with the observation of species-specific differences between galago and human Alu copies (Daniels et al. 1983).

Let S_w denote the number of differences between two copies from different sites from the same species. It follows from (13.6) that

$$E(S_w) = \theta_m(3/\theta + 1). \tag{13.14}$$

Hence, a plausible estimator of θ_m/θ, when θ is small, is simply $\hat{S}_w/3$ where $\hat{S}_w$ is the mean number of differences between copies at different sites in the same species. For the Alu data, the mean number of differences in the 42 possible comparisons (21 human pairwise comparisons and 21 chimpanzee comparisons) is 53.9. Thus our estimate of θ_m/θ is 53.9/3 = 18.0. One can also estimate θ_m/θ using the equation (13.6) in the obvious way. The total number of variable sites in the sample of seven human Alu copies is 138, and the estimate of θ_m/θ obtained from this observation is 14.6.

Let S_b denote the number of differences between two copies at the same site but from different species. It follows from (13.11) that

$$E(S_b) = \theta_m\left(t_0 + \frac{1}{1 + \theta}\right).$$

Thus we propose estimating the product on the right-hand side of this equation by $\hat{S}_b$, the mean number of differences between the human and chimpanzee copies at the same sites. The mean for the seven comparisons is 4.86.

One can use the two estimates just presented to estimate the lower bound for $R(7)$ given in (13.13). For θ small, the ratio $3E(S_b)/E(S_w)$ is approximately equal to $\theta(t_0 + 1)$, which is the quantity needed to calculate the lower bound for $R(7)$ in (13.13). For the Alu data, $3\hat{S}_b/\hat{S}_w = 3(4.86)/53.9 = 0.27$, and so we estimate that approximately $\exp(-\theta(t_0 + 1)) \simeq \exp(-.27) \simeq 76\%$ of elements present in humans are present in the same sites in chimpanzees. The lower bound for $R(7)$ is $\exp[-7(0.27)] = 0.15$. Thus the observation that all seven Alu copies are found in chimpanzees is not extremely unlikely, although the absence of one or more of the seven Alu copies would have been a more likely outcome. Our estimate of θ_m/θ depends on the assumption that the neutral mutation rate of the parent gene is the same as the neutral mutation rate of the copies dispersed throughout the genome. If the parent gene is subject to strong functional constraints that do not act on the dispersed Alu copies, then a much lower neutral mutation rate for the parent gene would result. If the neutral mutation rate of the parent gene is zero, then it is not difficult to show that $E(S_w) = \theta_m(2/\theta)$. Using the obvious estimator of θ_m/θ under this assumption, the estimate of the lower bound on $R(7)$ becomes 0.28, which is somewhat larger than before. Under either assumption, if fifteen or twenty more sites of human Alu copies were examined in the chimpanzee genome, then the parental gene model predicts that it is very improbable that all these locations in the chimpanzee genome would also be found to be occupied by Alu copies. Similarly, if one examines another primate not so closely related to humans, such as the orangutan, a significant fraction of the Alu copies should not be present in the same sites as in humans.

If one of the three parameters, θ_m, t_0, or θ, were known or could be estimated from other data, then we can estimate the other two parameters using the relationships given above. Very rough estimates of θ_m and t_0 are available. The divergence time of human and chimpanzee has been estimated to be approximately 5 million years. If $2N$ generations is small compared to 5 million years, we can estimate μ_m by $\hat{S}_b/2t_0 = 4.86/2(5 \times 10^6) = 4.86 \times 10^{-7}$ mutations per year or $4.86 \times 10^{-7}/265 = 1.8 \times 10^{-9}$ mutations per year per nucleotide site. This rate is quite close to other estimates of the neutral mutation rate (Li, Luo, and Wu 1985). Our estimate of θ_m/θ is 18.0, and so our estimate of μ, the deletion rate, is $\mu_m/18 = 4.86 \times 10^{-7}/18 = 2.7 \times 10^{-8}$ per Alu per year.

Under the parental gene model, we can also predict the mean divergence of the flanking direct repeats of Alu copies. We estimate that an Alu element, present in the human genome today, was inserted into its current site on average about $1/\mu \simeq 40$ million years ago. This time is two-thirds of the expected time back to the common ancestor of two copies at different sites ($\simeq 3/2\mu$). Since copies at different sites differ on average by $53.9/265 = 20\%$, we expect the mean divergences of the direct repeats created by Alu insertion to be approximately $20(2/3) = 13.3\%$. This is consistent with the estimates of Schmid and Shen (1985) who found that the direct repeats have diverged by about 12%.

Finally, in this section, we consider whether our model is consistent with the observation of species-specific differences between human and galago Alu sequences that have been examined. Species-specific differences are unlikely unless all the copies from at least one of the species have a common ancestor more recently than the divergence time of the species. All six of the type I Alu copies of galago sequenced by Daniels et al. (1983) are likely to have a common ancestor more recently than the galago-human divergence if all six elements were inserted more recently than the species divergence. The probability of this is $[1 - \exp(-\mu t_g)]^6$, where t_g is the time of divergence of galago and humans. Assuming that $t_g = 60$ million years, this probability is approximately 0.27. We conclude that the model is consistent with the observation of species-specific differences in the six galago copies of Alu.

Discussion

As discussed in the Introduction, Brookfield has shown that the Alu sequence data are not compatible with an equilibrium duplicative transposition model. The nonequilibrium properties of this model have been recently studied by Ohta (1986), who suggests that the nonequilibrium model fits much of the data well. We have analyzed a different model in which only one gene, the "parent" gene, is the source of new copies. This model is essentially the same as the model for processed pseudogenes studied by Walsh (1985). We have shown that the Alu sequence data from humans, chimpanzees, and galago are not inconsistent with the parental gene model. We point out that a potentially powerful test of the model is possible. If fifteen or twenty additional sites of Alu copies in humans are examined in the chimpanzee, the model predicts that some of the sites will lack Alu copies in the chimpanzee. Or, if the Alu sites of the α-globin region of humans are examined in a less closely related primate than the chimpanzee, such as the orangutan, then the model predicts that some of the sites will lack Alu copies.

An alternative model for the evolution of SINE families that has been proposed recently is that each SINE family was generated very rapidly (on an evolutionary time scale) at some time in the past, by the amplification of one or a few original copies (Deininger and Daniels 1986). Since that time, the copies have been stable, slowly diverging by the accumulation of neutral mutations and perhaps being excised occasionally. Few, if any, new copies were inserted since the rapid amplification. Under this amplification model there is essentially no turnover or homogenization of the family. Species-specific differences that are observed in some families are explained as being the result of separate amplification episodes in the different species. It is presumed that the source genes for the amplification in each species are somewhat diverged, resulting in the species-specific differences that are observed. Under this model, the expected number of differences between two copies at different sites is simply $2\mu_m t_a$, where t_a is the time since the expansion of the family. Obviously, in contrast to the parental gene model, the expected divergence of elements at different sites is not a function of the excision rate. We can estimate t_a by $\hat{S}_w/2\mu_m$, which for the Alu family is $53.9/(2 \times 4.86 \times 10^{-7}) \simeq 55$ million years.

The rapid amplification model and the parental gene model make different predictions about the amount of divergence of the direct repeats created at each end of inserted elements. Under the rapid amplification model, essentially all the elements were inserted at one time, and the divergence of the direct repeats is expected to be the same as the divergence of two elements, $2\mu_m t_a$. Under the parental gene model, the expected divergence of direct repeats is only two-thirds of the expected divergence of two elements, as shown in the previous section. Alu copies from different sites from humans are approximately 20% diverged. Therefore, we expect, under the rapid amplification model, that direct repeats will be approximately 20% diverged, and under the parental gene model we expect that they will be approximately $20(2/3) = 13.3\%$ diverged. Schmid and Shen (1986) estimate the divergence of direct repeats of Alu copies to be 12%, a value very close to the prediction of the parental gene model and quite distinct from the prediction of the rapid amplification model.

The two models also make different predictions about the variance of number of differences among elements from different sites. Under the rapid amplification model, all elements have a common ancestor at approximately the same time, and consequently under the infinite site model the variance of S_w is equal to the mean. Under the parental gene model, there is a large variance in the time back to the common ancestor of different elements, and so the variance of S_w is larger than the mean under an infinite site model. Thus it may be possible to distinguish the two models based on estimates of the variance of S_w.

The results we have presented are based on a model with a single parental gene. To illustrate the effects of having more than one parental gene, we consider a simple extension to n parental genes. It is assumed that each parental gene produces copies at the same rate. In this case, the mean divergence of two random copies from within a species can be written as

$$E(S_w) = \theta_m[3/\theta + 1/n + (1 - 1/n)E(T_p)],$$

where T_p is the time back to the most recent common ancestor of two randomly chosen parental genes. If n equals 1, this reduces, as we expect, to the earlier result given by equation 13.14. If $E(T_p)$ is large compared to $1/\theta$, then copies will show strong clustering into subfamilies. Copies from within a cluster will be little diverged, all being descended from the same parental gene. Copies from different clusters will be highly diverged, being descendants of different parental genes. If $E(T_p)$ is small compared to $1/\theta$, then the parental genes will be very similar to each other, and the divergence of copies will be essentially the same as with a single parental gene. If $E(T_p)$ is similar in magnitude to $1/\theta$, then further progress requires a model of the process that generates parental genes.

References

Brookfield, J.F.Y. 1985. A model for DNA sequence evolution within a transposable element family. *Genetics* 112: 393–408.

Daniels, G. R., and Deininger, P. R. 1985. Repeat sequence families derived from mammalian tRNA genes. *Nature* 317: 819–822.

Daniels, G. R.; Fox, G. M.; Loewensteiner, D. L.; Schmid, C. W.; Deininger, P. L. 1983. Species-specific homogeneity of the primate Alu family of repeated DNA sequences. *Nucleic Acid Res.* 11: 7579–7593.

Deininger, P. L., and Daniels, G. R. 1986. The recent evolution of mammalian repetitive DNA sequences. *Trends in Genetics* 2: 76–80.

Ewens, W. J. 1979. *Mathematical Population Genetics.* Springer-Verlag, Berlin.

Hudson, R. R., and Kaplan, N. L. 1986a. On the divergence of members of a transposable element family. *J. of Math. Biol.* 24: 207–215.

Hudson, R. R., and Kaplan, N. L. 1986b. On the divergence of alleles in nested subsamples from finite populations. *Genetics* 113: 1057–1076.

Langley, C. H.; Brookfield, J.F.Y.; and Kaplan, N. L. 1983. Transposable elements in Mendelian populations, I. A theory. *Genetics* 104: 457–471.

Li, W. H.; Luo, C. C.; and Wu, C. I. 1985. Evolution of DNA sequences. In *Molecular Evolutionary Genetics*, pp. 1–93 Ed. R. J. MacIntyre. Plenum, New York.

Ohta, T. 1986. Population genetics of an expanding family of mobile genetic elements. *Genetics* 113: 145–159.

Sawada, I.; Willard, C.; Shen, C.-K.J.; Chapman, B.; Wilson, A. C.; and Schmid, C. W. 1985. Evolution of Alu family repeats since the divergence of human and chimpanzee. *J. Molec. Evol.* 22: 316–322.

Schmid, C., and Shen, C.-K.J. 1985. The evolution of interspersed repetitive DNA sequences in mammals and other vertebrates. In *Molecular Evolutionary Genetics*, pp. 323–358. Ed. R. J. MacIntyre. Plenum, New York.

Singer, M. F. 1982. SINES and LINES: Highly repeated short and long interspersed sequences in mammalian genomes. *Cell* 28: 433–434.

Ullu, E., and Tschudi, C. 1984. Alu sequences are processed 7SL RNA genes. *Nature* 312: 171–172.

Ullu, E., and Weiner, A. M. 1985. Upstream sequences modulate the internal promoter of the human 7SL RNA gene. *Nature* 318: 371–374.

Walsh, J. B. 1985. How many processed pseudogenes are accumulated in a gene family? *Genetics* 110: 345–364.

Watterson, G. A. 1975. On the number of segregating sites in genetical models without recombination. *Theor. Pop. Biol.* 7: 256–276.

CHAPTER FOURTEEN

Statistics and Population Genetics

of the HLA System

Walter F. Bodmer and
Julia G. Bodmer

Introduction

Simple statistical 2×2 χ^2 analysis of reactions of groups of sera on a panel of cells from different individuals led to the definition of the HLA system. Positive associations between pairs of sera indicate common determinants, and so the first antigens were simply defined by the consensus of reactions of a set of sera, all or most pairs of which were significantly positively associated. This led Van Rood to describe the two-allele system he called 4a and 4b, later to become related to the HLA-B locus (Van Rood and Van Leeuwen 1963). Recognizing that alleles at a locus were negatively associated, Payne et al. (1964) described the alleles LA1 and LA2, now HLA-A1 and HLA-A2 of the HLA-A locus. Even though the experimental technique was unsophisticated and subject to relatively large experimental error, simple statistical analysis was all that was needed to bring some order out of an apparently chaotic set of serum reactions and so to lead to the discovery of the HLA system. These statistical approaches, with some added sophistication in the search for clusters of related sera and computer manipulation of the data, have remained at least until recently the mainstay of HLA serology, through which the majority of the HLA loci and their determinants have been defined. Since good sera are comparatively rare, collaboration has been very important in the development of the HLA system. This has taken the form of international workshops where reagents and methods have been compared with a major emphasis on the combined statistical analysis, especially in recent workshops, of very large bodies of data. The workshops were started by Amos in 1964, the ninth workshop was in 1984 (Albert 1984) and the tenth and most recent workshop took place in November 1987.

The extensive polymorphism of the HLA system was already clear from
the fact that highly polyspecific sera were produced by individuals who
had received multiple blood transfusions, and from the high frequency
with which foeto-maternal stimulation produced HLA antibodies, this
being the main source of sera used for the definition of the system. The
early workshops established that the major polymorphic determinants
were controlled by multiple alleles at two closely linked loci, HLA-A and
-B, and then identified a third, even more closely related and linked locus,
HLA-C. Cellular techniques using lymphocyte stimulation and analogous
statistical methodology to that used for the serological analysis identified
a series of determinants Dw, specific to B cells. These were later paralleled
by the serological determinants identified on B cells, HLA-DR. In recent
years techniques for making monoclonal antibodies and for cloning genes,
together with biochemical analysis, have revolutionized our knowledge of
the HLA system. Statistical and population genetic analyses are, however,
still necessary, and have if anything become more important as the complexity of the system, in terms of numbers of loci and of alleles, has increased. The aim of this paper is to provide a brief survey of the statistics,
population genetics, and evolutionary problems of the HLA system as it
is currently understood.

The HLA Region

A schematic overall map of the HLA region on chromosome 6, compared
with its mouse equivalent, H2, is shown in Figure 14.1. In addition to
the original HLA-A, B, C, now sometimes called Class I, determinants,
there are many other similar loci corresponding to QA and TL first discovered in the mouse H2 system. DNA sequence analysis shows that the
HLA-A, B, C products are very closely related to each other, as are the
genes within the QA and TL clusters. The difference in DNA sequence
between clusters approaches 30%, as compared to 5–10% differences
within clusters. Nevertheless, all of these genes are clearly associated with
$\beta 2$ microglobulin, coded for on a different chromosome. About half the
genes in the QA and TL region are nonfunctional pseudogenes. The function of the HLA-A, B, C products is to mediate the interaction between
killer T cells and their targets in the immune system, but there is so far
no real clue as to the function of the QA and TL genes whose tissue
distribution is much more restricted.

The HLA-D, or Class II, region codes for at least three subgroups of
determinants, DP, DQ and DR, each of which is formed by a combination
of two different polypeptides, α and β, all the relevant genes being encoded
within the HLA-D region. This contrasts with the lack of linkage between

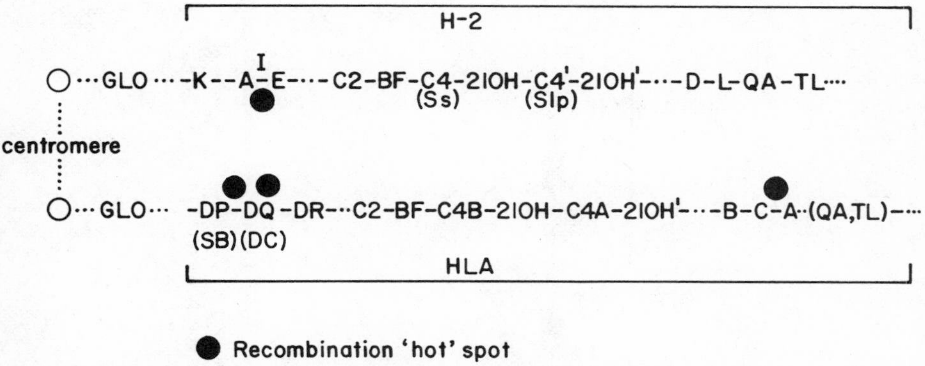

Figure 14.1. Schematic maps of the HLA and H2 regions. The regions are aligned to give the maximum correspondence of the sequences, leaving only H2-K out of line with the HLA sequence. HLA-A, B, C correspond to H2 DL and K and are the Class I genes. QA and TL are closely related but have different functions. H2 I-A corresponds to HLA-DQ and H2 I-E to HLA-DR. C2 and C4 are the second and fourth components of the classical complement system, and BF is factor B closely related to C2 but in the alternative pathway. 210H is the 21 hydroxylase gene deficient in congenital adrenal hypoplasia. GLO is glyoxalase, linked to both H2 and HLA. Ss and Slp are the original names for the mouse C4 genes in the H2 region and SB and DC are the former names for HLA-DP and -DQ, respectively. (Source: Bodmer et al. 1986. Recent data suggest that the correct orientation of the sequence C2 to 210H′ may be with C2 nearest to HLA-B and 210H′ to HLA-DR; Mueller et al. 1987.)

the genes for $\beta2$ microglobulin and the HLA-A, B, C determinants. Between these two sets of genes lie some of the genes for complement component, C2, factor B (BF), and two C4 genes, each closely associated with a gene for 21 hydroxylase, the enzyme that is deficient in congenital adrenal hyperplasia. The overall arrangement of the genes in mouse and man is remarkably similar, with only H2K, one of the mouse Class I genes, being out of line. A detailed molecular map of the HLA-D region and a comparison with its mouse equivalent, the H2-I region, is shown in Figure 14.2. This emphasizes the complexity that has been revealed by molecular genetic analysis. The DP and DQ regions contain pairs of α and β genes, presumably derived from each other by duplication. Only one set of DP genes is functional, while for DQ it is not known whether the second set of genes, DX, is expressed, though from their sequence the genes are potentially functional. In the DR region there is one α chain gene and three or more β chain genes, the number varying between individuals. The DO region contains an α and a β gene, but these may not correspond to each other. There is also evidence that a limited number of additional

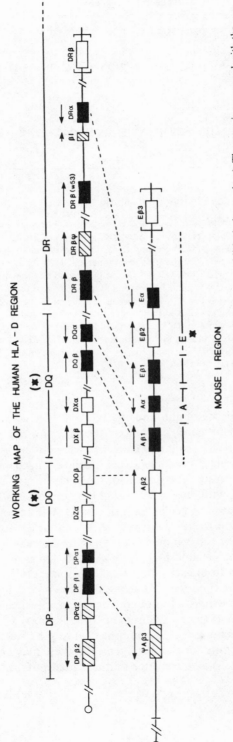

Figure 14.2. A schematic map of the HLA D region (based on Trowsdale et al. 1985 and further unpublished observations). The map is compared with the corresponding H2 I region. Filled-in rectangles indicate genes known to be expressed, hatched ones those known to be pseudogenes, and blanks those where expression is not yet clearly established. The arrows above indicate directions of transcription. The asterisks denote recombinational hot spots.

α and β genes may still be found within the region. As we will discuss later, some of the genes are much more polymorphic than others.

There is a clear correspondence between the detailed human and mouse maps, but there are also some striking differences. The mouse seems to lack a functional DP region and has only a vestigial pseudogene for a DP-like β chain. It has reasonable equivalents for DQ and DR, namely, I-A and I-E. Overall, the most striking difference is that the mouse I region is less complex and is physically more compact by a factor of at least two or three. (Overall summaries of the current state of the HLA region will be found in Albert 1984, and in Immunological Reviews, 1985, Nos. 84 and 85. A review of the use of monoclonal antibodies in the HLA system is also found in J. G. Bodmer and W. F. Bodmer 1984).

Gametic Association and Recombination Hot Spots

Gametic association (W. F. Bodmer and Payne 1965; sometimes also referred to as linkage disequilibrium) has long been a well-recognized feature of the population distribution of pairs of alleles of the different HLA loci. Indeed, it was gametic association that initially led to a confusion in the separate definition of the HLA-A, B, and C loci. Thus, positive associations between alleles may lead to positive associations between sera-recognizing determinants controlled by different loci, and this may be confused with the possibility that the two sera recognize the same determinant but with different additional activities. This problem was resolved firstly by more careful selection and clustering of sera, so that only those which were very highly associated were assigned to the same antigen cluster (gametic association being generally less strong than the association between two nearly monospecific sera recognizing the same determinant), and secondly by the fact that gametic associations are often different in different populations. These population differences allowed the separation of clusters of sera that identified associated antigens in one population, such as, for example A1 and B8 in Northern Europeans, which were much less associated in other populations, for example Caucasians from Southern Europe or India. Pairwise gametic associations in human populations are very largely determined by the the recombination fraction between the loci when this is very small, say, of the order of 0.2% or less. When the recombination fraction is larger than this and approaches 1%, gametic association may still persist but is less consistent between alleles at such loci. Thus, for example, alleles at the HLA-B and C loci show essentially the same patterns of gametic association worldwide, while the extent of gametic association between alleles at HLA-A and B is overall less and varies substantially from one population to another (see, e.g., W. F. Bodmer and J. G. Bodmer 1978).

A particular feature of the molecular analysis of the HLA and H2 regions is the identification of recombination "hot spots" that markedly influence the patterns of gametic association observed between pairs of loci within the HLA region. Recombination hot spots are positions where recombination is presumed to occur much more frequently than elsewhere. The most clearcut example for the HLA and H2 regions so far is the observation that in the H2-I region the majority of the recombinants have been localized to a 1.7 kb interval within a region that is at least 150 to 200 kb long (Steinmetz et al. 1982; Kobori et al. 1984). Recent molecular data clearly suggest that the relatively high recombination fractions between HLA-DP and DQ, and between HLA-A and C, probably also reflect the existence of one or more recombination hot spots in these intervals (see papers by Weissman et al. and Bell et al. in the *Cold Spring Harbor Symposium on the Molecular Genetics of* Homo Sapiens, 1987). A recombinant hot spot in the HLA DQ region is identified by a relative lack of gametic association between restriction fragment length polymorphisms (RFLP) at the DQ and DX loci. Thus, if the recombination fraction at a hot spot is of the order of, say, 0.5%, alleles at loci on either side of this hot spot will generally show little or no gametic association, while alleles at loci within the interval bounded by two recombinational hot spots are likely to show, on average, persistent and strong gametic associations. It is remarkable that classical linkage analysis coupled with an interpretation of patterns of gametic association and the argument that pairs of related duplicates, such as DX and DQ within the DQ region, are likely to be next to each other led to the present map of the HLA-D region, which has been fully confirmed by detailed molecular analysis. Thus, the fact that DQ and DR variants show strong gametic association, while DQ and DX do not, argues strongly for placing DX outside the DQ to DR interval. Since DR and complement alleles also show strong gametic association, DX cannot be between DR and the complement genes. This therefore establishes the sequence . . . DX . . . DQ . . . DR . . . . The consistency between the classical and the molecular analysis is reassuring but essential, for if there were a discrepancy, either one or the other analysis would have to be wrong.

It now seems likely that recombination, at least in mammals, is mainly localized to recombinational hot spots. These are probably small regions, perhaps no more than a thousand base pairs in length, where recombination occurs some five to ten times more often than elsewhere. Assuming the average hot spot gives rise to a recombination fraction of 0.5%, then with an estimated total human map length of about 2,500 centimorgans there would be about 5,000 hot spots in the genome overall. Since the HLA region is calculated, both by recombination fraction and molecular analysis, to be about one thousandth of the total human genome, this would give rise to an estimate of about five recombination hot spots within the HLA region, each separated by an average of 600,000 base pairs. Ascertainment bias means that it is likely that 0.5% is an overestimate for the recombination fraction with hot spots, and that the distribution of recombination fractions among hot spots is skewed toward lower values. A figure of 10,000 to 20,000 hot spots overall, implying 10 or 20 within the HLA region, may be likely. When loci are far apart in recombination fraction terms relative to the average recombination fraction within a hot spot, there is likely to be a reasonable correspondence between distance at the molecular level and the recombination fraction. Clearly, the smaller the recombination fraction is relative to that in a hot spot, the greater the heterogeneity in this relationship. For the HLA region, the total recombination fraction of 3% is consistent with it being one thousandth of the genome, and so one thousandth $\times$ 3 $\times$ 10^9, or 3 $\times$ 10^6 base pairs long. So far the region is known to code for some fifty products (including pseudogenes) corresponding to about 14,500 amino acids. This implies a coding ratio, namely, the proportion of the nucleotide sequence that codes for proteins, of about 1/70. (For a further discussion of some of these ideas, see W. F. Bodmer 1986b.)

Population Frequency Data

The value of the highly polymorphic HLA system for characterizing population interrelationships through analysis of gene frequency differences was recognized almost as soon as the system was discovered. Early field studies (see, e.g., J. G. Bodmer and W. F. Bodmer 1970) revealed marked differences in frequencies between different populations. The first, and from some points of view still the most comprehensive, analysis of the worldwide frequency distribution was achieved by the Fifth International Histocompatibility Testing Workshop held in 1972 (Dausset and Colombani 1972). The data obtained then revealed a number of features of the population distribution of HLA variants that have withstood the test of time. Since then, of course, more detailed and extensive analyses have been carried out. The number of alleles at the different loci and their level of polymorphism mean that phylogenetic analysis of the major human racial groups based on the HLA system alone is reasonably accurate, and provides essentially as much information as the classical blood group and enzyme electrophoretic markers. Analysis of frequency variation within European populations revealed a number of clines going from north to south, which are consistent with the hypothesis of Cavalli-Sforza and his coworkers of a migration of people from the center of agriculture in the Middle East outwards, starting some 9,000 years ago. Particularly striking in this respect is the cline in the haplotype *A1B8*, indicating a persistence of this gametic association for at least 5,000 to 10,000 years. This was one of the first lines of evidence used to support the existence of natural selection acting on the HLA system. The value of haplotype data for population comparisons was emphasized by J. G. Bodmer (1980) in her analysis of haplotype distributions in twenty-four predominantly European populations. Haplotype distribution has been analyzed more recently by Hedrick, Thomson, and Klitz (1986) to provide further evidence for selection that influences the population distribution of haplotypes showing persistent gametic associations.

The 1972 population data provided at least two further lines of evidence for the action of natural selection on the HLA system. The first came from comparing the expected pattern of homozygosity for different alleles in different populations at the various loci, with those expected under the assumption of neutrality (W. F. Bodmer, Cann, and Piazza 1973). This approach has also been extended to the analysis of more recent data by Hedrick, Thomson, and Klitz (1986). The second was the extraordinary pattern of HLA gene frequencies in American Indian populations, which have many fewer alleles, each with relatively high frequencies and so a markedly lower overall degree of heterozygosity. It was suggested by Cann, Bodmer, and Bodmer (1973) that this was consistent with a selective episode, possibly connected with diseases brought from the Old World, that restricted the HLA distribution to a much greater extent than that of other polymorphisms. A more recent analysis (J. G. Bodmer et al. 1987) confirms this very restricted distribution of haplotypes.

Analysis of HLA variation at the DNA level using RFLPs and more refined serological analysis using monoclonal antibodies are adding to the wealth of information that can now be collected to characterize populations with respect to their HLA distribution. Though many of the RFLPs show strong associations with clearly defined antigens, new variation is necessarily being discovered (see, e.g., papers in Albert 1984, and also Trowsdale et al. 1985, and J. G. Bodmer et al. 1987). Given enough effort, there is hardly any limit to the precision with which differences between populations can be defined. There are some notable gaps in the worldwide distribution of HLA frequencies, and indeed other polymorphic markers, especially in the transition from Europe to Asia across the Soviet Union. Interesting data is now being collected in China from north to south, suggesting marked differences between southern and northern Chinese, with the possibility of Caucasoid admixture perhaps connected with Mongoloid invasions in the north (Chen, pers. comm.).

HLA and Disease

The stimulus for the discovery of the HLA system was the search for a solution to transplant matching. The best evidence for the importance of HLA as a major histocompatibility (or tissue matching) system is the striking difference in survival of kidney and bone marrow grafts between HLA identical sibs, as compared to that across an HLA difference whether for Class I or Class II determinants. The effect of HLA matching on grafts exchanged between unrelated individuals is much less striking, though it is significant. The contrast between matching sibs and matching unrelated individuals is striking. Is it due to the fact that we do not yet know the determinants that most need to be matched, or is it because combinations of determinants are important? In either case, gametic association patterns must explain the difference between matching sibs and matching unrelated individuals. For if it is determinants not yet identified that matter, then if these showed a strong gametic association with already known determinants, population matching should be reasonably effective. Indeed, differences in the results of matching unrelated donors in the United States as compared to Europe can be attributed to the greater heterogeneity of the American population as a whole, which leads to a dilution of overall gametic association. If two or more determinants need to be matched at the same time for effective survival, then, again, if these are determined by alleles at loci between which there is expected to be an average strong gametic association, population matching should be relatively effective at least within comparatively homogeneous populations. In either case, however, if determinants that are important for graft survival are controlled by alleles at loci not showing a strong gametic association with those commonly typed, namely A, B, C, DQ, and DR, then sib matching will be expected to be much more effective than matching related individuals. The one possibility for this is the HLA-DP locus, since these determinants are known not to show a strong gametic association with alleles of the other HLA loci. Now that monoclonal antibodies and RFLPs are available for DP typing (see, e.g., Heyes et al. 1986), the suggestion that HLA-DP might be important for transplant matching can be examineá.

The demonstration by McDevitt and others that the mouse H2 system controlled genetic differences in immune response, and by Lilly that inherited differences in the incidence of virus induced leukemias were controlled by H2, stimulated the search for associations between HLA antigens and disease. The initial studies simply asked the question whether any antigen differed significantly in frequency between patients with a certain disease, and controls. This raised two obvious statistical problems. The first is the proper choice of controls, especially with respect to ethnic origin, since the frequencies of HLA types differ markedly from one population to another. The second is the fact that many HLA antigens are investigated in such studies, and so in assessing the significance of the association of a particular antigen with a disease, one must allow for the fact that this will be the most extreme difference among an appreciable number looked for. The simplest approximate statistical correction is to multiply the normal significance level by the number of comparisons made. (For an early discussion of these problems, see McDevitt and Landy 1972, and W. F. Bodmer 1972a).

Gametic association is important for the interpretation of HLA and disease association, since it explains how an association may be found with an antigen that is not itself the cause of a difference in an inherited susceptibility. The antigen may be controlled by an allele at a locus sufficiently closely linked to that which determines the disease susceptibility, for there to be a population association due to gametic association. Given the number of loci within the HLA region, patterns of disease association may be quite complex due to patterns of gametic association, even if there is only a single allele at one of the loci within the region that is determining susceptibility to the disease. Analysis of the frequency of antigens and corresponding genotypes in disease patients and controls can give estimates of disease gene frequencies and patterns of gametic association (see W. F. Bodmer and J. Bodmer 1974; Thomson and Bodmer 1977). Particular problems arise in assessing the significance of negative HLA antigen and disease associations when there are already significant positive associations (Thomson et al. 1985).

A direct search for linkage between HLA and susceptibility to a disease, using family data, should circumvent the problem of dependence on gametic association that is inherent in studying population frequencies of HLA antigens in patients and controls. Since, however, the HLA linked susceptibilities probably have a relatively low penetrance, conventional linkage analysis, even taking into account penetrance estimates, is very unreliable. This problem is overcome by the analysis of HLA distributions in affected sib pairs, which is a form of linkage analysis analogous to the classical Penrose sib pair approach, but which ignores unaffected individuals in a pedigree. These provide little information because they are a mixture of individuals without the susceptibility gene and those with the susceptibility gene but in whom it is not penetrant (see Thomson and Bodmer 1977; Motro and Thomson 1985). This approach to looking for linkage can, of course, be extended to any genetic marker, and also to any pedigree including two or more affected individuals, whatever their relationship. In this case, linkage is indicated by a distortion in the distribution of HLA or another marker among affected individuals in the pedigree, as compared to what would be expected from the normal pattern of Mendelian segregation. The expected distribution of markers in such complex pedigrees could be determined empirically by simulation, given data, for example, on the presumed genotypes of the input individuals, and population frequency data on the relevant genetic markers to take account of missing information. There is clearly, however, more scope for the theoretical analysis of marker distributions among an arbitrary set of individuals in a complex pedigree.

The first evidence for an HLA linkage with disease susceptibility obtained using this approach was for insulin-dependent diabetes mellitus (Cudworth and Woodrow 1975), and this has remained one of the most intensely studied HLA and disease associations. When it became clear that neither a simple recessive nor a dominant model could fit both the population and family data, "intermediate" models were proposed in which heterozygotes for a disease gene give rise to an increased susceptibility that is much less than for homozygotes (W. F. Bodmer 1980; Spielman, Baker, and Zmijewski 1980). The combination of population and family data can be used to estimate parameters, such as the disease gene frequency and penetrance, in an intermediate model from the combination of family and population data (see, e.g., Motro and Thomson 1985).

Further information can be obtained by comparing haplotype frequencies rather than single allele frequencies in patients and controls. Thus Winearls et al. (1984) showed that the gametic association between B15 and DR4 was much stronger in a group of diabetic patients than in controls from the same population. This suggests that neither B15 nor DR4 is involved, but perhaps another allele at a locus close to these but showing a gametic association with both B15 and DR4, and which is not found to the same extent in controls. An alternative explanation is that alleles at at least two loci are involved in controlling the disease susceptibility. The data are consistent with the suggestion that the disease susceptibility is controlled by the DQ and not the DR subregion. This would explain the relatively high frequency of *DR3/DR4* heterozygotes in diabetics by the suggestion that the association may be with particular heterozygous combinations of DQα and DQβ, both of which are polymorphic (Spielman et al. 1984).

Hodgkin's disease was the first example of an HLA and disease association, studied by Amiel (1967). The association is weak but significant because it has been so extensively investigated. Sib pair analysis within families has, however, very clearly indicated an HLA-linked effect. One explanation for this apparent discrepancy could be that the relevant gene is one that does not show major gametic association with the antigens that are usually typed for. Once again, as in the case of the transplant matching, this points to the possibility of an effect of HLA-DP, and this can now be investigated using the newer molecular techniques (see, e.g., W. F. Bodmer 1986a).

Evidence for Selection

An initial search for evidence of selection by looking for distortion of HLA segregation patterns in families was, perhaps not surprisingly, un-successful (Mattiuz et al. 1970). The association of HLA antigens with diseases and, by implication from the mouse data, with immune response differences, gives a simple rationale for how selection could act. In par-ticular, W. F. Bodmer (1972b), suggested that HLA variation associated with resistance to pathogens would give rise to a form of frequency-dependent selection in which particular new antigens have a selective advantage. Such a mechanism can readily give rise to the extensive poly-morphism observed for different loci of the HLA system, and remains perhaps the most plausible selective mechanism to explain the HLA polymorphism.

As already discussed, population distribution data provided initial evi-dence for the effects of selection, in particular through the distribution of particular haplotypes involving strong gametic associations and through differentials in levels of homozygosity between populations. The most striking evidence, however, for the action of natural selection is the fact that some of the genes in the HLA-D region, notably $DQ\alpha$, $DQ\beta$, and $DR\beta$, are very much more polymorphic than the other genes, notably than $DR\alpha$, which shows hardly any functional polymorphism. Both $DP\alpha$ and $DP\beta$ also show very limited levels of polymorphism, as do $DX\alpha$ and $DX\beta$, but in the latter case this could be connected with whether or not they are expressed. There is, furthermore, very significant heterogeneity in levels of polymorphism within these genes. Most of the polymorphism is found in the N-terminal regions of the chains and even within these there are localized areas of higher levels of variation. The data clearly suggest that it is these regions which have a major effect on immune response differences and so influence resistance to pathogens (see, e.g., Trowsdale et al. 1985). In the absence of natural selection, all the genes would be expected to show on average the same level of polymorphism, subject to overall differences in mutation rates. Differences between these very closely linked loci in mutation rates, and specifically in regions that are presumed to be functionally significant to the molecule, seem an extremely implausible explanation for the heterogeneity between loci in levels of polymorphism. Natural selection is the only plausible explanation, and the functional basis for it is clearly established through the role of these genes in controlling immune response and their observed associations with disease. We must emphasize that these disease associations are unlikely, in general, to have major postreproductive selective effects and must be thought of as the price to pay for the variation that has in the past been selected for as a protection against infectious diseases.

RFLPs have been detected particularly in association with the loci that are highly polymorphic by serological analysis. Since it is clear that the majority of the RFLPs lie outside the coding regions that are detected by the serological and cellular assays, the data suggest that there is an increased background level of variability in DNA sequences adjacent to polymorphic regions, as compared to nonpolymorphic regions. Once again, regional differences in mutation rate seem a most implausible mechanism to account for this observation. An alternative explanation is that these differences in variability are associated with patterns of selection for polymorphic variants that arise when new variants pull into the population associated variation that may have no functional significance. This would, of course, only happen for variations that show strong gametic association with the polymorphism that is being selected for. There are reasons for believing, as we will discuss later, that the generation of polymorphic variants itself may give rise to multiple differences from a single complex event at the DNA level. Since gametic association is a prerequisite for the hitchhiking of neutral variation associated with that being selected for, the region around a selected polymorphic region that shows an overall increased level of polymorphism should be bounded by the two nearest recombination hot spots on either side of the functionally important region. This explanation seems to fit the observed pattern of RFLPs very well. It might be interesting to check this suggestion by computational and theoretical analysis of appropriate population-genetic models.

Evolution

Sequence comparisons at the DNA level between the various HLA region genes can now provide an overall picture of their pattern of evolution. The Class I and Class II products each consist of closely related families of genes that are themselves distantly related to the immunoglobulins, as first suggested by W. F. Bodmer and Edelman many years before sequence data were available. Thus, the membrane proximal sequences of HLA-A,

B, C, and D products show 20–25% similarity with immunoglobulins and other members of the immunoglobulin "super family," including the sharing of certain common structural features. The HLA-A, B, C sequences are only slightly more similar to HLA-D than they are to immunoglobulins, and the difference between D region α and β chains is comparable to that between the Class I and Class II products. The similarities, though weak, suggest that all the genes for Class I and Class II products within the HLA region have a common origin by duplication. But the extent of differences suggests that the duplication events took place as long as 700 million years ago. Thus, in this case, duplicate genes have remained together within the same gene cluster over a very long period of time. Although in many cases the genes controlling different subunits of a heteropolymeric protein are unlinked, as in the case of $\beta2$ microglobulin and HLA-A, B, C, immunoglobulin heavy and light chains and the α and β chains of hemoglobin, this is not true for the HLA-D region α and β chains. Indeed, it was at first suggested that the HLA-D α and β chains might not both be coded for within the HLA region for this reason.

An evolutionary scheme for the HLA genes, especially the HLA-D region, based on percentages of shared sequence is shown in Figure 14.3. The pattern of evolution is a familiar one, starting from a single chain, possibly forming a homo-dimer, which duplicated and diverged to lead to diversification of function and more refined functional adaptation. This naturally suggests that there should be some functional differences, for example, between the various HLA-D subregion products. However, evolution is a messy patching-up process, and so clearcut functional differentiation between the products should not be expected and is not necessary to account for the divergence of the region through fine tuning of its adaptation to the needs for immune response.

There are considerable differences in sequence between some alleles, but these are localized to the N-terminal domains of the molecules. Thus, while DQ α alleles have up to 20% differences for the eighty-seven amino acid $\alpha1$ domain, differences elsewhere are no more than 4% or 5%. These are still somewhat higher than the differences between alleles at the much less polymorphic loci such as DP α, but this may be due to the phenomenon already discussed, namely, of a general increase ·in polymorphic levels around regions subject to selection. It seems likely that the differences outside the regions that are highly polymorphic reflect the rate of neutral substitution, and so these differences can be used to estimate the age of alleles. The data suggest that alleles may be 5–10 million years old in some cases, which predates the evolution of *Homo sapiens* but comes after the divergence of the hominids from the great apes (see, e.g., Bodmer et al. 1986).

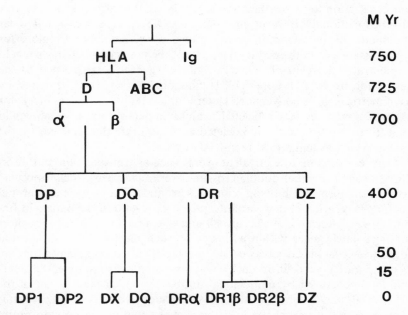

Figure 14.3. Evolutionary tree for the HLA D region and associated products. HLA and Ig refer to the primordial genes for the HLA and immunoglobulin system. D and ABC α and β refer similarly to the primordial genes for the Class II and Class I subregions and for the Class II subregion chains. Approximate divergence times are given in millions of years. (Source: Bodmer et al. 1986.)

Most of the sequence data obtained so far is from coding, and to some extent, from intervening sequences. Estimates of times of divergence from such sequences may clearly be subject to a considerable margin of error, especially for the more distant events. More precise estimates of divergence will come from comparisons of intronic and flanking region nucleotide sequences, which should correspond more nearly to the expected rates of evolution, assuming neutrality. These data are, however, laborious to obtain and not of such obvious functional interest. Nevertheless, more extensive sequence data will be obtained, which should then allow a much better assessment of the patterns of evolution of the HLA region genes, and in particular the times at which the various duplication events occurred.

It is already clear that in many cases there is much greater divergence in the intronic and flanking regions, as well as in the untranslated but transcribed 3' region, than in coding sequences. Sequence variation in the 3' untranslated region and in some parts of the coding sequence may reflect selection for diversification, and some of the intron sequences may also be functional. However, in general one would expect, especially near the middle of the intron and moving away into the flanking regions for some distance from the coding region, that the majority of changes would be neutral. The relatively high rate of divergence in introns led to the suggestion by W. F. Bodmer (1981) that some of this could be driven by asymmetrical gene conversion. Heteroduplex repair processes may, for example, favor repair of a deleted strand in a heteroduplex, which would lead to substitution of the nondeleted allele at a rate that depends on the rate at which asymmetrical repair takes place.

Gene conversion, or equivalent events such as transposition and intra-chromosomal unequal double crossing over, have been widely appealed to as mechanisms for moving sequences around within a gene cluster and so, for the creation of new variants. (See, e.g., several of the papers in Immunological Reviews, 1985, vols. 84 and 85). Transposition of sequences among related genes within a cluster by such mechanisms can account for sequence similarities between long-established adjacent genes that are of comparatively recent origin. These may explain, for example, the determinants recognized by monoclonal antibodies which react to all of the HLA-A, B, and C, or to all the HLA-D region molecules. Observations such as these suggested the hypothesis that some of the polymorphism within the HLA region could be with respect to control of expression of different genes (W. F. Bodmer 1973). This is still a plausible explanation for some of the polymorphic differences within the HLA region, but certainly does not apply for the majority. Transposition of sequences between genes, such as the different DR β genes, could account for serological cross-reactions between the products of alleles at different loci, as have been revealed by monoclonal antibodies. Since such a transposition event would have happened on one particular haplotype, the converted and original alleles will initially be on this same haplotype and so are likely to continue to show strong gametic association. That explains how there can be strong gametic association between cross-reacting determinants on different locus products.

Gene conversion and related events, by transposing a region from one gene into a corresponding region in another gene, are likely to create functionally effective variants, even when the donor gene for the event is a pseudogene. Such complex events may also, however, produce multiple changes at the DNA level, only some of which are functionally relevant. If the resulting new allele increases in frequency by selection, all the differences will be pulled into the population because of gametic association due to very close linkage. Although the frequency of gene conversionlike events may be comparatively low, the frequency with which they produce variants that could be selectively advantageous might be relatively high. Thus, a majority of functionally effective selected variants may be produced by these processes, rather than by conventional mutations. There may therefore be secondary selection pressures that favor situations which give rise to the potential for producing new, functionally effective variants by gene conversion or related mechanism, since a higher rate of production of such variants should allow more rapid and flexible selective response to novel pathogens. This is a way, for example, in which pseudogenes may be selectively advantageous, since they are potential donors for gene conversion to produce new variants. In general, the more related genes there are in a region, the greater the opportunity for production of new variants by gene conversionlike processes. Thus, perhaps the reason that the human HLA-D region is more complex than the mouse H2-I region is that the greater the life span of the species, the longer the exposure of an individual to a hostile pathogen-containing environment, and so the greater the need for a rapid and flexible response to novel pathogens. As a consequence there would be an advantage for more duplicate genes in the longer-lived human species than in the mouse.

The analogy between the function of the HLA Class I and Class II products and complement components provides a plausible explanation for selective mechanisms that could favor moving the complement genes into the HLA region. Thus, Porter (1985) suggested that complement variation may contribute to differential recognition of antigens in an analogous way to that postulated for HLA variations. A particular complement variant that was especially effective for a certain immune response associated with a pathogen might be advantageous in association with an HLA variant that conferred an advantageous immune response to the same pathogen. Such a synergistic interaction could have led to a selective advantage for a translocation or insertion event that brought the particular advantageous complement allele into the HLA region bearing the relevant HLA variant. This could then have led to the origin of complement genes within the HLA region (W. F. Bodmer 1986). There is so far no similar rationale for the presence of the 210H genes alongside the C4 genes within the HLA region. Perhaps when the C4 genes were transposed to the HLA region, they brought the 210H genes with them, simply as bystanders. Thus, in this case, the 210H genes may have hitchhiked their way into the HLA region by chance.

Conclusions

The development of the HLA system has provided a rich source of material for the application of statistical, population-genetic, and evolutionary ideas. The analysis of marker associations with disease, especially using the sib-pair approach, and the analysis of gametic associations in populations have been largely stimulated by work on the HLA system. Its complexity provides many challenges for an evolutionary understanding of gene clusters. The new information that has come from the applications of molecular techniques and the use of monoclonal antibodies, far from replacing the need for further statistical and population-genetic analysis, has created an even greater challenge for the population geneticist. There is a continuing need to relate the analysis to new observations, and to help with the problem of understanding the function of the HLA system and its involvement in disease associations.

Acknowledgment

Throughout most of our time at Stanford University, from 1962 to 1970, the statistical aspects of our work on the HLA system were supported by a joint grant with Sam Karlin. He has been a friend and collaborator for nearly a quarter of a century, and his help and stimulus made an important contribution to our early work on the HLA system.

References

Albert, E., ed. 1984. *Histocompatibility Testing 1984*. Springer-Verlag, Berlin, Heidelberg.

Amiel, J. L. 1967. Study of the leucocyte phenotypes in Hodgkin's disease. In *Histocompatibility Testing 1967*, pp. 79–81. Ed. E. S. Curtoni, P. L. Mattiuz, and M. R. Tosi. Munksgaard, Copenhagen.

Bodmer, J. G. 1980. The HLA system: The HLA-DR antigens and HLA haplotypes in 24 populations. In *Population Structure and Genetic Disorders*, pp. 211–238. Ed. A. W. Erikson. Academic Press, New York.

Bodmer, J. G., and Bodmer, W. F. 1970. Studies on African Pygmies, IV. A comparative study of the HL-A polymorphism of the Babinga Pygmies and other African and Caucasian populations. *Am J. Hum. Genet.* 22: 396–411.

Bodmer, J. G., and Bodmer, W. F. 1984. Monoclonal antibodies to HLA determinants. *Brit. Med. Bull.* 40: 276–282.

Bodmer, J. G.; Kennedy, L. J.; Lindsay, J.; and Wasick, A. M. 1987. Applications of serology and the ethnic distribution of three locus HLA haplotypes. *Brit. Med. Bull.* 43: 94–121.

Bodmer, W. F. 1972a. Association between HL-A type and specific disease entities. In *Genetic Control of Immune Responsiveness: Relationship to Disease Susceptibility*, pp. 331–365. Ed. Hugh McDevitt and Maurice Landy. Academic Press, New York.

Bodmer, W. F. 1972b. Evolutionary significance of the HLA system. *Nature* 237: 139–145.

Bodmer, W. F. 1973. A new genetic model for allelism at histocompatibility and other complex loci: Polymorphism for control of gene expression. *Transpl. Proc.* 5: 1471–1476.

Bodmer, W. F. 1980. The HLA system and disease. The Oliver Sharpey Lecture, 1979. *J. Roy. Coll. Physicians of Lond.* 14: 43–50.

Bodmer, W. F. 1981. Gene clusters, genome organization and complex phenotypes: When the sequence is known, what will it mean? *The Am. J. Hum. Genet.* 33: 664–682.

Bodmer, W. F. 1986a. HLA Today (An adapted version of the talk given at the ASHI Meeting in San Diego on 15 Oct. 1985). *Human Immunology* 17: 490–503.

Bodmer, W. F. 1986b. Human Genetics: The molecular challenge. In *Cold Spring Harbor Symposia of Quantitative Biology*, vol. 51, *Molecular Biology of Homo sapiens*, pp. 1–13. Cold Spring Harbor Laboratory, N.Y.

Bodmer, W. F., and Bodmer, J. G. 1978. Evolution and function of the HLA system. *Brit. Med. Bull.* 34: 309–316.

Bodmer, W. F., and Bodmer, J. 1974. The HL-A histocompatibility antigens and disease. Tenth Symposium in Advanced Medicine, Proceedings of Conference held at Royal College of Physicians, London, pp. 157–174. Ed. J.G.G. Ledingham. Pitman Medical Press, London.

Bodmer, W. F., and Payne, R. 1965. Theoretical consideration of leukocyte grouping using multispecific sera. In *Histocompatibility Testing 1965*, pp. 141–149. Munksgaard, Copenhagen.

Bodmer, W. F.; Cann, H.; and Piazza, A. 1973. Differential genetic variability among polymorphisms as an indicator of natural selection. In *Histocompatibility Testing 1972*, pp. 753–767. Munksgaard, Copenhagen.

Bodmer, W. F.; Trowsdale, J.; Young, J.; and Bodmer, J. G. 1986. Gene clusters and the evolution of the major histocompatibility system. *Phil. Trans. Roy. Soc. Lond. B* 312: 303–315.

Cann, H.; Bodmer, J. G.; and Bodmer, W. F. 1973. The HL-A polymorphism in Mayan Indians of San Juan La Laguna, Guatemala. In *Histocompatibility Testing 1972*, pp. 367–375. Munksgaard, Copenhagen.

Cudworth, A. G., and Woodrow, J. 1975. Evidence for HLA-linked genes in "juvenile" diabetes mellitus. *Brit. Med. J.* 3: 133–135.

Dausset, J., and Colombani, J., eds. 1973. *Histocompatibility Testing 1972*. Munksgaard, Copenhagen.

Hedrick, P. W.; Thomson, G.; and Klitz, W. 1986. Evolutionary Genetics: HLA as an exemplary system. In *Evolutionary Processes and Theory*, pp. 583–606. Ed. S. Karlin and E. Nevo. Academic Press, New York.

Heyes, J.; Austin, P.; Bodmer, J.; Bodmer, W. F.; Madrigal, A.; Mazzilli, M.; and Trowsdale, J. 1986. Monoclonal antibodies to HLA-DP-transfected mouse L cells. *Proc. Natl. Acad. Sci. USA* 83: 3417–3421.

Immunological Reviews. 1985. Nos. 84, 85.

Kobori, J. A.; Winot, A.; McNicholas, J.; and Hood, L. 1984. Molecular characterization of the recombination region of six murine major histocompatibility complex (MHC) 1 region recombinants. *J. Molec. Cell. Immunol.* 1: 125–131.

McDevitt, H. O., and Landy, M., eds. 1972. Genetic organization of H-2 and its relationship to the Ir and MLC genes. In *Genetic Control of Immune Responsiveness: Relationship to Disease Susceptibility*, pp. 73–142, 338–365. Academic Press, New York, London.

Mattiuz, P. L.; Ihde, D.; Piazza, A.; Ceppellini, R.; and Bodmer, W. F. 1970. New approaches to the population genetic and segregation analysis of the HL-A system. In *Histocompatibility Testing 1970*, pp. 193–205. Munksgaard, Copenhagen.

Motro, M., and Thomson, G. 1985. The affected sib method, I. Statistical features of the affected sib-pair method. *Genetics* 110: 525–538.

Mueller, U.; Stephan, D.; Philippsen, P.; and Steinmetz, M. 1987. Orientation and molecular map position of the complement genes in the mouse MHC. *The EMBO Journal* 6: 369–373.

Payne, R.; Tripp, M.; Weigle, J.; Bodmer, W.; and Bodmer, J. 1964. A new leukocyte isoantigen system in man. *Cold Spring Harbor Symp. on Quantitative Biology* 29: 285–295. Cold Spring Harbor Laboratory, N.Y.

Porter, R. R. 1983. Complement polymorphism and the major histocompatibility complex and associated diseases: A speculation. *Mol. Biol. Med.* 1: 161–168.

Spielman, R. S.; Baker, L.; and Zmijewski, C. M. 1980. Gene dosage and susceptibility to insulin dependent diabetes. *Ann. Hum. Gen.* 44: 135–150.

Spielman, R. S.; Lee, J.; Bodmer, W. F.; Bodmer, J. G.; and Trowsdale, J. 1984. Six HLA-D α chain genes on human chromosome 6: Polymorphisms and associations of DC α-related sequences with DR types. *Proc. Natl. Acad. Sci. USA* 81: 3461–3465.

Steinmetz, M.; Minard K.; Korvath, S.; McNicholas, J.; Strelinger, J.; Wake, C.; Long, E.; Mach, B.; and Hood, L. 1982. A molecular map of the immune response region from the major histocompatibility complex of the mouse. *Nature* 300: 35–42.

Thomson G., and Bodmer, W. F. 1977. The genetic analysis of HLA and disease associations. In *HLA and Disease*, pp. 84–93. Ed. J. Dausset and A. Svejgaard. Munksgaard, Copenhagen.

Thomson, G.; Nicholas, F. W.; Bodmer, W. F.; O'Neill, M. E.; Hedrick, P. W.; and Hudes, E. 1985. Analysis of negative and multiple HLA antigen disease associations. *Tissue Antigens* 26: 293–306.

Trowsdale, J.; Young, J.A.T.; Kelly, A. P.; Austin, P.; Carson, S.; Meunier, H.; So, A.; Ehrlich, H. A.; Spielman, R. S.; Bodmer, J.; and Bodmer, W. F. 1985. Structure, sequence and polymorphism in the HLA-D region products. *Immunol. Rev.* 84: 136–173.

Van Rood, J. J., and Van Leeuwen, A. 1963. Leukocyte grouping: A method and its application. *J. Clin. Invest.* 42: 1382–1390.

Winearls, B. C.; Bodmer, J. G.; Bodmer, W. F.; Bottazzo, G. F.; McNally, J.; Mann, J. I.; Thorogood, M.; Smith, M. A.; and Baum, J. D. 1984. A family study of the association between insulin dependent diabetes-mellitus, autoantibodies and the HLA system. *Tissue Antigens* 24: 234–246.

LIST OF CONTRIBUTORS

Julia G. Bodmer, Imperial Cancer Research Fund Laboratories, P.O. Box 123, Lincoln's Inn Fields, London WC2A 3PX, England

Walter F. Bodmer, Imperial Cancer Research Fund Laboratories, P.O. Box 123, Lincoln's Inn Fields, London WC2A 3PX, England

Luigi L. Cavalli-Sforza, Department of Genetics, Stanford University School of Medicine, Stanford, Calif. 94305-5120, USA

Freddy Bugge Christiansen, Department of Ecology and Genetics, University of Aarhus, DK-8000 Aarhus C, Denmark

C. Clark Cockerham, Department of Statistics, North Carolina State University, Raleigh, N.C. 27650-8203, USA

Warren J. Ewens, Department of Mathematics, Monash University, Clayton, Victoria 3168, Australia

Marcus W. Feldman, Department of Biological Sciences, Stanford University, Stanford, Calif. 94305-5020, USA

John H. Gillespie, Department of Genetics, University of California, Davis, Calif. 95616, USA

Richard R. Hudson, National Institutes of Environmental Health Sciences, P.O. Box 12233, Research Triangle Park, N.C. 27709, USA

Norman L. Kaplan, National Institutes of Environmental Health Sciences, P.O. Box 12233, Research Triangle Park, N.C. 27709, USA

Sabin Lessard, Département de Mathématiques et de Statistique, Université de Montréal, Montréal, Quebec H3C 3J7, Canada

Uri Liberman, Department of Statistics, Tel Aviv University, Tel-Aviv, Israel

Michael E. N. Majerus, Department of Genetics, University of Cambridge, Downing Street, Cambridge CB2 3EH, England

Peter O'Donald, Department of Genetics, University of Cambridge, Downing Street, Cambridge CB2 3EH, England

Jonathan Roughgarden, Department of Biological Sciences, Stanford University, Stanford, Calif. 94305-5020, USA

Simon Tavaré, Department of Mathematics, University of Utah, Salt Lake City, Utah 84112, USA

Marcy K. Uyenoyama, Department of Zoology, Duke University, Durham, N.C. 27706, USA

Geoffrey A. Watterson, Department of Mathematics, Monash University, Clayton, Victoria 3168, Australia

Bruce S. Weir, Department of Statistics, North Carolina State University, P.O. Box 5457, Raleigh, N.C. 27650, USA

INDEX